College Geometry

second edition

COLLEGE GEOMETRY

Frank C. Denney
David M. Minor
Chabot College

PRENTICE HALL, ENGLEWOOD CLIFFS, NEW JERSEY 07632

Library of Congress Cataloging-in-Publication Data

Denney, Frank C.
College geometry / Frank C. Denney, David M. Minor. — 2nd ed.
p. cm.
Rev. ed. of: College geometry / David M. Minor. 1972.
Includes index.
ISBN 0-13-142142-5
1. Geometry. I. Minor, David M. II. Minor, David M. College geometry. III. Title.
QA445.D46 1991 90-21584
516—dc20 CIP

Editorial/production supervision: Debra Wechsler and Valerie Zaborski
Cover and interior design: Donna M. Wickes
Prepress buyer: Paula Massenaro
Manufacturing buyer: Lori Bulwin
Acquisitions editor: Priscilla McGeehon
Photo researcher: Anna Block

A Division of Simon and Schuster
Englewood Cliffs, New Jersey 07632

Printed in the United States of America

10 9 8 7 6 5 4 3 2 1

ISBN 0-13-142142-5

Prentice-Hall International (UK) Limited, *London*
Prentice-Hall of Australia Pty. Limited, *Sydney*
Prentice-Hall Canada Inc., *Toronto*
Prentice-Hall Hispanoamericana, S.A., *Mexico*
Prentice-Hall of India Private Limited, *New Delhi*
Prentice-Hall of Japan, Inc., *Tokyo*
Simon & Schuster Asia Pte. Ltd., *Singapore*
Editora Prentice-Hall do Brasil, Ltda., *Rio de Janeiro*

Contents

Preface

Over the years since publication of the first edition, we have had the opportunity to see what happens to students who manage to enter courses in trigonometry and calculus without a firm foundation in geometry. The results have been, for the most part, disastrous. "They know not what they do not know." Geometry is not easy to teach, nor is it easy to learn, but we believe that there is no substitute for the immersion in methods of logical thinking that geometric proofs provide. There seem to be no other place in the mathematics curriculum where there is time for this kind of concentration. Mere statement of facts and memorization of formulas is not enough.

Our subject matter is primarily the geometry of Euclid. We believe that every person can profit from a close acquaintance with this historic, traditional approach. We do not mean to deny the importance of and the great impact upon mathematical and scientific thought of non-Euclidean geometries, translations, network theory, and the fractals and chaos theory of Mandelbrot. The assertion that geometry is a dead subject is absolutely contradicted by recent progress in the field. But these important topics are worth studying in their own time. For a one-semester or one-quarter course we feel the need to concentrate on plane geometry.

That we include an early introduction to coordinate geometry and trigonometry may seem a contradiction to the previous statement, but we believe that it is not. The objective for the inclusion of both topics is reinforcement and application. Since a separate course in analytic geometry has all but disappeared from the mathematics curriculum, there is seldom enough time to spend on this important topic. Success in trigonometry is heavily dependent on basic geometric facts. The current emphasis on trigonometric functions of real numbers may be more appropriate for calculus, but all too often the student is left wondering what all the bother is about when lacking the assurance provided by the historical triangle measurement approach.

Our text has been called fairly rigorous, and we appreciate that. We would have liked to have it more so but did not want to make it too formidable for students taking the course for the first time. The level of rigor for this course can be set by the teacher. It is easier for an individual instructor to "lighten up" a more rigorous text than it is to tighten up a less rigorous one.

The second edition differs from the first primarily in the distribution of geometric constructions throughout the text rather than placing them in a separate chapter. We have also formed a separate chapter on perpendicular and parallel lines, since these concepts are so important in their own right.

We regard the chapter on space geometry as a mere introduction—some instructors may not find time to include it—but with the background provided in the previous chapters, it should be possible to cover the material fairly quickly.

We wish to acknowledge the encouragement for undertaking this revision given to us over the years by our colleague Eric Nelson of Chabot College. We also appreciate the helpful reviews provided by:

Glen Jacobs, Greenville Technical College.
James McKinney, California State Polytechnic University.
Larry Mischel, Cuesta Community College.
Richard C. Petersen, California State University at Fresno.
Robert G. Russell, West Valley College.
Eric R. Stietzel, Foothill College.
Jean A. Vrechek, Sacramento City College.
Edward T. Walsh (Tom), City College of San Francisco.

The assistance provided by the staff of Prentice Hall, notably Priscilla McGeehon, Debra Wechsler, and Valerie Zaborski, enabled us to complete our task most efficiently.

EUCLID, a teacher at the University in Alexandria, Egypt, about 300 B.C., collected the geometric facts known to the ancient Greeks and put them into organized form in his work *Elements*. According to the historian Stobaeus, when a student after learning the first theorem asked Euclid "But what advantage shall I get by learning these things?" the great teacher called to his assistant and said "Give him threepence, since he must needs make profit out of what he learns." (*Brown Brothers*)

1

Introduction

There is no branch of mathematics however abstract which may not someday be applied to phenomena of the real world.

NIKOLAI IVANOVICH LOBACHEVSKI (1793–1856) Russian mathematician.

1-1 SOME APPLICATIONS

Over 2000 years ago, a Greek mathematician named Eratosthenes was able to calculate the circumference of the earth to a figure that differs from the most recent ones by only several hundred miles. Considering that it was some 1700 years later before Columbus "proved" that the world was round, this was indeed a remarkable accomplishment.

The student of geometry will find Eratosthenes' method to be quite simple once the basic ideas involved have been learned. He measured the length of the shadow cast by a pillar in the town of Alexandria, Egypt, and used the known distance to the town of Syene (about 501 miles) and the fact that at the time he measured the shadow in Alexandria, the sun shone directly to the bottom of a deep well in Syene. Figure 1-1 shows the relationship of sun to earth, of shadow to pillar. The assumption that the sun's rays are essentially parallel is correct, for all practical purposes, as might be verified by noticing the sun's rays shining through a venetian blind.

We use this example to show some of the important features of our study of geometry. It will be *proved* that the angles at Q, the center of the earth, and P, the top of the pillar, are equal in measure. Eratosthenes knew that the ratio of the degree measure of the angle at Q to 360° is the same as the ratio of the Syene–Alexandria distance to the circumference of the earth. Thus

$$\frac{7.2^\circ}{360^\circ} = \frac{501}{C}$$

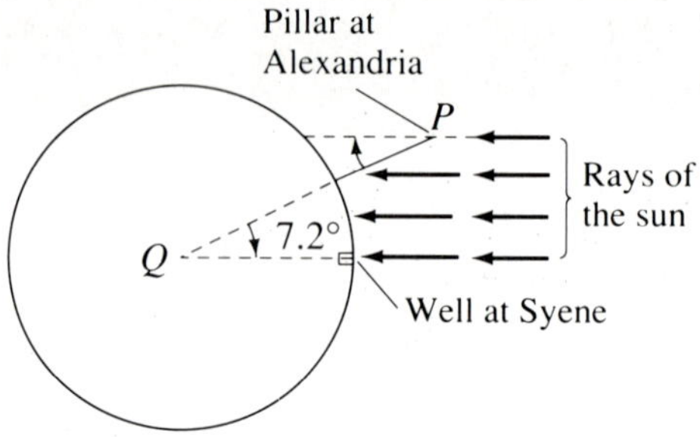

Figure 1-1

and when we solve this proportion for C, we find $C = 25050$ miles. (The latest measurement is approximately 24090 miles.) Although the implications of the ideas we present may not always seem so immediately applicable, taken together they can produce some far-reaching and extremely useful results. Whenever possible we will apply our findings in some practical way.

It is further appropriate to note concerning our first example that the word "geometry" means "earth measurement." Applications of geometry however, are not limited to earth measurement alone. The circuit diagram of Fig. 1-2 can be analyzed by means of the vector diagram of Fig. 1-3. By means of a geometric principle which was used by the Egyptians and proved by the Greek mathematician Pythagoras, we know that if lines OX and OY form a 90° angle, $E^2 = (IR)^2 + (IX_L)^2$, or that $E = \sqrt{(IR)^2 + (IX_L)^2}$, so we can calculate $E = \sqrt{(50)^2 + (120)^2} = \sqrt{2500 + 14400} = \sqrt{16900} = 130$.

The Egyptians are thought to have used a special case of this principle in order to resurvey the fields after the flooding of the Nile River. By using a rope with knots spaced equally like the one pictured in Figure 1-4, a 90° angle was formed at point O. Note that $5^2 = 3^2 + 4^2$. This relationship among the three

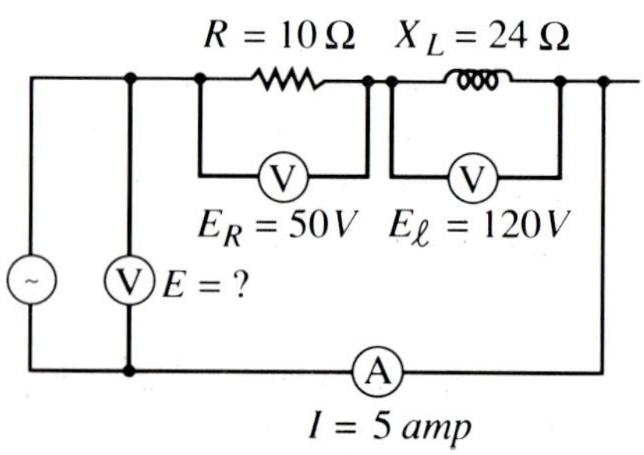

Figure 1-2

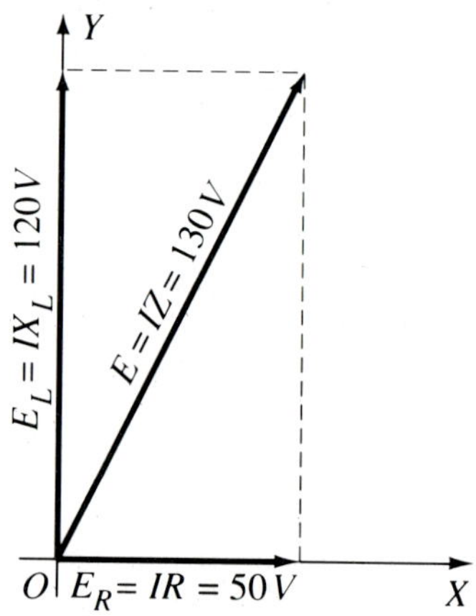

Figure 1-3

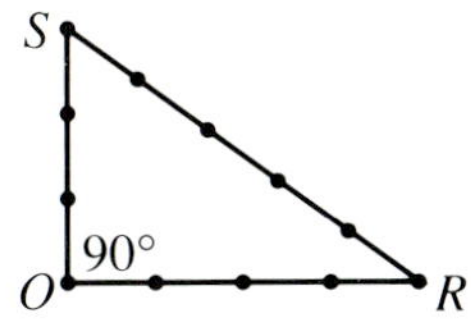

Figure 1-4

sides of a right triangle is one of the most important ones in all mathematics and its applications are widespread.

Although our examples have demonstrated some practical measuring applications of geometry, our purpose will also be to develop patterns of thought and to provide background for further work in mathematics. A knowledge of geometric facts is essential to the study of trigonometry, analytic geometry, and the calculus. A significant number of the mathematics questions on the GMAT (Graduate Management Aptitude Test), which is required for acceptance to many MBA (master of business administration) programs, are geometry problems. The test writers are not so much interested in knowledge of geometric facts, however, as they are in the ability to reason well.

Exercise Set 1.1

1. **(a)** Use a compass to draw a circle with a radius of 4 in. Assume for a moment that we do not know how to find its circumference.
   (b) Cut a piece of string 3 in. long, and lay it along the edge of the circle.
   (c) Mark the endpoints of the string on the circle and draw lines from each of these points to the center of the circle.
   (d) Use a protractor to measure the angle formed by the two radii.
   (e) Use Eratosthenes' proportion to find the circumference ($A/360° = 3$ in.$/C$, if A is the measure of the angle at the center of the circle).
   (f) Check by using the formula $C = 2\pi r$, with $\pi = 3.14$ and r the length of the radius. (Your answer is fine if it is within 10 percent of this one. Your measurement of the string is probably as accurate as Eratosthenes' measurement of the distance from Alexandria to Syene.)
2. Repeat Exercise 1, this time using a 3-in. radius and the 3-in. string.
3. You can construct a triangle with sides of given lengths by the following method. Suppose we want a triangle with sides of 4, 7, and 9. Draw a part of a line AQ, longer than 9 units, such as shown in the figure below. Mark off a distance of 9 units with your compass, placing the point at A and marking through a point on the line AQ, which we will call B. With the compass points 7 units apart draw a section of a circle above AB

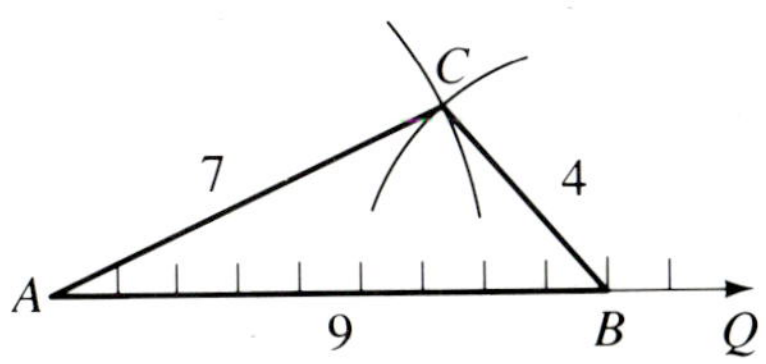

with A as the center. Now with the compass points 4 units apart and using B as the center draw part of a circle above AB so that it intersects the section drawn previously. Call the point of intersection C. Connect C with A and with B and the triangle is completed. For the following exercises, you may want to use graph paper which is *4* squares per inch; otherwise use a scale of $\frac{1}{4}$ in. equals 1 unit. Draw a triangle with sides

(a) 6, 8, and 10 units
(b) 6, 7, and 9 units
(c) 8, 10, and 13 units
(d) 5, 12, and 13 units

4. **(a)** Which of the triangles in Exercise 3 appear to be right triangles?
(b) Use the Pythagorean relation to determine which are actually right triangles. (If a triangle is a right triangle, the square of the length of the longest side is equal to the sum of the squares of the lengths of the other two sides.)

5. Predict which of the following sets of numbers would represent lengths of sides of a right triangle and draw the triangles for those which are

(a) 4, 5, 6
(b) 8, 15, 17
(c) 9, 11, 13
(d) 3.5, 12, 12.5

1-2 ESSENTIALS OF A LOGICAL SYSTEM

In discussing Eratosthenes' problem at the beginning of the chapter we mentioned that the rays of the sun arrive at the earth in nearly parallel lines. This conclusion may be reached by the process of *inductive reasoning,* which is essential to much of our scientific advancement.

You found when looking at the triangles in Exercise 3 of the first exercise set that the right triangles all have sides such that the square of the length of the longest side is equal to the sum of the squares of the lengths of the two shorter sides. If you were to measure the circumferences and diameters of a large number of circles you might observe that the length of the circumference is always a little more than three times the length of the diameter. In a problem in the next chapter you will be asked to find a formula for determining the number of lines which can be drawn through n distinct points, after actually drawing lines through two, three, and four points. These are examples of inductive reasoning, or generalizing from a series of specific observations.

The early Egyptians and Greeks arrived at many significant conclusions in this way, and a large body of inductively derived geometric facts helped the Egyptians to build the pyramids and other colossal structures as well as to survey and rebuild after the flooding of the Nile. Later, Greek mathematicians organized the facts, drew conclusions, and arrived at a system of geometry which remained virtually unchanged for over 2000 years.

The method used by these later Greek mathematicians is known as *deductive reasoning.* In this method one begins with known or assumed "facts" and deduces other facts which follow as a logical consequence of the original ones. As an example of the difference between inductive and deductive reasoning, consider Fig. 1-5. Assume that lines l_1 and l_2 are parallel. If we measure the numbered angles with a protractor, it becomes apparent that the measures of angles 1, 3, 5, and 7 are equal to each other, as are 2, 4, 6, and 8. We also find that angles 1′, 3′, 5′, and 7′ all have equal measures, as do 2′, 4′, 6′, and 8′. From this we might conclude that any time we draw a line that intersects two parallel lines, the measures of the angles formed will be equal in the same way. This is inductive reasoning. Our conclusion happens to be true, but not just because our two observations confirm it. Increasing the number of observations would strengthen our belief that this should always be true, but that alone is not enough to prove it is true. For example, suppose that we claim that the number $k = n^2 - n - 15$ is positive only if n is greater than 5. When we test this claim by substituting 6, 7,

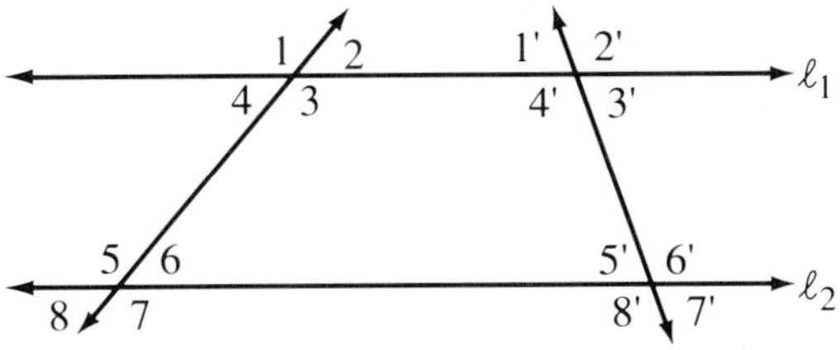

Figure 1-5

and 8, we find that k is positive. When we substitute 4, 3, 2, 1, we find that k is negative. (Try it!) We might now consider our claim to be true. But suppose we try $n = -4$. What happens?

Referring to Fig. 1-6, let us assume that we have already proved the following: (1) When two lines intersect, the pairs of opposite angles formed—such as the pair 1 and 3 and the pair 2 and 4 (these are called *vertical angles*)—have equal measures; (2) when two parallel lines are intersected by a third line, the pairs of angles situated as 4 and 6 (between the parallel lines and on opposite sides of the line) have equal measures, which means that 3 and 5 also have equal measures. We prove these relationships in a later chapter. The point is that facts (1) and (2) are sufficient to prove that the measures of angles 1 and 7 are equal, since the measure of 1 equals the measure of 3, the measure of 3 equals that of 5, and the

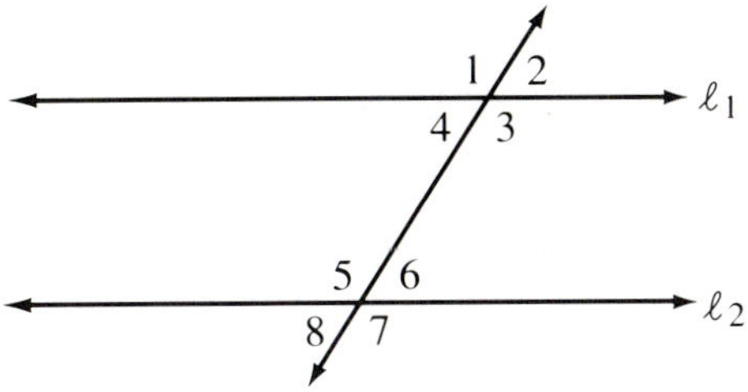

Figure 1-6

measure of 5 equals that of 7. This is an example of deductive reasoning. We knew only part of the facts but deduced the others from that which was known. Now we are able to use these facts to prove others.

Basic Connectives

Before we continue, let us establish some ground rules for our work. Our study of geometry is based upon *statements* about things such as points and lines, together with drawings of various combinations of these points and lines. We need to agree upon certain word meanings in order to be understood. Certain simple statements will be made, such as "A is a square" or "A is a rectangle," and then we will combine these statements to form more complicated ones using the connectives *and, or,* and *if . . . , then. . . .* For example, we might write "A is a square and A is a rectangle"; "A is a square or A is a rectangle"; and "If A is a square, then A is a rectangle."

We will also call the disclaimer *it is false that . . .* a connective, even though it does not connect two statements but rather is written before a given statement.

Truth and Falsity of Statements

In working with statements we will be concerned with whether they are true or false. If the letters P and Q stand for statements, then let us agree on the following points.

1. The connected statement "*P and Q*" is true only when P and Q are both true and is false otherwise. For example, "This page is shaped like a rectangle and the page you are reading is page 6" is true. "This page is shaped like a rectangle and the page you are reading is page 153" is false. "The page you are reading is page 153 and this book weighs 50 pounds" is also false. In formal logic the statement "P and Q" is called the *conjunction* of P and Q.
2. The connected statement "*P or Q*" is true whenever either P, or Q, or both P and Q are true. It is false only when P and Q are both false. Thus "This page is shaped like a rectangle or the page you are reading is page 153" is true, and so is "This page is shaped like a rectangle or the page you are reading is page 6." The statement "The page you are reading is page 153 or this book weighs 50 pounds" is false. Note that we do consider the statement "*P or Q*" as true when *both* P and Q are true. Some people use the word "or" to mean "either . . . or, but not *both*," as in "I will eat lunch here or I will go home to eat." To say "Nine is less than ten or nine is less than twenty" is a true statement, however, since both parts are true. It is this interpretation of the word *or* which we will use throughout the book. The statement "P or Q" is called the *disjunction* of P and Q.
3. If the statement "P" is true, then its *negation*, "*not* (P)" or "It is false that P" is false. If "P" is *false*, however, then "not (P)" is true. For example,

the statement "It is false that this page is shaped like a rectangle" is false. The statement "It is false that the page you are reading is page 153" is true.

Implications and Related Statements

The fourth connective to be used, perhaps most frequently in the study of mathematics, is the *implication* "If P, then Q." In this statement we will call P the *hypothesis* and Q the *conclusion*. An implication is considered to be false when the hypothesis P is true and the conclusion Q is false, but it is true under every other condition of P and Q. Other ways in which we will find implications expressed are "If P, Q," "Q if P," and "P only if Q."

Consider the statement "If you buy a Maxicar, you will average better than 30 miles per gallon." Suppose that you do buy a Maxicar. If it averages better than 30 miles per gallon, then you would consider the statement to be true. If it does not average better than 30 miles per gallon, the statement would be false. Now if you do *not* buy a Maxicar, it makes no difference whether it averages better than 30 miles per gallon or not, since you haven't put the statement to the test. Note that the only condition that shows the statement to be false is the second one, in which P is true and Q is false. Of course the implications we will be most interested in now will be mathematical ones, but the study of *logic* can also be applied to reasoning in everyday life.

For convenience let us use the symbolism $P \rightarrow Q$ to mean "If P, then Q." We can then write three additional statements related to the original implication:

1. $P \rightarrow Q$, the implication
2. $Q \rightarrow P$, the *converse* of the implication $P \rightarrow Q$
3. not $(Q) \rightarrow$ not (P), the *contrapositive* of $P \rightarrow Q$
4. not $(P) \rightarrow$ not (Q), the *inverse* of $P \rightarrow Q$

Truth Tables

If two mathematical statements are simultaneously true or false—that is, if statement X is true only when Y is true and if X is false only when Y is false—then X and Y are said to be *equivalent* statements. To illustrate this let us use what is known as a *truth table* for $P \rightarrow Q$. On the left we write the "truth value" of the statements P and Q; on the right, the truth value for the implication $P \rightarrow Q$, as defined above.

P	Q	$P \rightarrow Q$
T	T	T
T	F	F
F	T	T
F	F	T

The truth table for the contrapositive is given below. Remember that when P is true, not (P) is false and when P is false, not (P) is true.

P	Q	not (Q)	not (P)	not (Q) $\rightarrow$ not (P)
T	T	F	F	T
T	F	T	F	F
F	T	F	T	T
F	F	T	T	T

In the contrapositive the hypothesis is not (Q) and the conclusion is not (P), so the only case for which this implication is false is found in the second line, where we see that not (Q) is true and not (P) is false. Comparing this with the truth table for the original implication, we see that the two are equivalent.

The truth table for $Q \rightarrow P$ is:

P	Q	$Q \rightarrow P$
T	T	T
T	F	T
F	T	F
F	F	T

Here we see that when P is true and Q is false the implication $P \rightarrow Q$ is false but the converse is true, whereas when P is false and Q is true the implication is true and the converse is false. Thus an implication and its converse are not equivalent statements.

If an implication $P \rightarrow Q$ and its converse $Q \rightarrow P$ are both true, then we can say "P if and only if Q" and use the symbol $P \leftrightarrow Q$: for example, "Two distinct lines are parallel if and only if they do not intersect."

A logical statement that is always true is called a *law of logic*. Consider the statement "$P \rightarrow$ (P or Q)." The truth table for this statement is as follows.

P	Q	P or Q	$P \rightarrow$ (P or Q)
T	T	T	T
T	F	T	T
F	T	T	T
F	F	F	T

This law is sometimes known as the *Law of Addition*.

Examples

1. Write the converse, inverse and contrapositive of the implication "If an engine has four cylinders, then it needs four sparkplugs."

Solution Converse: If an engine needs four sparkplugs, then it has four cylinders. Inverse: If an engine does not have four cylinders, then it does not need four sparkplugs.

Contrapositive: If an engine does not need four sparkplugs, then it does not have four cylinders.

2. Write the truth table for the statement "*P* and *Q*."

Solution Write the first two columns showing the possible truth values of *P* as related to *Q*. The third column follows because "*P* and *Q*" is true only when both are true, according to our definition.

P	*Q*	*P* and *Q*
T	T	T
T	F	F
F	T	F
F	F	F

3. By means of a truth table show that the statement "not (*P* and *Q*)" is equivalent to "not (*P*) or not (*Q*)."

Solution Again, write the first two columns of truth values for *P* and *Q*. The column headed by (1) is the truth table found in the previous example. Each value in the next column (2) is the opposite of the corresponding value in (1), since one is the negation of the other. Columns (3) and (4) are the negations of the original truth values for *P* and *Q*. Column (5) is the disjunction of the values in columns (3) and (4), since the statement is false only when both are false. Comparing columns (2) and (5), we see that they are identical, hence the two given statements are equivalent.

		(1)	(2)	(3)	(4)	(5)
P	*Q*	*P* and *Q*	not (*P* and *Q*)	not (*P*)	not (*Q*)	not (*P*) or not (*Q*)
T	T	T	F	F	F	F
T	F	F	T	F	T	T
F	T	F	T	T	F	T
F	F	F	T	T	T	T
			↑			↑

◁

Exercise Set 1.2

1. Write the converse, inverse, and contrapositive of each of the following implications:
   (a) If the lamp is burning, then the current is on.
   (b) If $X = 3$, then $X^2 = 9$.
   (c) If you don't use Fluordent, then you will have cavities.
2. Write the truth table for the statement "*P* or *Q*."
3. By means of a truth table show that the statement "not (*P*) or *Q*" is equivalent to "$P \to Q$."
4. By means of a truth table show that "not (*P* or *Q*)" is equivalent to "not (*P*) and not (*Q*)."

5. By means of a truth table show that "not $(P \rightarrow Q)$" is equivalent to "P and not (Q)."
6. Are an implication and its inverse equivalent? Are the converse and inverse of a given implication equivalent? Use truth tables to decide.
7. If an implication $P \rightarrow Q$ and its converse $Q \rightarrow P$ are both true, then we can say "P if and only if Q," and use the symbol $P \leftrightarrow Q$. Complete the following truth table:

P	Q	$P \rightarrow Q$	$Q \rightarrow P$	$(P \rightarrow Q)$ and $(Q \rightarrow P)$

8. Write an original true implication for which the converse is false.
9. Write an original true implication for which the converse is also true.
10. By means of a truth table show that $[(P \rightarrow Q)$ and $P] \rightarrow Q$ is a law of logic. (This is known as the Law of Detachment.)
11. Use the set of truth values below to show that $[(P \rightarrow Q)$ and $(Q \rightarrow R)] \rightarrow (P \rightarrow R)$ is a law of logic. (This is the Law of Syllogism.)

P	Q	R
T	T	T
T	T	F
T	F	T
T	F	F
F	T	T
F	T	F
F	F	T
F	F	F

1-3 METHODS OF PROOF

The Law of Detachment, which states "If $P \rightarrow Q$ is true and P is true, then it follows that Q is true" (refer to Exercise 10 in Exercise Set 1.2), will be one of the laws of logic most frequently used in this text. Suppose, for example, we later find the following: "If two sides of a triangle are equal in length, then the triangle is isosceles." Now if we should observe that in a triangle RST, sides RS, ST, and TR are 2 in., 3.5 in., and 2 in. in length, respectively, it follows that triangle RST must be isosceles. Any property that an isosceles triangle has must also be a property of triangle RST.

The Law of Syllogism, which states "If $P \rightarrow Q$ is true and $Q \rightarrow R$ is true, then $P \rightarrow R$ is true" (refer to Exercise 11 in Exercise set 1.2), together with the Law of Detachment, enable us to put together some extremely complex arguments. If we know that "If a triangle is isosceles, then it has two angles of the same size," then triangle RST of the previous paragraph must have two angles of the same size.

The Law of Disjunctive Syllogism states "If (P or Q) is true and we find that P is false, then Q is true." Suppose the student-body president says "The dance will be held in the gymnasium or the cafeteria," but the gymnasium floor must be resurfaced on the night of the dance. Then the dance will be held in the cafeteria.

Exercise Set 1.3

1. Show by means of a truth table that $[(P \text{ or } Q) \text{ and not } (P)] \rightarrow Q$ is a law of logic.
2. Suppose a program says "The artist will play a tune by George Gershwin or a tune by Cole Porter." The artist does not play a tune by Cole Porter. What composer's work does the artist play? What law of logic did you use to reach your conclusion?
3. Consider the following: "If Max mows the lawn, he'll take Cathy to the movies. This is a true statement, because Max's father said he could use his car tonight if Max mows the lawn. Max said he'd take Cathy to the movies tonight if he could borrow his father's car." What law of logic was used in this argument? Identify the different parts of the argument by letters—that is, P is the statement "Max mows the lawn," and so on.
4. Show that $(P \text{ and } Q) \rightarrow P$ is a law of logic by means of truth tables. This is sometimes known as the Law of Simplification.
5. Consider the following argument: "A card is either red or black. If the card is red, then it is a diamond or a heart. If the card is black, then it is a club or a spade. This card is not black and it is not a heart." What is this card, and what laws of logic did you use to deduce the answer?
6. Recall that an implication and its contrapositive are equivalent statements, and consider the following: "If you don't have a permit, then you can't park in this lot. If you have a permit, you must be a student at this school. Louise's car is parked in the lot." What may we conclude about Louise, and what laws of logic are used to reach the conclusion?

2

Points, Lines, and Planes

In times gone by, mathematicians tried to define the terms point *and* line. *Pythagoras defined a point as "a monad having position," Plato, as "the beginning of a line," and Euclid, as "that which has no part." Some early definitions of a line were "the flux of a point" or "length without breadth."*

2-1 POINTS AND LINES

The study of geometry has much to do with the ideas of points and lines. Rather than try to *define* these terms, we will leave them *undefined* and base later definitions on them. There are many objects in the world around us which represent the idea of a point as we intend to use it; a pencil mark on a piece of paper, a raindrop on a window, or a position at which a spider crosses its web are only a few examples. A line, or at least part of it, might be represented by the corner of a building where two walls meet, or by a mark made on paper when a pencil is drawn along a straightedge. To show that a line is meant to continue without limit in each direction, we place arrowheads on each end, as in Fig. 2-1. In the future, whenever we mention *line* we will mean a *straight* line unless specified otherwise. There are figures such as curved lines and "broken lines," as shown in Fig. 2-2, but when we want to talk about these we will identify them.

Figure 2-1

Figure 2-2

Our basic terms point and line are the first of the *geometric figures* that we will encounter.

Definition 2-1 **A geometric figure** is a nonempty set of points.

Postulates in Geometry

A *postulate* is an assumption that is considered to be true without proof. Postulates are generally associated with geometry, although not necessarily so. Sometimes a postulate that we state will be obvious, sometimes not. The discussion in the next paragraph will lead to our first postulate.

We can always draw a representation of part of a line by locating two points on a piece of paper, laying a straightedge between them, then connecting the points. Now suppose we locate two points A and B, draw a line, and then draw another line through the same points A and B. Of course the same line appears, if we have truly used the same two points.

Postulate 2-1(a) For any two distinct points there is exactly one line containing them.

A related postulate is the following.

Postulate 2-1(b) Every line contains at least two distinct points.

Meaning of the Word Between

As we continue our discussion we will frequently refer to a point B that is "between" two points A and C. We need to agree on exactly what that means to us because outside of mathematics the word can be used differently. We might, for example, say that the black ball is between the red and the white balls. In the situations in the top row of Fig. 2-3 it is easy to decide the truth or falsity of the statement "B lies between R and W." In the bottom row it is somewhat more difficult to decide.

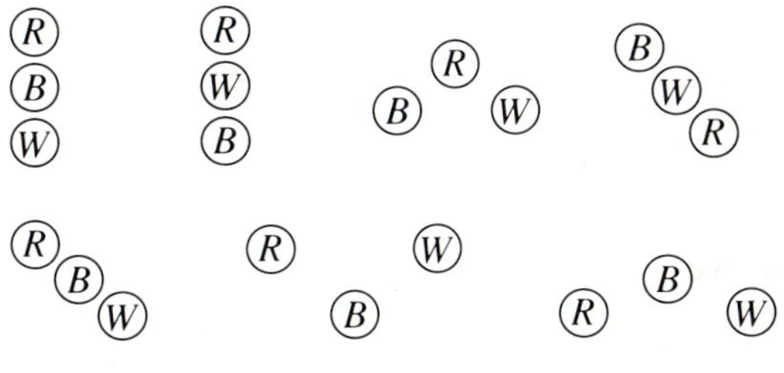

Figure 2-3

We choose not to define between explicitly; however, if we say that point B is between points A and C (B is distinct from A and C), we may visualize a relationship similar to one of those shown in Fig. 2-4.

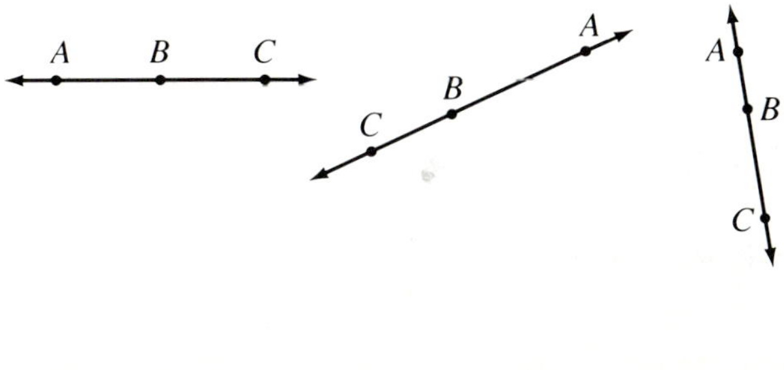

Figure 2-4

In written form *A-B-C* indicates that *B* lies on the line *AC* and is read, "*B* lies between *A* and *C*." Exercise Set 2.1 will help us to develop common intuitive feelings for these terms.

Exercise Set 2.1

1. List five physical representations of
  (a) points (bullet hole, town on map, etc.)
  (b) lines or parts of lines (edge of paper, phone line, etc.)

2. What is the difference between a physical representation of a point and a geometric point?

3. How many points are there on a line?

4. Draw two distinct lines ℓ_1 and ℓ_2 intersecting at a point *P*.

5. On a line ℓ locate a point *S* and two points *R* and *T* on the same side of *S*.

6. On a line ℓ locate a point *S* and two points *R* and *T* on opposite sides of *S*.

7. Draw a figure consisting of three distinct points *A*, *B*, and *C*.
  (a) Draw a point *D* between *A* and *B*.
  (b) Draw a point *E* between *B* and *C*.
  (c) Draw a point *F* between *C* and *A*.

8. Draw a figure consisting of four distinct points *A*, *B*, *C*, and *D*.
  (a) What is the minimum number of lines that can be determined?
  (b) What is the maximum number of lines that can be determined?

9. **(a)** How many distinct lines are determined by one point?
  (b) How many distinct lines can be drawn through one point?
  (c) How many lines are determined by two distinct points?
  (d) How many lines are determined by three distinct points?
  (e) Does it make any difference in part (d) how the points are situated?

10. If no three of the points are collinear (see page 32)
  (a) how many lines are determined by four points (taken two at a time)?
  (b) how many lines are determined by five points?
  ***(c)** how many lines are determined by *n* points?

11. In each of the following, draw a figure with three points *A*, *B*, and *C* such that
  (a) *A-B-C* **(b)** *B-A-C* **(c)** *A-C-B*

12. Draw points *R* and *S* about two inches apart.
  (a) Draw *T* such that *R-T-S*.
  (b) Draw *U* such that *T-U-S*. Is *T* between *R* and *U*?
  (c) Draw *V* such that *R-V-T*. Is *V* between *R* and *S*?
  (d) How many points do we now have between *R* and *S*? How many more points can we place between *R* and *S*?

13. Draw two distinct points *A* and *B*. Now draw the figure that will contain *A* and *B* and all the points between them.

14. Postulate 2-1(a) has many practical applications, such as in sighting a rifle and in setting fence posts along a straight line. To set the fence posts we need only to locate the end

* The more difficult problems are indicated by an asterisk.

posts, then sight between them to check the alignment of the others. Give at least one additional example of a practical application of this postulate.

15. A news photograph might be considered to be a geometric figure, since it is made up of thousands of "points." Take a close look at a photograph from your local newspaper. A television picture is composed of a sequence of "lines" and "points." Consult a reference book on television to see how the television image is produced.

2-2 SEGMENTS, RAYS, AND HALF-LINES

Segments

A line is infinite in extent and we usually deal with only a limited portion of it. Our discussion of betweenness and the previous exercise set suggest Post. 2-2.

Postulate 2-2 (*Betweenness Postulate*) Given any two distinct points A and B, there exists at least one point between them.

If we call the point whose existence is guaranteed by Post. 2-2 point C, then of course new pairs of points are determined. The new pairs, AC and CB, have at least one point between each of them. In Fig. 2-5, we see that C lies between A and B, there is a point D between A and C and point E is between C and B.

Figure 2-5

If the process were to be continued, there would appear to be an infinite number of points between any two distinct points A and B. We will call this figure composed of points A and B and *all* the points between them a *segment*.

Definition 2-2 A **segment** is the geometric figure containing two distinct points and the set of all points between them.

There are several ways to indicate that we are talking about segments. We can denote the segments with endpoints A and B as segment AB or simply $\overline{AB}$. We can also use a combination of set notation and the betweenness notation A-X-B, in which X represents the set of points between A and B. Using the latter we might even redefine segment AB as

$$\overline{AB} = \{A, B\} \cup \{X | A\text{-}X\text{-}B\},$$

or "segment AB is the set of points A and B union the set of points X such that X is between A and B." (See Appendix II for an explanation if this notation is not familiar to you.)

A visual representation of segment AB is simply as shown in Fig. 2-6.

Figure 2-6

Exercise Set 2.2

1. List five physical representations of a line segment. Examples: Edge of book, cutting edge of wood plane, etc.
2. **(a)** Draw two points R and S and the segment determined by them.
   (b) Draw a point T between R and S. List the segments formed.
3. Draw a segment AB with C between A and B.
   (a) Draw point D between A and C, then draw E between A and D.
   (b) Draw point F between A and E. (How long could this process continue, theoretically?)
   (c) How many segments were determined in parts (a) and (b)? (Count carefully, since $\overline{DB}$ is a segment too.)
   (d) Can you place a point next to A? That is, can you place a point X next to A such that no other point can be placed between them?
   (e) If the point A is removed from $\overline{AB}$ will the set of points still have two endpoints? Why?
4. Draw a segment AB. Then draw the following.
   (a) Draw a point X_1 such that $A\text{-}B\text{-}X_1$.
   (b) Draw X_2 such that $B\text{-}X_1\text{-}X_2$, then X_3 such that $X_1\text{-}X_2\text{-}X_3$.
   (c) How long could this operation be continued, theoretically?
   (d) Draw a representation of the set consisting of $\overline{AB}$ and the union of all possible X_n's described above. What is the resulting figure?
5. Draw a segment AB. Then draw the following.
   (a) Draw a point Y_1 such that $Y_1\text{-}A\text{-}B$.
   (b) Draw Y_2 such that $Y_2\text{-}Y_1\text{-}A$, then Y_3 such that $Y_3\text{-}Y_2\text{-}Y_1$.
   (c) Continue the above process using three more selected points. How long could this operation be continued, theoretically?
   (d) Draw a representation of the set consisting of $\overline{AB}$ and the union of all possible Y_n's described above. What is the resulting figure?
6. Redraw the figures resulting in 4(d) and 5(d) to show

$$\overline{AB} \cup \{X|A\text{-}B\text{-}X\} \cup \{Y|Y\text{-}A\text{-}B\}$$

Rays

From the previous exercise we have seen that for each distinct pair of points R and S, there is a point T that lies "beyond" S, as illustrated in Fig. 2-7. The point T is not in segment RS. This suggests the next postulate.

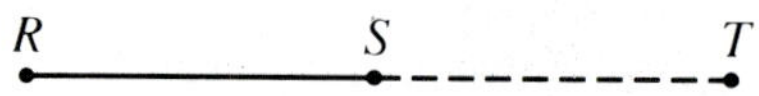

Figure 2-7

Postulate 2-3 Given any two distinct points R and S, there exists a point T such that $R\text{-}S\text{-}T$.

Postulate 2-3 extends our notion of segment somewhat, since the point T determines a segment RT, and in fact, by Post. 2-3, there is another point X_1, beyond T, and X_2 beyond X_1, and X_n beyond X_{n-1}, so that we have an infinite set of points beyond S. When we combine the segment with this new infinite set of points, the result is a figure called a *ray*.

Definition 2-3 A **ray** is the figure formed by a segment and the set of all points beyond exactly one endpoint, but not both. The ray with **endpoint** R and the set of points X beyond S will be called ray RS and denoted by $\overrightarrow{RS}$. $\overline{RS}$ and the set of points X beyond R will be called ray SR and written $\overrightarrow{SR}$.

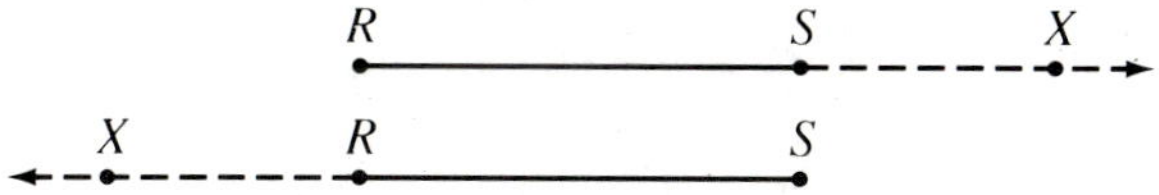

Figure 2-8

Thus $\overrightarrow{RS} = \overline{RS} \cup \{X|R\text{-}S\text{-}X\}$ and $\overrightarrow{SR} = \overline{RS} \cup \{X|X\text{-}R\text{-}S\}$, as in Fig. 2-8.

Notation for Line

Since Post. 2-3 permits us to extend a segment in either direction, the segment RS could produce both $\overrightarrow{RS}$ and $\overrightarrow{SR}$. The union of these two figures is a line, and we can denote line RS by $\overleftrightarrow{RS}$, as in Fig. 2-9.

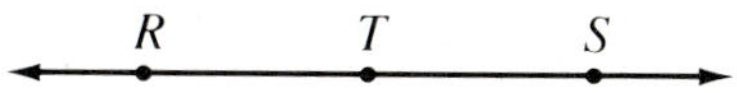

Figure 2-9

Definition 2-4 If T is a point on $\overleftrightarrow{RS}$ such that $R\text{-}T\text{-}S$, then $\overrightarrow{TS}$ and $\overrightarrow{TR}$ are called **opposite rays.** (Note that $\overrightarrow{TS} \cup \overrightarrow{TR} = \overleftrightarrow{RS}$.)

Half-lines

It is sometimes convenient to refer to a geometric figure that is similar to a ray except that it lacks the endpoint.

Definition 2-5 A **half-line** is a ray excluding its endpoint.

If T is on $\overleftrightarrow{RS}$ and $R\text{-}T\text{-}S$, then half lines may be denoted by $\overset{\circ}{\overrightarrow{TS}}$ or $\overset{\circ}{\overrightarrow{TR}}$, as in Fig. 2-10.

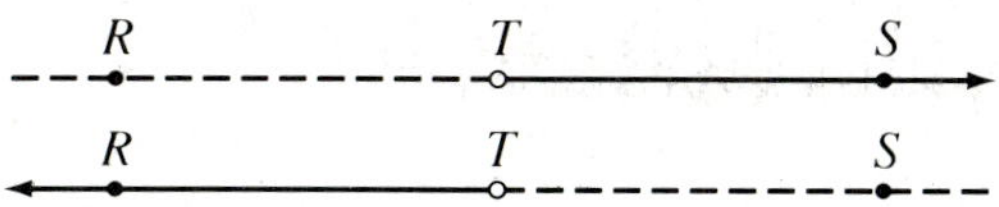

Figure 2-10

Exercise Set 2.3

1. Draw three points A, B, and C not on the same line.
   (a) Draw $\overrightarrow{AB}$, $\overrightarrow{BC}$, and $\overrightarrow{CA}$.
   (b) Using another three points A, B, and C situated as in part (a), draw $\overrightarrow{BA}$, $\overrightarrow{CB}$, and $\overrightarrow{AC}$.
2. On a line AB draw a point C distinct from A and B.
   (a) List the distinct line segments that have been determined.
   (b) List the rays that have been determined.

3. Is $\overline{AB}$ the same set of points as $\overline{BA}$?
Is $\overrightarrow{AB}$ the same set of points as $\overrightarrow{BA}$?
Explain.

4. **(a)** What geometric figures would the notation $\overset{\circ\quad}{\overline{AB}}$ and $\overset{\quad\circ}{\overline{RS}}$ suggest? Describe and then draw them.
   (b) What geometric figure would the notation $\overset{\circ\;\;\circ}{\overline{PQ}}$ suggest? Describe and then draw it.
   (c) The figures in part (a) are sometimes called *half-open intervals,* whereas those in (b) are called *open intervals*. What would a line segment be likely to be called in this sense?

5. **(a)** Is the union of two half lines necessarily a line?
   (b) *Can* the union be a line? (Illustrate.)

6. **(a)** Is the union of two rays necessarily a line?
   (b) *Can* the union be a line? (Illustrate.)

7. Draw a line *AB* and a point *C* such that *A-B-C*. Describe the following figures or sets of points in a simpler way. (Example: $\overrightarrow{AB} \cup \overrightarrow{BC} = \overrightarrow{AB}$.)

   (a) $\overrightarrow{AB} \cap \overrightarrow{BC}$ **(b)** $\overrightarrow{AB} \cup \overrightarrow{AB}$
   (c) $\overset{\circ}{\overrightarrow{BC}} \cup \overset{\circ}{\overrightarrow{CB}}$ **(d)** $\overset{\circ}{\overrightarrow{BC}} \cup \overrightarrow{CB}$
   (e) $\overline{AB} \cap \overline{BC}$ **(f)** $\overset{\circ}{\overrightarrow{BC}} \cap \overset{\circ}{\overrightarrow{BA}}$
   (g) $\overrightarrow{AC} \cap \overrightarrow{CA}$ **(h)** $A \cup \overset{\circ}{\overrightarrow{AB}}$

8. Draw a line *RS* and a point *T* such that *R-T-S*. Sketch the set of points indicated below. Example: $\overrightarrow{RS} \cap \overset{\circ}{\overrightarrow{TR}}$. Solution:

   (a) $\overrightarrow{RS} \cap \overset{\circ}{\overrightarrow{SR}}$ **(b)** $R \cup \overset{\circ}{\overrightarrow{TS}}$
   (c) $\overset{\circ}{\overrightarrow{TS}} \cap \overset{\circ}{\overrightarrow{TR}}$ **(d)** $\overleftrightarrow{RS} \cap \overset{\circ}{\overrightarrow{RT}}$
   (e) $\overset{\circ}{\overrightarrow{RT}} \cap \overset{\circ}{\overrightarrow{ST}}$

When you can measure what you are talking about and express it in numbers, you know something about it.

LORD KELVIN (1824–1907) British physicist.

2-3 THE REAL NUMBER LINE

From the beginning of our study of arithmetic we have used the set of *natural numbers* $N = \{1, 2, 3, 4, 5, \ldots\}$. This is an infinite set, since there is no "largest" natural number. Consider a horizontal line $\overleftrightarrow{OU}$, such that *U* lies to the right of *O*. If we associate the number 1 with the point *U*, we may then lay off lengths equal to the length of $\overline{OU}$ to the right of *U* so that each endpoint is associated with the successive natural numbers in the set. See Fig. 2-11.

If we lay off lengths equal to the length of $\overline{OU}$ to the left of *U*, we may identify the endpoints by -1, -2, -3, etc., and if we associate the number zero

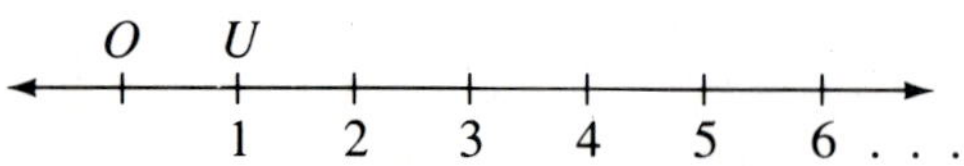

Figure 2-11

with the point O of $\overline{OU}$ we have a representation of set $J = \{0, \pm 1, \pm 2, \pm 3, \ldots\}$ which is known as the set of *integers,* as illustrated in Fig. 2-12. The union of $\{0\}$ and N, or $\{0, 1, 2, 3, 4, 5, \ldots\}$ is usually called the set of *whole numbers.*

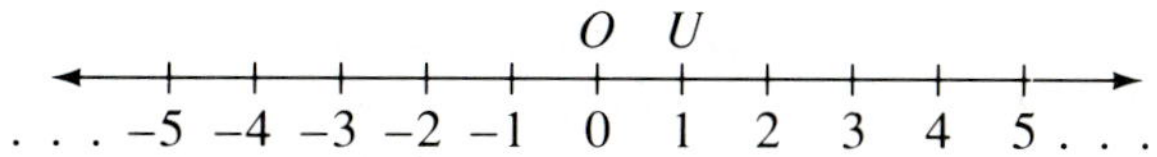

Figure 2-12

We have seen that to each integer we may correspond a point on a line. Postulate 2-2 on betweenness tells us that there is a point between each pair of points, and in this way we can indicate correspondence of the set of *rational numbers* to points on a line. Rational numbers are those numbers which can be expressed in the form p/q, where p and q are integers but q is not equal to zero. Thus $6 = \frac{6}{1}$, $-4 = \frac{-4}{1}$, $\frac{3}{5}$, $\frac{-3}{4}$, $\frac{10}{3}$, $\frac{12}{-5}$, and $5.6 = \frac{56}{10}$ are rational numbers, all of which can be located on the number line. See Fig. 2-13. If a and b are different rational numbers, then there are two points A and B on the number line that correspond to them. There is always a number $c = (a + b)/2$ between a and b and a point C that corresponds to c. Considering A and C, then, we can find point D such that $d = (a + c)/2$ (Fig. 2-14). Since this process can be continued indefinitely, this shows that there are infinitely many rational numbers between a and b.

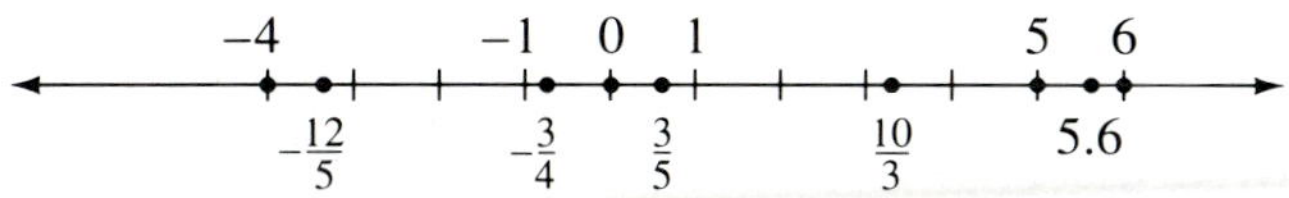

Figure 2-13

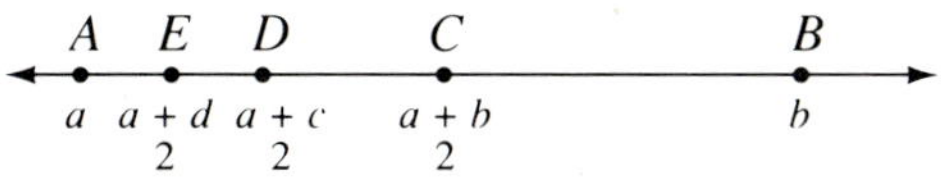

Figure 2-14

In Chapter 5 we will see how to locate any rational number on the number line. Although we have considered a large number of points on the line we have by no means exhausted all of the possibilities. The number $\sqrt{2}$ cannot be written in the form p/q (see Appendix III), yet there is a point that corresponds to it, as well as points that correspond to $\sqrt{3}$, $\sqrt{5}$, π, etc. These numbers that cannot be expressed in the form p/q, where p and q are integers and $q \neq 0$, are the *irrational numbers.* The set of irrational numbers together with the rational numbers form the set R of *real numbers.*

Now the correspondence between the set of real numbers and the set of points on a line is complete; that is, for every point on a line there is a real number and for every real number there is a point on the line. This line will sometimes be referred to as the *real number line.* To formalize this we state the following postulate.

Postulate 2-4 (*Ruler Postulate*) The points on a line can be associated with the set of real numbers so that

(a) to every point there corresponds a unique real number, and
(b) to every real number there corresponds one and only one point.

This in turn leads to a definition that will be of considerable use to us as we progress.

Definition 2-6 The number corresponding to each point on a line is called the **coordinate** of the point, and the correspondence is called a **coordinate system.**

In Fig. 2-15 the coordinate of A is -3, of C is 1, of E is 3.5, etc. This is the usual arrangement of numbers on the real number line, although it could be reversed, or the line could be vertical, or it could be neither horizontal nor vertical. Following the usual custom, however, we can agree that for real numbers x and y with x less than y, point X lies to the left of point Y (Fig. 2-16). Symbolically we denote this as $x < y$ or $y > x$.

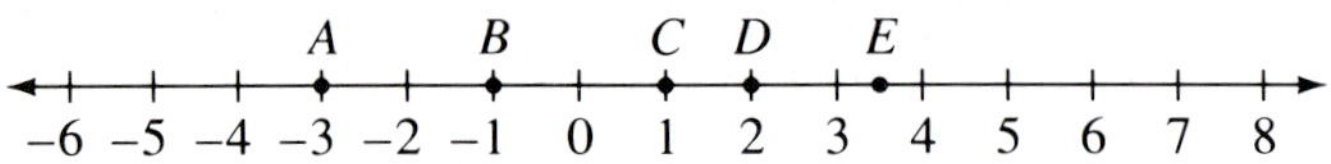

Figure 2-15

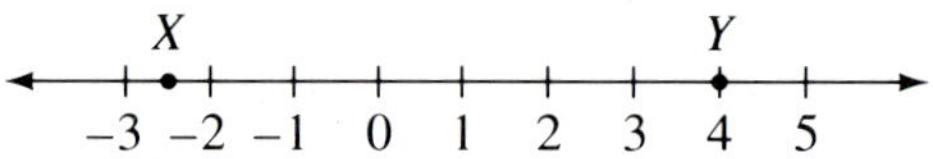

Figure 2-16

Thus we can see that every negative number is *less than* any nonnegative number, since each negative number lies to the left of each nonnegative number. For example, $-2567 < 0$. You have more than likely met *inequality* relations such as these previously, but if further study or review is needed, refer to Appendix I.

In order to have a common method of drawing points on a line, we will agree that $\{x|a < x < b\}$ refers to the set of numbers between a and b but not including a or b, and the graph will look like Fig. 2-17.

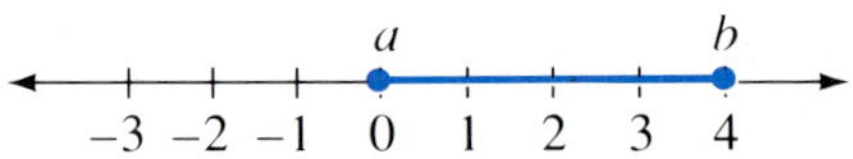

Figure 2-17

If a and b are to be included we will write $\{x|a \le x \le b\}$, and the endpoints will be filled in on the graph, as in Fig. 2-18.

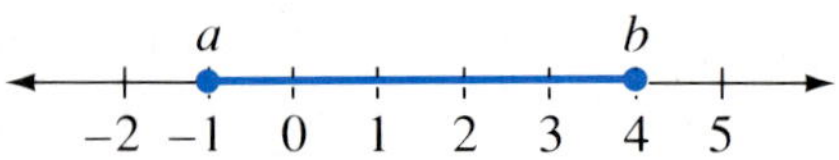

Figure 2-18

Another convention which is generally agreed upon is that *every positive number has exactly one positive square root*. If we write $x = \sqrt{a}$, $(a > 0)$, we are indicating that x is a positive number, the positive square root of a. The following statements are true: $5 = \sqrt{25}$, $\frac{2}{3} = \sqrt{\frac{4}{9}}$, $\sqrt{169} = 13$. These are false: $-5 = \sqrt{25}$, $-6 = \sqrt{36}$, $\sqrt{16} = \pm 4$.

Exercise Set 2.4

1. Arrange the points corresponding to each of the following sets of numbers on a number line in their appropriate order and label them with their coordinates.
   (a) $\{3, 7, 5, -2\}$
   (b) $\{-1.7, -.03, +2.1, .43\}$
   (c) $\{\frac{7}{8}, 1\frac{1}{4}, \frac{2}{3}, -1\frac{1}{5}\}$

2. Locate the points corresponding to the following pairs of numbers on a number line. Then replace each question mark with the proper symbol, $<$ or $>$.
   (a) 3 ? 5 **(b)** −2 ? 7
   (c) −1 ? −4 **(d)** 6 ? 2
   (e) −4 ? −2

3. Find a replacement for each variable which will make each of the following a true statement; then plot the corresponding points on a number line:
   (a) $r > s$ (both r and s positive)
   (b) $r > s$ (both r and s negative)
   (c) $r > s$ (r positive and s negative)
   (d) $r < s$ (both r and s negative)
   (e) $r < s$ (both r and s positive)

4. If a, b, and c are positive numbers and $a > b$, which of the following statements are always true? (Refer to Appendix I if necessary. Give a counterexample for those that are not always true.)
   (a) $b < c$ **(b)** $a - b > 0$
   (c) $a - 4 < b - 4$ **(d)** $a/2 < c$
   (e) $a/b > 1$

5. How many coordinates of the form a/b are there between 0 and 1 if $a \in \{1, 2, 3, 4, \ldots, n, \ldots\}$ and $b = a + 1$?
   List the first ten elements of the set of coordinates.

6. Locate two points B and C on a number line and associate the numbers −5 with B and 2 with C. Use the same scale.
   (a) What number would correspond to C if the coordinate of B were changed to 0?
   (b) What would B become if the C coordinate were 0?
   (c) If the C coordinate is $2\frac{3}{4}$, what is the B coordinate?
   (d) If the B coordinate is $-3\frac{2}{3}$, what is the C coordinate?
   (e) What would the B coordinate become if the C coordinate were 4.2?

7. If a, b, and c are coordinates of points on a line such that $c > a$ and $b < a$, which point is between the other two?

8. If x is an integer, draw each of the following over the part of the number line from −5 through +10. (Use a different number line for each part.)
   (a) $\{x|x > 2\}$ **(b)** $\{x|-4 < x < 3\}$
   (c) $\{x|5 \le x \le 8\}$ **(d)** $\{x|-3 < x < 2\}$

9. If x is a real number, draw the graph of each of the following over the part of the number line from −5 through +5. (Use a different number line for each part.)
   (a) $\{x|x < -1\}$ **(b)** $\{x|x \ge 0\}$
   (c) $\{x|-1 < x < 3\}$ **(d)** $\{x|0 \le x < 5\}$

10. Tell which point in each of the following pairs lies to the right of the other, and how much greater its coordinate is than that of the other.
(a) $x_1 = 3, x_2 = 5$
(b) $x_1 = -3, x_2 = 2$
(c) $x_1 = -4, x_2 = -8$
(d) $x_1 = 3, x_2 = 0$
(e) $x_1 = 3, x_2 = -4$

11. Locate on a number line the point corresponding to each of the following coordinates.
(a) $\sqrt{36}$
(b) $\sqrt{49}$
(c) $\sqrt{8}$
(d) $\sqrt{3}$
(e) $3\sqrt{2}$
(f) $2\sqrt{2}$
(g) $4\sqrt{3}$
(h) $\sqrt{64}$
(i) $\sqrt{12}$
(j) $\sqrt{48}$

12. How many coordinates of the form $(a + x_n)/2$ are there between two points A and B with coordinates $a = 2$ and $b = x_1 = 3$, if $x_2 = (2 + x_1)/2 = (2 + 3)/2 = 2\frac{1}{2}$, $x_3 = (2 + x_2)/2$, $x_4 = (2 + x_3)/2$, etc.? List the first five coordinates.

13. If x is a real number, draw the graph of each of the following over the section of the real number line from -5 through $+5$. (Use a different number line for each graph.)
(a) $\{x|x > 3\} \cup \{x|x < 0\}$
(b) $\{x|x > -1\} \cap \{x|x \leq 5\}$
(c) $\{x|-3 \leq x \leq 4\} \cup \{x|x = 5\}$
(d) $\{x|-4 \leq x \leq 2\} \cap \{x|-3 \leq x \leq 3\}$

In fact everything that can be known has number, for it is not possible to conceive of or to know anything that has not.
PHILALOUS OF CROTON (500 B.C.)
Philosopher

2-4 DISTANCE

When we consider two distinct points on a line, we can always associate the zero point of a set of coordinates with one of them in such a way that the second of the two points corresponds to a positive number. In Fig. 2-19, for example, we see that if B is to the right of A we have the usual arrangement. On the other hand, if B is to the left of A we could reverse the number line so that B is still positive, as in Fig. 2-20.

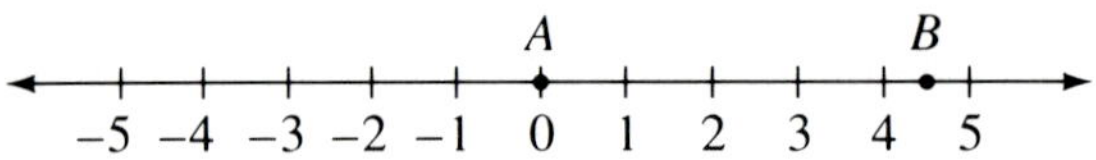

Figure 2-19

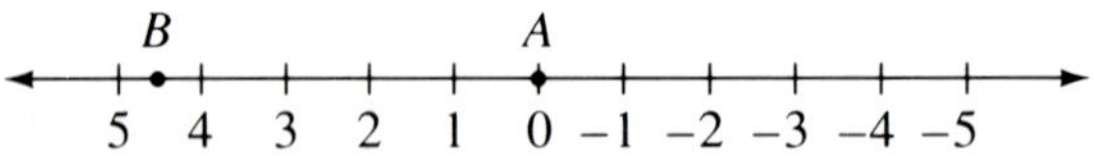

Figure 2-20

Postulate 2-5 (*Ruler Placement Postulate*) For any two points on a line, we may assign zero to one of them and a positive coordinate to the other.

Since the choice of a unit of measure is arbitrary, depending upon the situation and custom (distance between cities in miles or kilometers, not in inches or centimeters), we will refer merely to "units" of distance rather than to feet, inches, etc., except in some practical applications later on. Hence all our theorems will hold true for any given unit of measure.

Postulate 2-6 (*The Distance Postulate*) To every pair of distinct points we can assign a unique positive number.

Definition 2-7 The **distance** between two points is the positive number given by the Distance Postulate or zero if they are the same point. The distance between points A and B may be denoted as $m\overline{AB}$, read "measure of segment AB." The measure of a segment is called its **length.**

Absolute Value

When three distinct points are involved we can express the distances between them on the number line as follows:
If A-B-C (Fig. 2-21), we can let A have the coordinate 0 and write $m\overline{AB} = 3$, $m\overline{AC} = 8$.

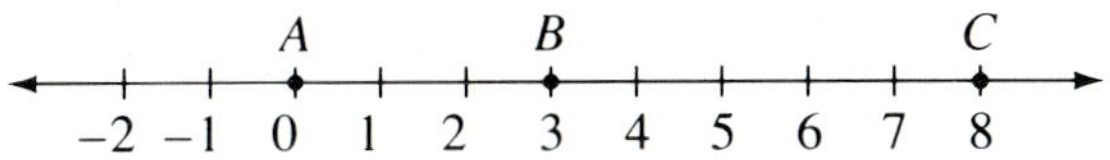

Figure 2-21

Using the same three points A, B and C with the zero coordinate at B (Fig. 2-22) we see that $m\overline{BC} = 5$, but we would need to renumber the line to find $m\overline{BA}$.

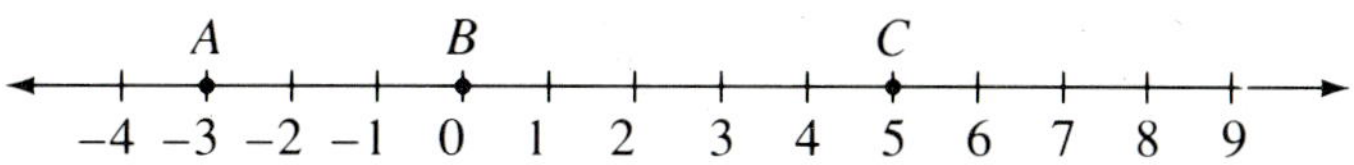

Figure 2-22

The distance from A to B is the same as the distance from B to A according to our definition. In order to avoid reversing the number line, we can denote the distance of any point from the origin (zero) by the *absolute value* of the coordinate of the point. Since $-(-3) = 3$, $-(-\frac{5}{2}) = \frac{5}{2}$, etc., we can recall the definition generally used in algebra.

Definition 2-8 Absolute value of a number:
For every real number a, $|a| = a$, if a is positive ($a > 0$).
$|a| = 0$, if a is zero ($a = 0$).
$|a| = -a$, if a is negative ($a < 0$).

Thus $m\overline{BC} = |5| = 5$ and $m\overline{BA} = |-3| = -(-3) = 3$. In Fig. 2-23, we see that $m\overline{RS} = |\frac{9}{4}| = \frac{9}{4}$, $m\overline{RQ} = |\pi| = \pi$, $m\overline{RT} = |-\frac{5}{2}| = -(-\frac{5}{2}) = \frac{5}{2}$, $m\overline{RU} = |5| = 5$, $m\overline{RV} = |-4| = -(-4) = 4$.

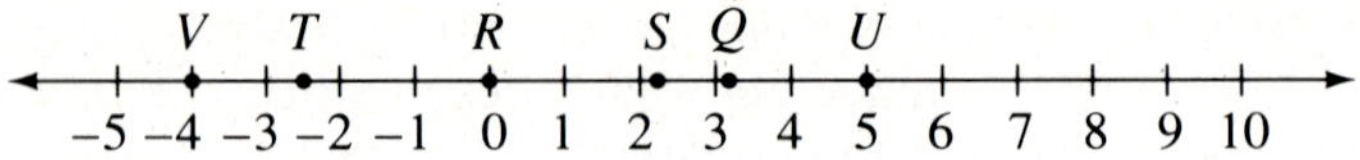

Figure 2-23

From Fig. 2-22 it can be determined that $m\overline{AC}$ is 8 units, since $m\overline{BA} = 3$, $m\overline{BC} = 5$, and $m\overline{AC} = m\overline{BA} + m\overline{BC}$. On the number line of Fig. 2-24 we can observe that

$m\overline{GJ}$ is from 1 to 4, or 3 units.
$m\overline{HK}$ is from 2 to 5, or 3 units.
$m\overline{JE}$ is from 4 to -1, and is 5 units.

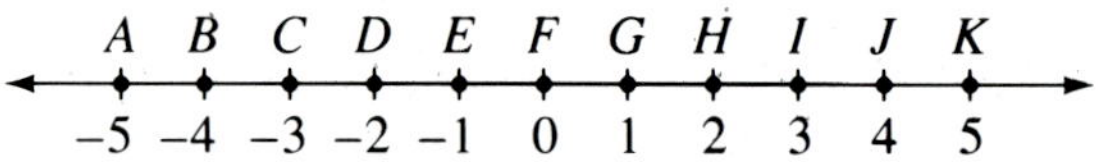

Figure 2-24

The distance between the points relates to the difference of the corresponding numbers—i.e., $3 - 1 = 2$ but $1 - 3 = -2$—so in order to be consistent with Def. 2-7, we can state the following.

Definition 2-9 The **distance between two points** is the absolute value of the difference between the numbers corresponding to the points. That is, $m\overline{AB} = |a - b| = |b - a|$.

Referring again to Fig. 2-24, we find $m\overline{BG} = |-4 - 1| = |-5| = 5$, $m\overline{HC} = |2 - (-3)| = |5| = 5$, and $m\overline{GJ} = |1 - 4| = |-3| = 3$.

Example

Given the following points on a number line, find the distance from

(a) A to D
(b) C to B
(c) E to D
(d) E to B
(e) B to A

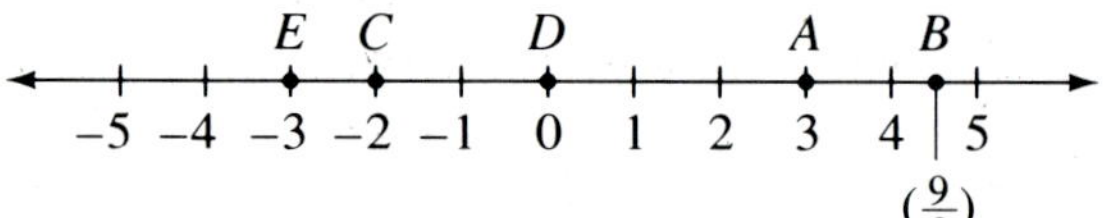

Solution By the definition of the distance between two points (Def. 2-9), we have

(a) $m\overline{AD} = |3 - 0| = |3| = 3$
(b) $m\overline{CB} = |-2 - 4\frac{1}{2}| = |-6\frac{1}{2}| = 6\frac{1}{2}$
(c) $m\overline{ED} = |-3 - 0| = |-3| = 3$
(d) $m\overline{EB} = |-3 - 4\frac{1}{2}| = |-7\frac{1}{2}| = 7\frac{1}{2}$
(e) $m\overline{BA} = |4\frac{1}{2} - 3| = |1\frac{1}{2}| = 1\frac{1}{2}$ ◁

Exercise Set 2.5

1. Which of the following statements are always true? (x, y, and n are real numbers.)

   (a) $|7| = 7$
   (b) $|-5| = 5$
   (c) $|4 - 7| = -3$
   (d) $|3 - 2| = 1$
   (e) $|5 - 14| = -(-9)$
   (f) $|x| = x$
   (g) $|-x| = x$
   (h) $|y^2| = y^2$
   (i) $|6 + n| = |n + 6|$
   (j) $|3 - x| = |x - 3|$

2. Find the distance from the origin to each point named on the line below.

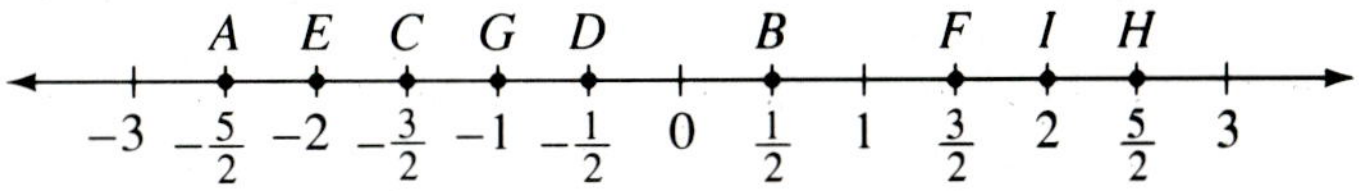

3. If x is a real number, draw each of the following sets of points on a separate number line:

   (a) $\{x|\ |x| = 3\}$
   (b) $\{x|\ |x| < 3\}$
   (c) $\{x|\ |x| > 3\}$
   (d) $\{x|\ |x| \leq 3\}$

4. Find the distance between two points with the following coordinates:

   (a) 0 and 7
   (b) 4 and 0
   (c) -3 and 0
   (d) 0 and $-4\frac{1}{2}$
   (e) -4 and -7
   (f) -6 and -2
   (g) -3 and 3
   (h) a and b
   (i) $(a + 1)$ and $-(a + 1)$
   (j) $(x + y)$ and $(x - y)$

5. Simplify each expression:

   (a) $|4 + 7|$
   (b) $|3 - 8|$
   (c) $|-3 + 5|$
   (d) $|-2 - 3|$
   (e) $|-4 - (-5)|$
   (f) $|(-n) - n|$
   (g) $|x - (-x)|$
   (h) $|(-a) - (-a)|$

6. Two different coordinate systems are on the same number line. The points A, B, and C are assigned values as follows: For the first system A is 4, C is 7. In the second system A is -3 and B is 2. What is the coordinate of B in the first system? Of C in the second system? Which point is between the other two?

7. Three points P, Q, and R lie on the same line but not necessarily in that order. If P is 8 units from Q and R is 12 units from Q, how many ways are there of arranging the points? (Illustrate.)

8. Consider the line AC with A-B-C, such that $m\overline{AB} = m\overline{BC} = 6$ and the coordinate of B is 3. Find the coordinates of A and C.

*9. *Prove*: If A, B, and C are three points on a line with coordinates a, b, and c, respectively, such that $a < b < c$, then B lies between A and C.

Congruence of Segments

As we discuss segments and their measures, we want to distinguish carefully between the set of points—$\overline{AB}$, for instance—and their measure, $m\overline{AB}$. If we write $\overline{AB} = \overline{CD}$, that implies $\overline{AB}$ and $\overline{CD}$ are the same set of points. When $m\overline{AB} = m\overline{CD}$, however, it is not necessarily the case that $\overline{AB} = \overline{CD}$. If $m\overline{AB} = m\overline{CD}$, the segments are said to be *congruent*.

Definition 2-10 Any two **segments** are **congruent** ($\cong$) if and only if their measures are equal.

Therefore, if $m\overline{AB} = m\overline{CD}$, we say that $\overline{AB} \cong \overline{CD}$; and if $\overline{AB} \cong \overline{CD}$, then $m\overline{AB} = m\overline{CD}$. Since the length of a segment is a real number, the *equivalence relations* for equality of real numbers hold (Appendix I). That is, for every a, b, $c \in R$ it is true that

1. $a = a$ (Reflexive Axiom)
2. If $a = b$, then $b = a$ (Symmetric Axiom)
3. If $a = b$ and $b = c$, then $a = c$ (Transitive Axiom)

From these and Def. 2-10 it can be shown that congruence of segments is also an equivalence relation, and we state this as a theorem.

Theorem 2-1 For every segment AB, CD, and EF

(a) $\overline{AB} \cong \overline{AB}$ (Congruence of segments is reflexive.)
(b) If $\overline{AB} \cong \overline{CD}$, then $\overline{CD} \cong \overline{AB}$ (Congruence of segments is symmetric.)
(c) If $\overline{AB} \cong \overline{CD}$ and $\overline{CD} \cong \overline{EF}$, then $\overline{AB} \cong \overline{EF}$ (Congruence of segments is transitive.)

Let us show that Theorem 2-1(b) is true. Proofs of the other two parts are similar.

Theorem 2-1(b) For every segment AB, CD if $\overline{AB} \cong \overline{CD}$, then $\overline{CD} \cong \overline{AB}$.

Given: $\overline{AB} \cong \overline{CD}$
Prove: $\overline{CD} \cong \overline{AB}$

Proof

Statements	*Reasons*
1. $\overline{AB} \cong \overline{CD}$	1. Given
2. $m\overline{AB} = m\overline{CD}$	2. If two segments are congruent, then their measures are equal (Def. 2-10)
3. $m\overline{CD} = m\overline{AB}$	3. Symmetric Axiom of Equality for real numbers
4. $\overline{CD} \cong \overline{AB}$	4. If two segments are equal in measure, they are congruent (Def. 2-10)

As our geometric figures become more complicated, we will need to "add" and "subtract" parts of segments. The following postulates and theorems will enable us to do this.

Postulate 2-7 If C is a point on $\overline{AB}$ such that A-C-B, then $m\overline{AC} + m\overline{CB} = m\overline{AB}$.

Theorems 2-2 and 2-3 follow as a direct consequence of Post. 2-7.

Theorem 2-2 If A-B-C and R-S-T such that $\overline{AB} \cong \overline{RS}$ and $\overline{BC} \cong \overline{ST}$, then $\overline{AC} \cong \overline{RT}$.

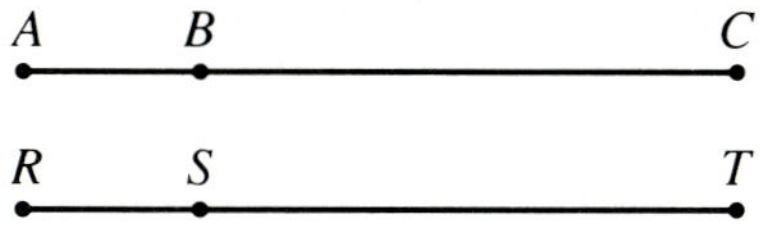

Proof

Statements	*Reasons*
1. $\overline{AB} \cong \overline{RS}$, $\overline{BC} \cong \overline{ST}$	1. Given
2. $m\overline{AB} = m\overline{RS}$ and $m\overline{BC} = m\overline{ST}$	2. Definition of congruent segments (Def. 2-10)
3. $m\overline{AB} + m\overline{BC} = m\overline{RS} + m\overline{ST}$	3. Addition Law for Equals (Appendix I)
4. $m\overline{AB} + m\overline{BC} = m\overline{AC}$; $m\overline{RS} + m\overline{ST} = m\overline{RT}$	4. Postulate 2-7
*5. $\therefore m\overline{AC} = m\overline{RT}$	5. Substitution Axiom of Equality for real numbers
6. $\therefore \overline{AC} \cong \overline{RT}$	6. Definition of congruent segments. (Def. 2-10)

The next theorem can be proved in a similar fashion and is left as an exercise for the reader.

Theorem 2-3 If A-B-C and R-S-T such that $\overline{AC} \cong \overline{RT}$ and $\overline{AB} \cong \overline{RS}$, then $\overline{BC} \cong \overline{ST}$.

The Ruler Placement Postulate (Post. 2-5) together with the Distance Postulate (Post. 2-6) lead us to another useful theorem.

Theorem 2-4 (*Segment Construction*) For any line AB and segment CD there exists a point P on ray AB such that $m\overline{CD} = m\overline{AP}$ and $\overline{CD} \cong \overline{AP}$.

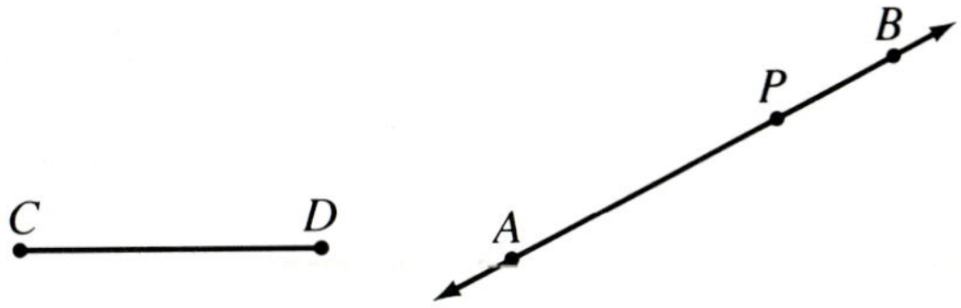

* $\therefore$ symbolically represents "therefore."

Proof

Statements	*Reasons*
1. Let m$\overline{CD} = x$, a positive number	1. Distance Postulate (Post. 2-6) and definition of distance (Def. 2-7)
2. Let AB be a line and choose the coordinate system such that A corresponds to 0 and B corresponds to b, a positive number	2. Ruler Placement Postulate (Post. 2-5)
3. There exists some point P on $\overleftrightarrow{AB}$ corresponding to x	3. Ruler Postulate (Post. 2-4)
4. $\therefore$m$\overline{AP} = x$	4. Distance Postulate and definition of distance
5. $\therefore$m$\overline{CD}$ = m$\overline{AP}$	5. Transitive Axiom of Equality for real numbers
6. $\therefore\overline{CD} \cong \overline{AP}$	6. Definition of congruent segments (Def. 2-10)

To illustrate the use of some of these theorems, consider the following.

Example

A, B, C and D are points on a line such that $\overline{AC} \cong \overline{BD}$, A-B-C and B-C-D. Show that $\overline{AB} \cong \overline{CD}$.

Solution (Draw the figure.) Since $\overline{BC} \cong \overline{BC}$ according to Theorem 2-1(a), and it is given that $\overline{AC} \cong \overline{BD}$, it follows that $\overline{AB} \cong \overline{CD}$ by Theorem 2-3. ◁

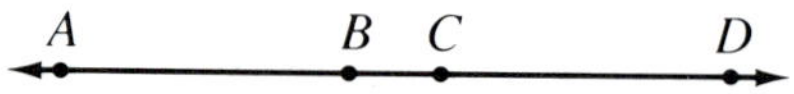

Midpoint of a Segment

A special point of interest in discussing segments will be defined as follows.

Definition 2-11 If a segment AB contains a point P such that m$\overline{AP}$ = m$\overline{PB}$, then P is the **midpoint** of $\overline{AB}$. A distinct figure that passes through the midpoint of a segment is said to **bisect** the segment. Such a figure is called the **bisector** of the segment.

In Fig. 2-25 $\overleftrightarrow{PQ}$ is said to bisect (be the bisector of) $\overline{AB}$ if P is the midpoint of $\overline{AB}$.

Note that since m$\overline{AB}$ is equal to a unique real number n, m$\overline{AP}$ = m$\overline{PB}$ = $n/2$. Now we can prove another theorem.

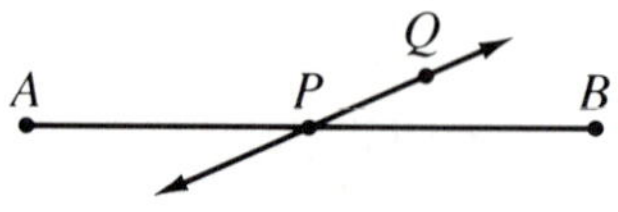

Figure 2-25

Theorem 2-5 Every segment has exactly one midpoint.
Given: $\overline{AB}$ with midpoint P
Prove: Midpoint P is unique

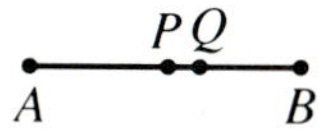

Assume that two distinct points P and Q are midpoints of $\overline{AB}$.

Proof

Statements	*Reasons*
1. $m\overline{AB} = m\overline{AP} + m\overline{PB}$ and $m\overline{AB} = m\overline{AQ} + m\overline{QB}$	1. Postulate 2-7
2. $m\overline{AP} = m\overline{PB}$ and $m\overline{AQ} = m\overline{QB}$	2. Definition of midpoint (Def. 2-11)
3. Hence $2m\overline{AP} = m\overline{AB}$ and $2m\overline{AQ} = m\overline{AB}$	3. Substitution Axiom of Equality for real numbers
4. $2m\overline{AP} = 2m\overline{AQ}$ or $m\overline{AP} = m\overline{AQ}$	4. Transitive Axiom of Equality for real numbers
5. $\therefore \overline{AP} \cong \overline{AQ}$	5. Definition of congruent segments (Def. 2-10)
6. Hence $P = Q$, a contradiction of the assumption	6. Ruler Postulate (Post. 2-4). To each point there corresponds a *unique* real number

Thus we have shown that $\overline{AB}$ has exactly one midpoint.

The proof of Theorem 2-5 was an *indirect proof*. We met the first such proof at the end of Chapter 1. This time we assumed something that we did not think was correct in order to demonstrate that the desired conclusion was the only possible one. In the following exercise set, you will be asked to give reasons for statements in some direct proofs. You may wonder how to decide which method to use, what statements to make, and in what order to make them. Actually there are many different possibilities that may be correct, so at this time you should read the proofs in the text carefully to determine why each statement is necessary, and perhaps state the proof again in your own words. Then try to complete a proof on your own. Some theorems are easy and obvious; some are lengthy and difficult. In either case they provide a challenge, an excellent opportunity for you to use your mental processes, so that you will not only gain a number of facts from your study of geometry but also develop a way of thinking that should be helpful in other areas. The proofs of the next two theorems will also be left as exercises.

Theorem 2-6 P is the midpoint of $\overline{AB}$ if and only if $\overline{AP} \cong \overline{PB}$ and A-P-B.

(Remember that the statement "P if and only if Q" means that both "If P, then Q" and "If Q, then P" are true. We must prove both.)

Theorem 2-7 The segments formed when congruent segments are bisected are congruent.

Exercise Set 2.6

Give reasons for each of the statements in the following proofs of theorems. You may need to use some of the axioms and theorems listed in Appendix I.

1. Theorem 2-3. If A-B-C and R-S-T such that $\overline{AC} \cong \overline{RT}$ and $\overline{AB} \cong \overline{RS}$, then $\overline{BC} \cong \overline{ST}$.
Given: A-B-C, R-S-T, $\overline{AC} \cong \overline{RT}$, $\overline{AB} \cong \overline{RS}$
Prove: $\overline{BC} \cong \overline{ST}$

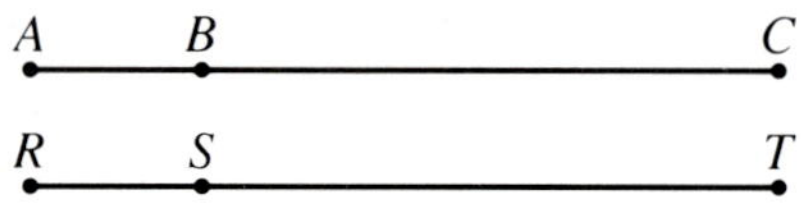

Proof

Statements	*Reasons*
1. A-B-C, R-S-T, $\overline{AC} \cong \overline{RT}$ and $\overline{AB} \cong \overline{RS}$	1. (Why?)
2. $m\overline{AC} = m\overline{RT}$ and $m\overline{AB} = m\overline{RS}$	2. (Why?)
3. $m\overline{AB} + m\overline{BC} = m\overline{AC}$ and $m\overline{RS} + m\overline{ST} = m\overline{RT}$	3. (Why?)
4. $\therefore m\overline{AB} + m\overline{BC} = m\overline{RS} + m\overline{ST}$	4. (Why?)
5. $m\overline{BC} = m\overline{ST}$	5. (Why?)
6. $\therefore \overline{BC} \cong \overline{ST}$	6. (Why?)

2. Theorem 2-6. P is the midpoint of $\overline{AB}$ if and only if $\overline{AP} \cong \overline{PB}$ and A-P-B.
(1) If $\overline{AP} \cong \overline{PB}$ and A-P-B, then P is the midpoint of $\overline{AB}$.
Given: $\overline{AP} \cong \overline{PB}$ and A-P-B
Prove: P is the midpoint of $\overline{AB}$

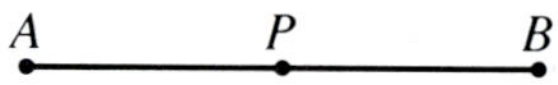

Proof

Statements	*Reasons*
1. $\overline{AP} \cong \overline{PB}$ and A-P-B	1. Given
2. $m\overline{AP} = m\overline{PB}$	2. (Why?)
3. $\therefore P$ is the midpoint of $\overline{AB}$	3. (Why?)

(2) If P is the midpoint of $\overline{AB}$, then $\overline{AP} \cong \overline{PB}$ and A-P-B.
Given: P is the midpoint of $\overline{AB}$
Prove: $\overline{AP} \cong \overline{PB}$ and A-P-B

Proof

Statements	*Reasons*
1. P is the midpoint of $\overline{AB}$	1. Given
2. $m\overline{AP} = m\overline{PB}$, A-P-B	2. (Why?)
3. $\overline{AP} \cong \overline{PB}$	3. (Why?)

3. In Theorem 2-6 is it necessary that we state *A-P-B*? Could $\overline{AP}$ and $\overline{PB}$ be congruent and *P not* lie on $\overline{AB}$? Illustrate.

4. Theorem 2-7. The segments formed when congruent segments are bisected are congruent.

   Given: $\overline{AB} \cong \overline{CD}$. *P* bisects $\overline{AB}$ and *Q* bisects $\overline{CD}$
   Prove: $\overline{AP} \cong \overline{PB} \cong \overline{CQ} \cong \overline{QD}$

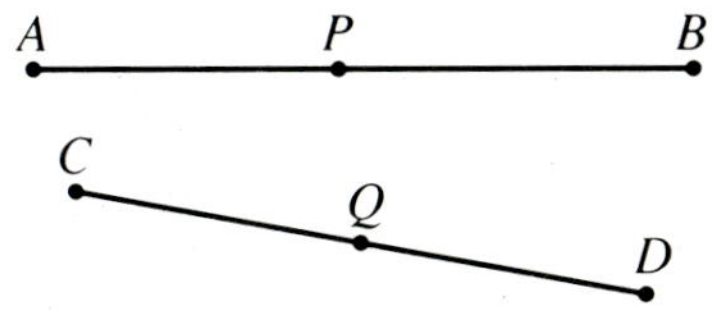

Proof

Statements	*Reasons*
1. *P* bisects $\overline{AB}$ and *Q* bisects $\overline{CD}$	1. (Why?)
2. $m\overline{AP} = m\overline{PB} = \frac{1}{2}m\overline{AB}$ and $m\overline{CQ} = m\overline{QD} = \frac{1}{2}m\overline{CD}$	2. (Why?)
3. $\overline{AB} \cong \overline{CD}$	3. (Why?)
4. $m\overline{AB} = m\overline{CD}$	4. (Why?)
5. $\frac{1}{2}m\overline{AB} = \frac{1}{2}m\overline{CD}$	5. (Why?)
6. $\therefore m\overline{AP} = m\overline{PB} = m\overline{CQ} = m\overline{QD}$	6. (Why?)
7. $\overline{AP} \cong \overline{PB} \cong \overline{CQ} \cong \overline{QD}$	7. (Why?)

5. Suppose *M* is the bisector of $\overline{RS}$ and *N* is the bisector of $\overline{TU}$. Is it necessary that $\overline{RM} \cong \overline{TN}$?

6. If *AB* is a segment, then $\overline{AB} \cong \overline{BA}$.

   Given: *AB* is a segment
   Prove: $\overline{AB} \cong \overline{BA}$

Proof

Statements	*Reasons*
1. $m\overline{AB} = \lvert a - b \rvert$	1. (Why?)
2. $m\overline{BA} = \lvert b - a \rvert$	2. (Why?)
3. $\lvert a - b \rvert = \lvert b - a \rvert$	3. (Why?)
4. $m\overline{AB} = m\overline{BA}$	4. (Why?)
5. $\overline{AB} \cong \overline{BA}$	5. (Why?)

2-5 PLANES, HALF-PLANES, AND CONVEX SETS

Since we have defined geometric figures in terms of sets of points, we can speak of a point being an element or member *of* that figure or as a point *in* the figure. That is, we may write "$P \in \ell$" to mean that a point *P* belongs to the line ℓ or "$P \in \overline{AB}$" to say that *P* belongs to the segment *AB*. We may also find it convenient to write "point *P* is on ℓ" to indicate *P* is an element of the set of points that we call line ℓ.

Points that lie on the same line are said to be *collinear*. In Fig. 2-26 we see that points A, B, and C are collinear, whereas A, B, and D are *noncollinear* points because they do not lie on the same line.

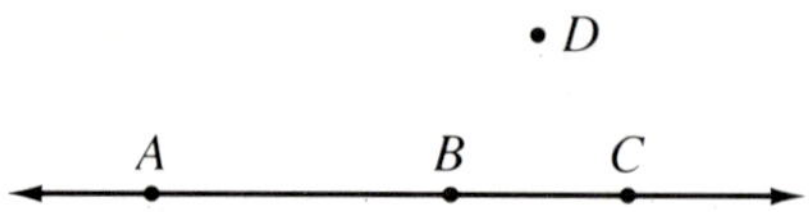

Figure 2-26

Planes

If we now consider three noncollinear points R, S, and T, the lines RS, RT, and ST are determined. If we draw a large number of lines from R intersecting $\overleftrightarrow{ST}$, from S intersecting $\overleftrightarrow{RT}$, and from T intersecting $\overleftrightarrow{RS}$, the resulting figure looks something like Fig. 2-27.

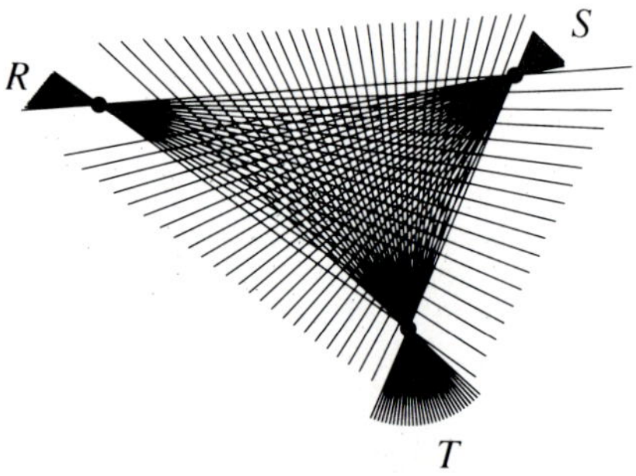

Figure 2-27

If we were to continue drawing lines, we would find that the figure which is the union of all such lines is what we commonly call a plane. Note that each line contains one of the three noncollinear points and a point of the line connecting the other two given points; that is, for each such line the two determining points are dependent on the *three* given points.

Postulate 2-8 A plane is uniquely determined by three noncollinear points.

Definition 2-12 The set of points in a plane are said to be **coplanar.**

The following theorem will be useful in our later work. How many planes are determined by two different lines that intersect?

Theorem 2-8 Two different lines intersect in at most one point.

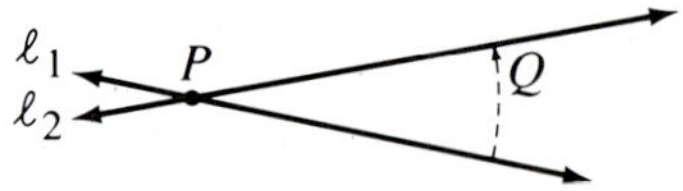

Proof Assume that either (1) the lines intersect in more than one point, two of which are P and Q, or (2) the lines intersect in one point P or not at all. Since there are no further possibilities to consider, let us determine what happens under assumption (1): If lines ℓ_1 and ℓ_2 intersect at two distinct points P and Q, then $\ell_1 = \ell_2$, since Post. 2-1(a) states that through any two distinct points one and only one straight line can be drawn. This contradicts the statement that the lines are different, and assumption (2) must be correct.

Half-Planes and Convex Sets

A point separates a line into two half-lines. Similarly a line may separate a plane into two half-planes. In Fig. 2-28 if we call the half-planes $\mathcal{H}_1$ and $\mathcal{H}_2$, then $\mathcal{H}_1$ is a set of points such that any two points P and R of $\mathcal{H}_1$ determine a segment PR with all its points in $\mathcal{H}_1$, whereas $\mathcal{H}_2$ is a set of points such that any two points Q and S of $\mathcal{H}_2$ determine a segment QS with all its points in $\mathcal{H}_2$. Sets having this property are *convex sets*.

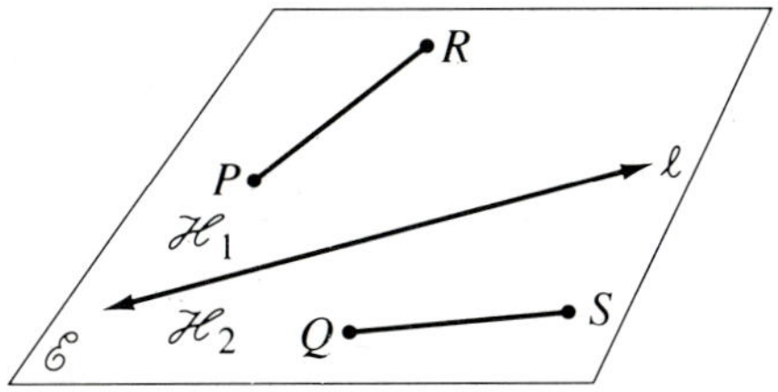

Figure 2-28

Definition 2-13 A set is called a **convex set** if, for every two distinct points A and B in the set, the segment AB lies entirely in the set.

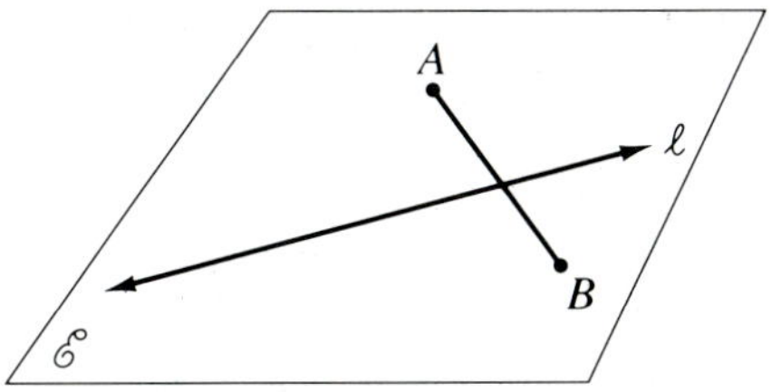

Figure 2-29

In Fig. 2-29 points A and B are in the distinct sets determined by line ℓ and segment AB intersects ℓ. From the preceding we can see the meaning of the following postulate.

Postulate 2-9 (Plane Separation Postulate) A line ℓ contained in a plane $\mathcal{E}$ separates the plane into three convex sets $\mathcal{H}_1$, $\mathcal{H}_2$ and ℓ such that if $A \in \mathcal{H}_1$ and $B \in \mathcal{H}_2$ then $\overline{AB}$ intersects the line ℓ.

Note that the *union* of $\mathcal{H}_1$, $\mathcal{H}_2$, and ℓ is $\mathcal{E}$ ($\mathcal{H}_1 \cup \mathcal{H}_2 \cup \ell = \mathcal{E}$) and the *intersection* of $\mathcal{H}_1$, $\mathcal{H}_2$, and ℓ is the empty set ($\mathcal{H}_1 \cap \mathcal{H}_2 \cap \ell = \emptyset$).

Definition 2-14 The sets $\mathcal{H}_1$ and $\mathcal{H}_2$ of Post. 2-9 are called **half-planes,** and ℓ is said to be their common **edge**.

Referring to Fig. 2-29, A and B are said to be on *opposite sides* of ℓ if $\overline{AB}$ intersects ℓ in exactly one point. A and B are said to be on the *same side* of ℓ if $\overline{AB} \cap \ell = \emptyset$.

Example

Is each of the following a convex set?

(a) a figure consisting of exactly two distinct points
(b) a ray
(c) a half-line
(d) a plane with one point removed
(e) the figure formed by the union of the set of points between each pair of three noncollinear points.

Solution

(a) No. By definition a convex set exists only if every point on the *segment* joining two points of the given set is entirely in the (convex) set. Thus if A and B are the two points, any point X such that A-X-B is not in the set.
(b) Yes. Consider $\overrightarrow{RS}$. Any pair of points in the set $\overrightarrow{RS} \cup \{X|R\text{-}S\text{-}X\}$ determines a segment entirely within $\overrightarrow{RS}$.
(c) Yes. Same argument as above, since a half-line is simply a ray with its endpoint removed.
(d) No. Many pairs of points in the resulting figure will be such that the segment joining them contains a point (the one removed) that is not in the figure.
(e) No. Noncollinear points A, B, and C determine $\overline{AB}$, $\overline{BC}$, and $\overline{CA}$; but some points D and E exist such that A-D-B and B-E-C, and segment DE does not belong to the set $\overline{AB} \cup \overline{BC} \cup \overline{AC}$. ◁

Exercise Set 2.7

1. Explain why a three-legged stool or a tripod is more likely to be "steady" than a four-legged chair or table.
2. **(a)** Is a line a convex set? Explain.
   (b) If one point is removed from a line do the remaining points form a convex set? Explain.
   (c) Do two intersecting lines form a convex set? Explain.
   (d) Is the graph of $\{x|-2 < x < 3\}$ a convex set? Explain.
3. If $\mathcal{H}_1$ and $\mathcal{H}_2$ are coplanar and:
   (a) If $\mathcal{H}_1$ and $\mathcal{H}_2$ have the same edge, is $\mathcal{H}_1 \cup \mathcal{H}_2$ a convex set? Explain.
   (b) If the edge of $\mathcal{H}_1$ intersects the edge of $\mathcal{H}_2$ in exactly one point, is the set convex? Explain.
4. Consider a point P on a line AB.
   (a) P separates $\overleftrightarrow{AB}$ into how many sets? List them. Are they convex sets?
   (b) If P is removed from $\overleftrightarrow{AB}$, list the sets that remain. What are they called? Are they convex sets?

5. As a consequence of Post. 2-8 we can denote a plane by three noncollinear points, i.e., plane ABC, plane RST, etc. If A, B, C, and D are not coplanar, how many planes are determined? Name them.

2-6 GEOMETRIC CONSTRUCTIONS

A *construction* in geometry is strictly confined to figures that can be drawn using only a compass and *straightedge,* which can best be described as a ruler with no markings on it (Fig. 2-30). Although there are many devices available for making precise drawings, the early Greek mathematicians were committed to keeping things simple and enjoyed the challenge of creating complicated constructions with just these two devices.

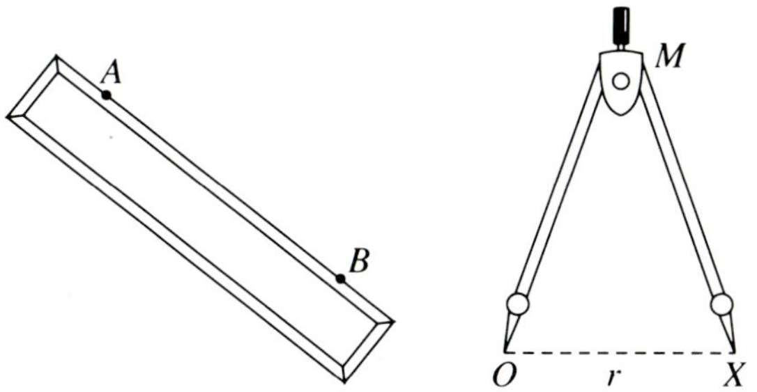

Figure 2-30

We met the Segment Construction Theorem (Theorem 2-4) in the previous section. With the straightedge we can draw (part of) a line through two given points. We assume that the compass consists of two legs $\overline{MO}$ and $\overline{MX}$ joined at a point M so that the distance between points O and X can be set and a circle drawn with center at any point C and radius $m\overline{OX} = r$. Let us assume we can draw another circle with center at another point D and radius $m\overline{OX}$.

Construction 1: Construction of congruent segments.

Suppose we have a segment CD and a ray AB and would like to find a point P on $\overrightarrow{AB}$ such that $\overline{AP} \cong \overline{CD}$:

1. Place point O of the compass on point C of the segment and place point X of the compass on point D of the segment (this is the *compass setting*).
2. Without changing the compass setting, place point O of the compass at point A on the ray and draw an arc of a circle with radius $m\overline{OX}$ intersecting the circle at point P.

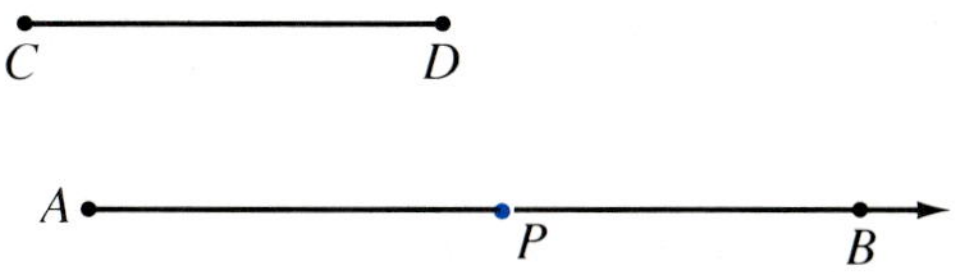

In order to make your first exercise in constructions more interesting, we introduce a figure called the *ellipse,* which is defined to be a set of points, the sum of whose distances from two fixed points is constant. In the following figure, the two fixed points, called *foci,* are designated F_1 and F_2. The distances of the foci from point P on the ellipse are d_1 and d_2, respectively. If d_1 is 5 and d_2 is 10, then the constant sum is 15. For other points on the ellipse, the distances might be 3 and 12, 7 and 8, or any other pair of numbers for which the sum is 15. The following figure is an ellipse, and the method of constructing four of the points is shown.

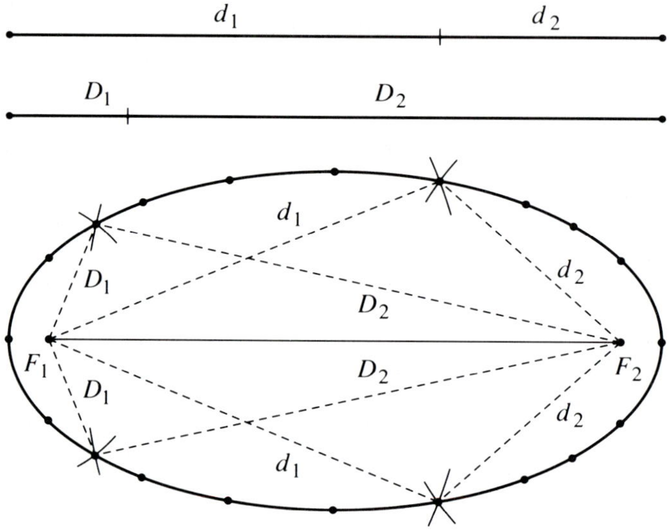

Exercise Set 2.8

1. On a line ℓ_1 draw a segment PQ approximately 3 in. long and a point R such that P-R-Q.
   (a) Construct $\overline{RS} \cong \overline{PQ}$ so that R-Q-S.
   (b) Construct $\overline{RT} \cong \overline{PQ}$ so that R-P-T.
   (c) Draw any line ℓ_2 such that $\ell_2 \neq \ell_1$ and $R \in \ell_2$.
   Construct $\overline{RK} \cong \overline{PQ}$ so that $K \in \ell_2$.
2. On a line ℓ draw a segment AB approximately 1 in. long. Construct $\overline{BC} \cong \overline{AB}$ so that A-B-C, $\overline{CD} \cong \overline{AB}$ so that B-C-D, and $\overline{DE} \cong \overline{AB}$ so that C-D-E. If $m\overline{AB} = a$, $m\overline{AE} = ?$
3. Use the scale given below to construct an ellipse with foci 12 units apart and with points for which the sum of the distances from the foci is 15 units. Use all natural-number values of d_1 from 2 through 13. (This may take some time and some patience, but if you do it, you'll be one in a million!)

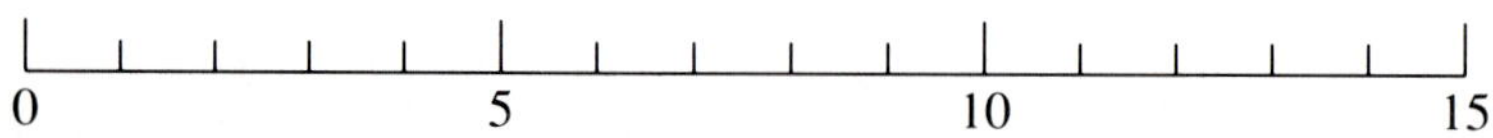

4. A *hyperbola* is a set of points, the difference of whose distances from two fixed points is constant. Part of a hyperbola, which has two branches, is shown below. Use the scale

given previously with foci 6 units apart and common difference between foci and points on the hyperbola $4\frac{1}{2}$ units to construct a hyperbola.

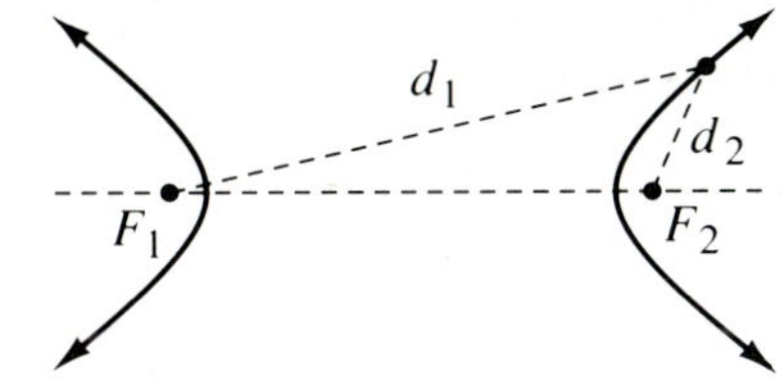

REVIEW EXERCISES

Indicate which of the following statements are always true according to the definitions, axioms, postulates, and theorems developed in Chapter 2. If the statement is not always true, give a reason or drawing to show why it is not.

Example: 1. Any three points are always collinear.
False

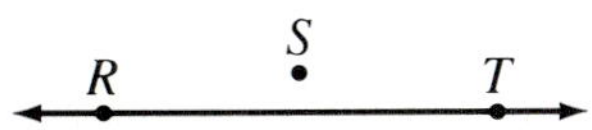

2. Any three points are always coplanar.
True

1. Any segment has exactly one midpoint.
2. Every ray has exactly one midpoint.
3. If $\overline{AB} = \overline{CD}$, then m$\overline{AB}$ = m$\overline{CD}$.
4. If $\overline{AB} \cong \overline{CD}$, then $\overline{AB} = \overline{CD}$.
5. If M is the midpoint of $\overline{RS}$ and P is the midpoint of $\overline{XY}$, then $\overline{RM} \cong \overline{PX}$.
6. If the coordinates of A, B, C, and D are -2, 1, 2, and 5, respectively, then m$\overline{AC}$ = m$\overline{BD}$.
7. If $|n| < 5$, then n cannot be a negative number.
8. If A-D-B and C-D-B, then A-C-B.
9. If $\overline{AP} \cong \overline{PB}$, then P is the midpoint of $\overline{AB}$.
10. If P is the midpoint of $\overline{AB}$, then $\overline{AP} \cong \overline{PB}$.

Complete each of the following statements:

11. A ________________ is the figure formed by $\overline{AB}$ and the set of all points X such that A-B-X.
12. $\overset{\circ\!\!\rightarrow}{AB} \cup \overline{AB}$ = ________.
13. $\overline{AB} \cap \overrightarrow{AB}$ = ________.
14. Four points, no three of which are collinear, determine ________ straight lines.
15. The set of points of a ray, excluding its endpoint, is called a ________________.

16. Which of the following is *not* a geometric figure?
(a) A ray
(b) A convex set
(c) $\overrightarrow{RS}$
(d) plane XYZ
(e) $m(\overline{RS})$

17. Which one of the following is *not* a convex set?
(a) line
(b) plane
(c) segment
(d) ray
(e) point

18. If two segments are congruent, then they have the same
(a) endpoints
(b) measure
(c) points
(d) triangles
(e) equivalence

Draw the graph of each of the following sets of real numbers on a separate number line.

19. $\{x | x \leq -1\}$

20. $\{x | \, |x| > 2\}$

21. $\{x | -2 < x < 3\}$

22. The points shown below are on the same line. List all the points contained in each of the following sets:
(a) $\overrightarrow{CE}$
(b) $\overrightarrow{DC}$
(c) $\overleftrightarrow{EF}$
(d) $\overline{BD}$
(e) $\overline{AD} \cap \overline{BE}$

3

Angles and Proofs of Theorems

Geometry will draw the soul toward truth and create the spirit of Philosophy.

PLATO (429–347 B.C.) Greek philosopher.

In Chapter 2 we were concerned with intersecting lines and rays, but we paid no particular attention to the figure that was formed when they intersected. Now we will investigate these figures more thoroughly. We also met proofs in the previous chapter and the reader was asked to complete a few of them. By the end of the present chapter the idea of proof should be quite familiar.

3-1 ANGLES

Definition 3-1 An **angle** is the set of points that is the union of two rays that have the same endpoint. The rays are called the **sides** of the angle, and the common endpoint is the **vertex** of the angle. If its sides are opposite rays, the angle is a **straight angle.**

In Fig. 3-1(a), the angle is formed by $\overrightarrow{BA}$ and $\overrightarrow{BC}$. The vertex is at B. We will denote this angle as $\angle ABC$, using the symbol $\angle$ to represent an angle, A and C the points on the rays, and B the vertex of the angle. In Fig. 3-1(b) we have $\angle XYZ$, $\angle PYZ$ and $\angle XYP$. Figure 3-1(c) shows an angle that can be denoted by $\angle UVW$, $\angle MVN$, $\angle MVW$, or $\angle UVN$. In both (a) and (c) we might have named the angle by its vertex letter alone: $\angle ABC = \angle B$, $\angle UVW = \angle V$. Note that when three letters are used, the vertex letter is between the letters of the points on the sides. Could we use just one vertex letter for the angle(s) in Fig. 3-1(b)?

Other acceptable methods for labeling angles are shown in Fig. 3-2. There should be no confusion as to which is $\angle 1$ or $\angle 2$, $\angle a$ or $\angle b$.

According to our definition the figure in Fig. 3-3(a) would not be an angle

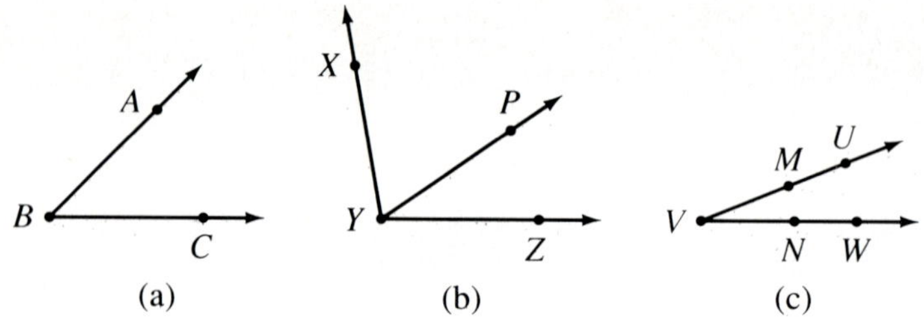

Figure 3-1

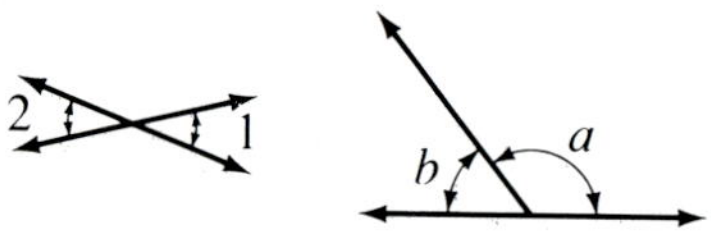

Figure 3-2

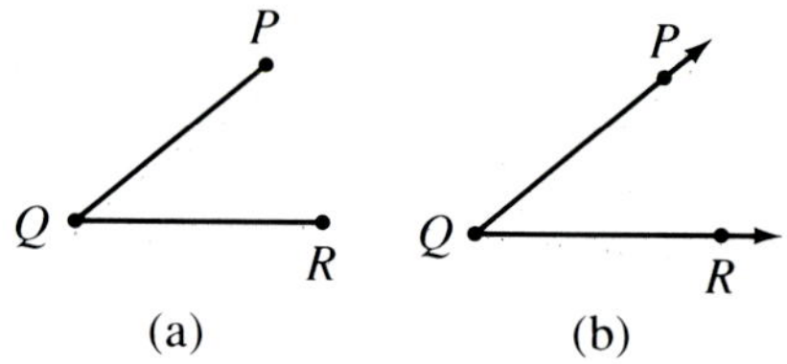

Figure 3-3

because it is made up of segments. Since $\overline{QP}$ and $\overline{QR}$ are subsets of $\overrightarrow{QP}$ and $\overrightarrow{QR}$, however, we will agree that two segments with a common endpoint may determine an angle.

Exterior and Interior of an Angle

In Fig. 3-4 we see an angle *ABC* and two points *P* and *Q* that lie in the same plane but not on the angle. Under these conditions we say that *P* lies in the interior of the angle and *Q* lies in the exterior of the angle.

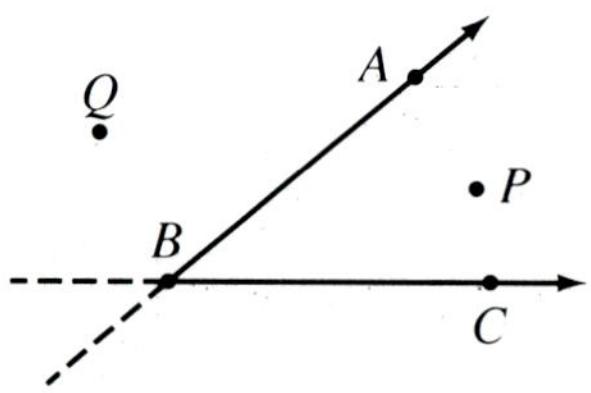

Figure 3-4

Definition 3-2(a) A point *P* lies in the **interior** of the angle *ABC* if (1) *P* and *A* are on the same side of the line *BC* and (2) *P* and *C* are on the same side of line *AB*. The **exterior** of angle *ABC* is the set of points that do not lie on the angle or in its interior.

Following are alternate definitions of interior and exterior.

Definition 3-2(b) A point *P* lies in the **interior** of angle *ABC* (*A*, *B*, and *C* are not collinear) if and only if there are points $M \in \overrightarrow{BA}$ and $N \in \overrightarrow{BC}$ such that *M-P-N*. A point *Q* lies in the **exterior** of an angle if it does not lie on the angle or in the interior of the angle.

From the preceding it appears that all angles other than straight angles separate the plane into three disjoint sets:

1. the angle itself
2. the interior of the angle (a convex set)
3. the exterior of the angle

Note that if P is in the interior and Q is in the exterior, it will always be true that $\overline{PQ}$ intersects the angle.

Exercise Set 3.1

Exercises 1–3 refer to the following figure. Angle ABC consists of $\overrightarrow{BA}$ and $\overrightarrow{BC}$, which are determined by $\overleftrightarrow{BA}$ and $\overleftrightarrow{BC}$.

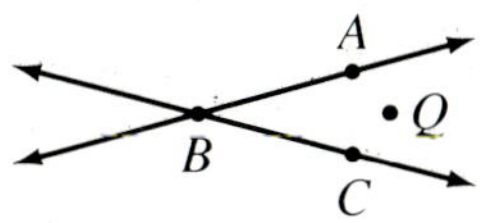

1. The point Q is on the same side of $\overleftrightarrow{BC}$ as point ______.
2. The point C lies on the same side of $\overleftrightarrow{AB}$ as point ______.
3. Q is in the ______ of $\angle ABC$.

Exercises 4–7 refer to the following figure.

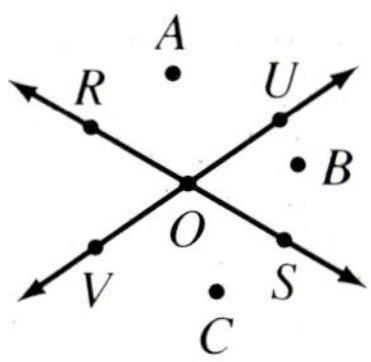

4. Name the angles in the figure.
5. The point A is interior to what angle(s)?
6. The point C is interior to what angle(s)?
7. Name the points exterior to $\angle ROV$.
8. **(a)** Name the angles in the figure below.
   (b) Which angles could you name using the vertex letter only?

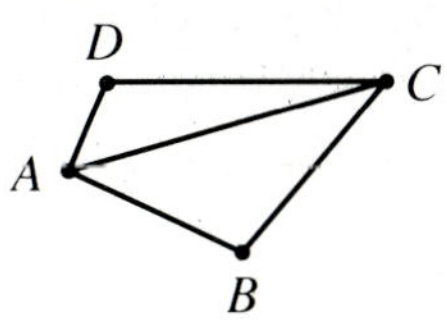

9. Does a straight angle have an interior and an exterior? Look again at the definitions and explain your answer.

10. Name all the angles in the following figure.
  (a) with vertex *B*
  (b) with vertex *C*
  (c) with vertex *F*
  (d) that may be named by the letter at the vertex only
  (e) that have *E* as a vertex

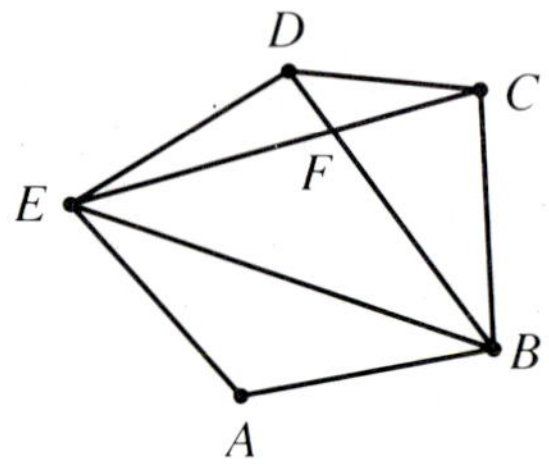

11. Use the following figure.
  (a) Name the points of the figure interior to the angle.
  (b) Name the points of the figure on the angle.
  (c) Name the points of the figure exterior to the angle.

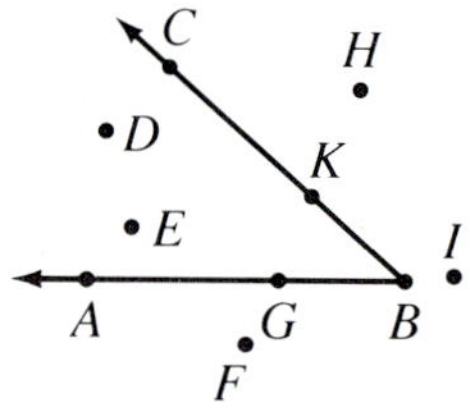

12. In the figure for Exercise 11, is the interior of the angle a convex set? Is the exterior of the angle a convex set?

13. How does a straight angle separate a plane? Does this separation differ from that of a nonstraight angle? Explain.

14. Complete the following statements.
  (a) A point *Q* lies in the interior of an angle if and only if ________.
  (b) A point *Q* lies in the exterior of an angle if and only if ________.

15. If *A* and *B* are exterior to an angle, does it follow that $\overline{AB}$ is in the exterior of the angle? Explain.

3-2 MEASUREMENT OF ANGLES

When we considered the measure of a segment, we assigned an arbitrary *unit* of length and compared other segments to this unit. In a similar manner we will assign an arbitrary unit for the angle, which we will call a degree.

Definition 3-3(a) A **degree** is the measure of an angle that is $\frac{1}{180}$ of a straight angle.

An instrument commonly used to measure angles is the *protractor.* Most protractors are divided into degree units.

If we place a protractor on line *AB* as illustrated in Fig. 3-5, it will lie in one of the half planes formed by $\overleftrightarrow{AB}$ and we can read the measure of some of the angles in this half plane which have their vertex at *A*.

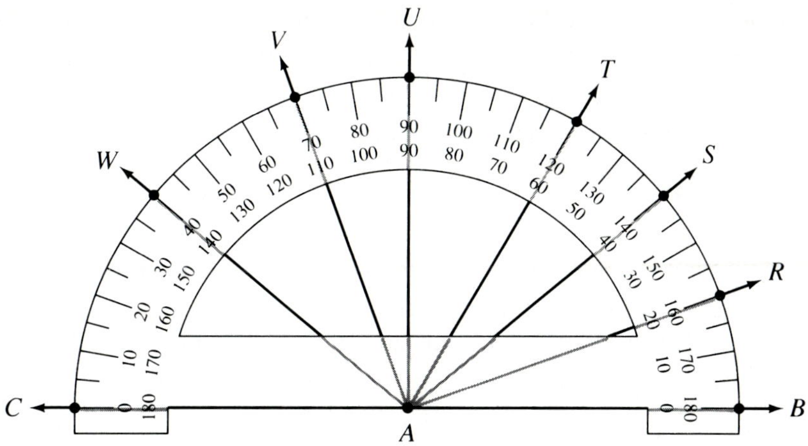

Figure 3-5

Definition 3-3(b) The number of degrees contained in an angle is called its **measure.** As in the measure of length of segments we will write m$\angle X$ and say "the measure of angle X."

In Fig. 3-5 we see that m$\angle BAR = 20$, m$\angle BAS = 40$, and m$\angle BAW = 140$. Note that m$\angle RAS = 40 - 20$ or m$\angle BAS -$ m$\angle BAR$, whereas m$\angle TAV =$ m$\angle BAV -$ m$\angle BAT = 110 - 60 = 50$. As shown in Fig. 3-6, since m$\angle BAR = 20$, we call $\angle BAR$ a 20-degree angle and use the degree symbol (°) to indicate the measurement of the angle.

If two rays coincide, the measure of the angle formed will be 0. The interior of an angle of 0° is the empty set. (See Appendix II)

Postulate 3-1 (*Angle Measurement Postulate*) To every angle there corresponds a real number a such that $0 \leq a \leq 180$.

Postulate 3-2 (*Angle Construction Postulate*) Let $\overleftrightarrow{AB}$ be a line in a plane $\mathscr{E}$. For every number $a(0 < a < 180)$ there is exactly one ray AR with R in $\mathscr{H}_1$ such that m$\angle RAB = a$.

For example, see Fig. 3-7, where $a = 135$.

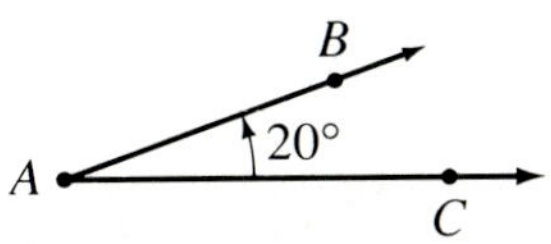

Figure 3-6

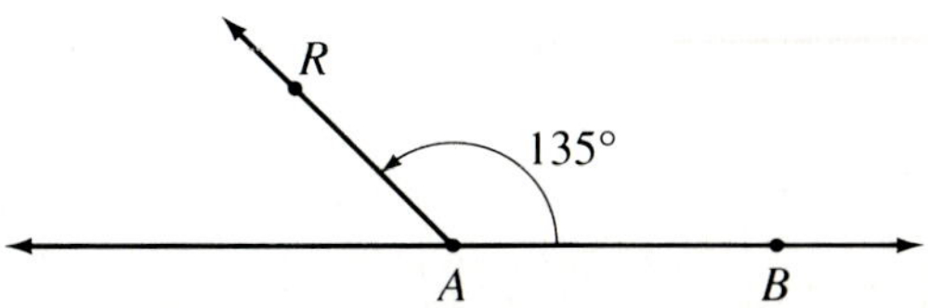

Figure 3-7

Postulate 3-3 (*Angle Addition Postulate*) If P is in the interior of an $\angle ABC$, or if P is a point not on a straight angle ABC, then $m\angle ABC = m\angle ABP + m\angle PBC$, or $m\angle ABP = m\angle ABC - m\angle PBC$. See Fig. 3-8.

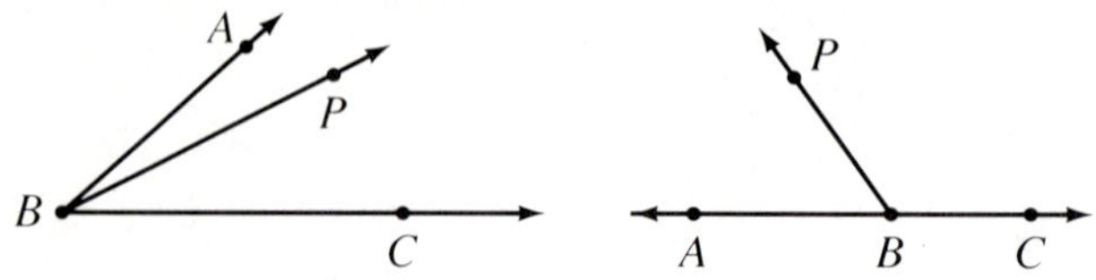

Figure 3-8

In Fig. 3-9 we see that if $m\angle ABP = 17$ and $m\angle PBC = 21$, $m\angle ABC = 21 + 17 = 38$. Also if $m\angle ROS = 42$ and $m\angle TOS = 12$, then $m\angle TOR = 42 - 12 = 30$.

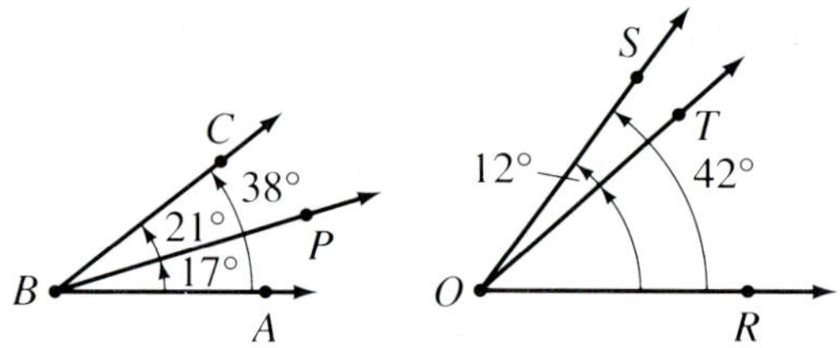

Figure 3-9

Exercise Set 3.2

1. Using a protractor measure

(a) $\angle XOR$ **(b)** $\angle ROC$
(c) $\angle SOE$ **(d)** $\angle SOT$
(e) $\angle TOH$

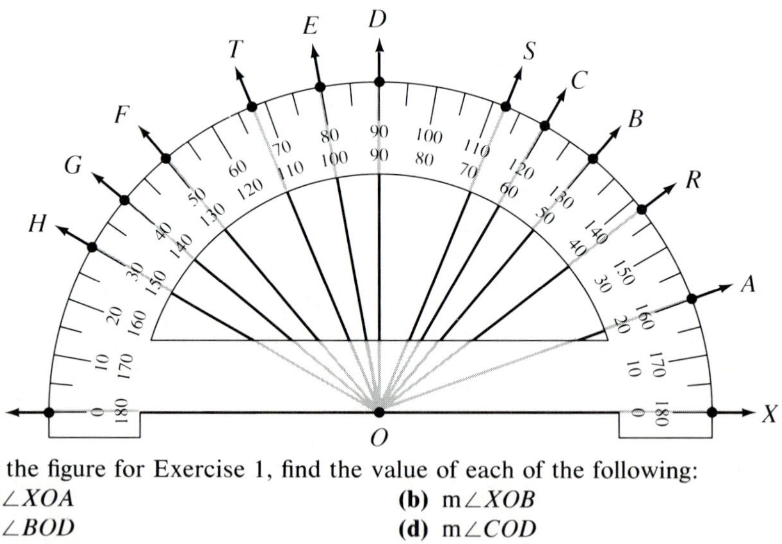

2. Using the figure for Exercise 1, find the value of each of the following:

(a) $m\angle XOA$ **(b)** $m\angle XOB$
(c) $m\angle BOD$ **(d)** $m\angle COD$
(e) $m\angle AOE$ **(f)** $m\angle EOH$

(g) m$\angle AOR$ + m$\angle ROC$
(h) m$\angle BOD$ + m$\angle DOF$
(i) m$\angle XOD$ + m$\angle DOF$
(j) m$\angle HOE$ + m$\angle EOC$
(k) m$\angle XOX$

3. Without using a protractor, *estimate* the measurement of the following angles (to the nearest 5°).

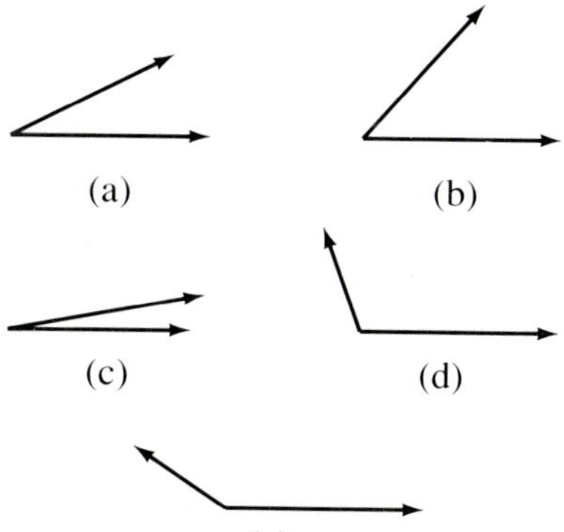

4. Measure the angles in Exercise 3 with a protractor and compare your answers to your estimates.

5. Using only a straightedge, draw a set of angles whose measurement will approximate angles of 30°, 45°, 60°, 90°, 120°, 135°, 150°.

6. Measure the angles you drew in Exercise 5 and record the measurements.

7. In the figure below
(a) m$\angle RPS$ + m$\angle SPT$ = m$\angle$______
(b) m$\angle RPT$ + m$\angle TPU$ = m$\angle$______
(c) m$\angle RPU$ − m$\angle TPU$ = m$\angle$______
(d) m$\angle UPS$ − m$\angle TPS$ = m$\angle$______
(e) m$\angle UPT$ + m$\angle TPS$ + m$\angle SPR$ = m$\angle$______
(f) m$\angle VPT$ + m$\angle TPS$ = m$\angle$______
(g) m$\angle VPU$ + m$\angle TPU$ = m$\angle$______

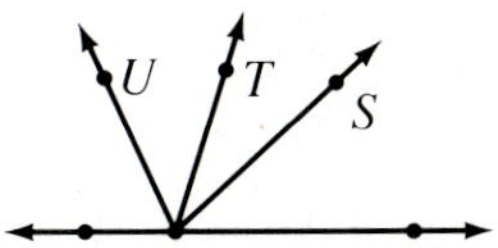

8. **(a)** Draw $\overleftrightarrow{AB}$ and $\overleftrightarrow{CD}$ intersecting at E.
(b) Find m$\angle AED$ and m$\angle DEB$, m$\angle BEC$ and m$\angle AEC$ with your protractor.
(c) Which pairs appear to have the same measure?

9. **(a)** Draw noncollinear points P, Q, and R, then $\overline{PQ}$, $\overline{QR}$, and $\overline{RP}$.
(b) Find m$\angle P$, m$\angle Q$, and m$\angle R$ with your protractor.
(c) Find s if s = m$\angle P$ + m$\angle Q$ + m$\angle R$.

10. **(a)** Draw four points L, M, N, and O, as in the following figure, and then draw $\overline{LM}$, $\overline{MN}$, $\overline{ON}$, and $\overline{OL}$.

(b) Find $m\angle L$, $m\angle M$, $m\angle N$, and $m\angle O$ with your protractor.

(c) Find t if $t = m\angle L + m\angle M + m\angle N + m\angle O$.

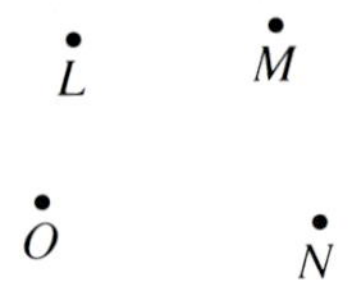

11. **(a)** In the figure drawn in Exercise 10, draw $\overline{MO}$.

(b) Find $m\angle LOM$ and $m\angle LMO$ with your protractor.

(c) Find x if $x = m\angle L + m\angle LOM + m\angle LMO$.

(d) Find $m\angle NOM$ and $m\angle NMO$ with your protractor.

(e) Find y if $y = m\angle N + m\angle NOM + m\angle NMO$.

Bearings: The Azimuth Circle

Angles are used to locate the relative position of objects in many interesting ways. Perhaps the easiest is the *azimuth circle,* which is used by the armed forces, the forest service and in air and marine navigation. The direction north is associated with 0°, east with 90°, south with 180°, and west with 270°, as indicated in Fig. 3-10.

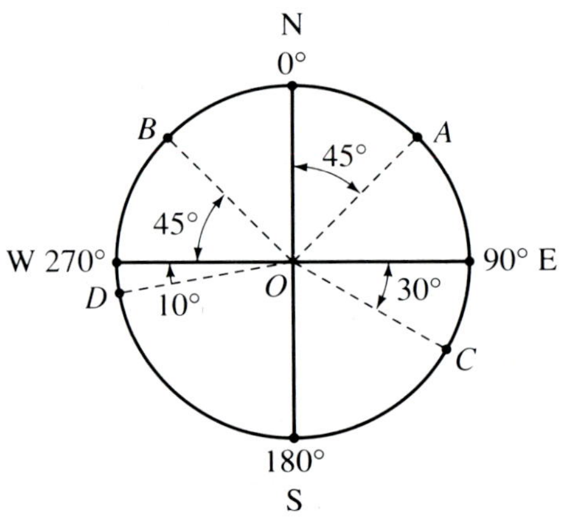

Figure 3-10

Although we have restricted our discussion of angles to those having measures from 0 through 180, the *bearing* of a point X from the point O is the angle NOX if X is east of line ON and it is $180 + m\angle SOX$ if X is west of line ON. Thus the bearings from O of points A, B, C, and D in Figure 3-10 are 45°, 315°, 120° and 260°.

Exercise Set 3.3

1. In the following figure give the bearing of each of the points A, B, C, D, F, and G.

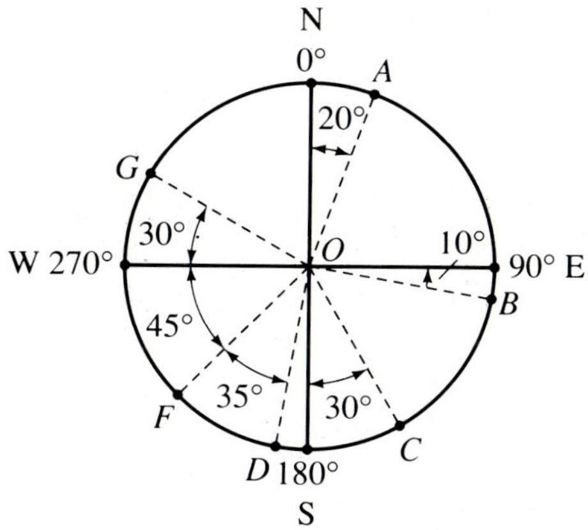

2. A radar operator observes a ship at a bearing of 40° and a helicopter at a bearing of 280°.
   (a) What is the measurement of the angle between the two?
   (b) If the ship is 10 mi away and the helicopter is 14 mi away from the radar, use a scale of $\frac{1}{4}$ in. = 1 mi to draw the relative positions of the radar set, the ship, and the helicopter.
   (c) If the radar set is at point O, the ship at S and the helicopter at H, use a protractor to measure angles OHS and OSH.

3-3 SPECIAL ANGLES AND RELATIONSHIPS

For further discussion of angles, we will need to classify them in terms of their measure.

Definition 3-4 Given that a is the measure of an angle BAC (Refer to Fig. 3-11)
(a) the angle is an **acute angle** if $0 < a < 90$,
(b) it is a **right angle** if $a = 90$,
(c) it is an **obtuse angle** if $90 < a < 180$,
(d) it is a **straight angle** if $a = 180$.

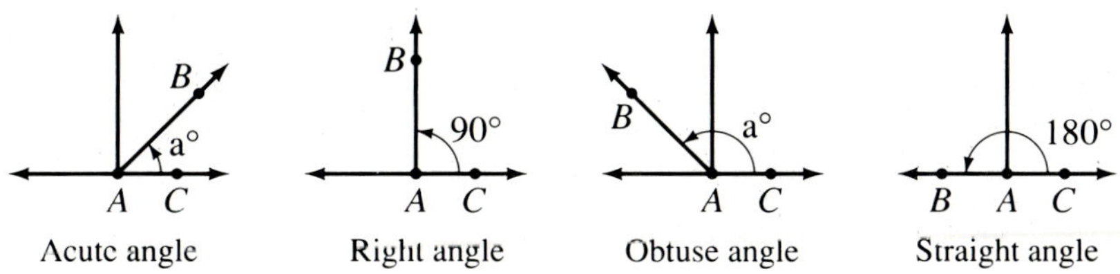

Figure 3-11

Definition 3-5 **Adjacent angles** are two angles that have the same vertex and a common side such that the intersection of their interiors is the empty set. The two sides that are not common to both of two adjacent angles are said to be the **exterior sides** of the angles.

In Fig. 3-12 $\angle ABC$ and $\angle CBD$ have the common side $\overrightarrow{BC}$ and vertex B. Sides $\overrightarrow{BD}$ and $\overrightarrow{BA}$ are the exterior sides.

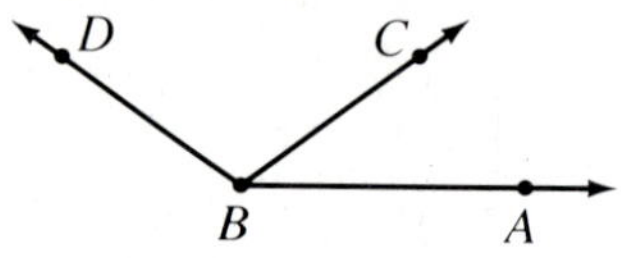

Figure 3-12

Definition 3-6 Two angles are **supplementary** if the sum of their measures is 180. Each such angle is said to be the **supplement** of the other.

It is not required by the definition that the angles be adjacent, so all the pairs shown in Fig. 3-13 are supplements of each other.

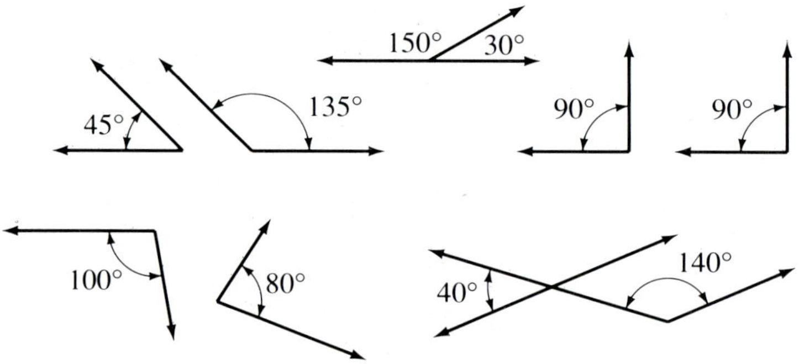

Figure 3-13

Definition 3-7 Two angles are said to be **complementary** if the sum of their measures is 90. Each such angle is said to be the **complement** of the other.

If two adjacent angles are complementary their exterior sides form a right angle, as shown in Fig. 3-14(b). We use the symbol ⌝ to indicate a right angle. Angle RST in Fig. 3-14(c) is a right angle.

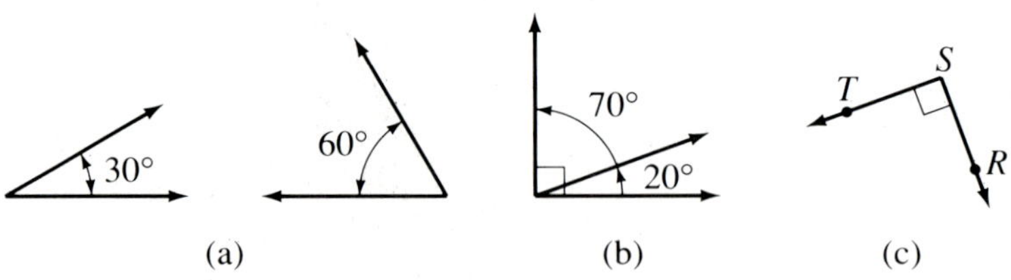

Figure 3-14

Example

(a) If $\angle A$ and $\angle B$ are supplementary and $m\angle A$ is twice $m\angle B$, find the measure of each.

(b) If m$\angle A$ is 20 more than the measure of its complement, what is the measure of $\angle A$?

Solution

(a) m$\angle A$ + m$\angle B$ = 180 and m$\angle A$ = 2m$\angle B$, $\therefore$ 2m$\angle B$ + m$\angle B$ = 180; 3m$\angle B$ = 180, m$\angle B$ = 60, m$\angle A$ = 180 − 60 = 120.

(b) Let m$\angle A = x$, m$\angle B = y$, and $x + y = 90$.
$\therefore x = y + 20$ and by substitution we have $y + 20 + y = 90$, $2y = 70$, $y = 35$. Since m$\angle A = x = y + 20$, it follows that m$\angle a = 55$. ◁

Exercise Set 3.4

1. In the following figure, name two angles that are
   (a) acute
   (b) obtuse
   (c) right
   (d) straight
   (e) adjacent

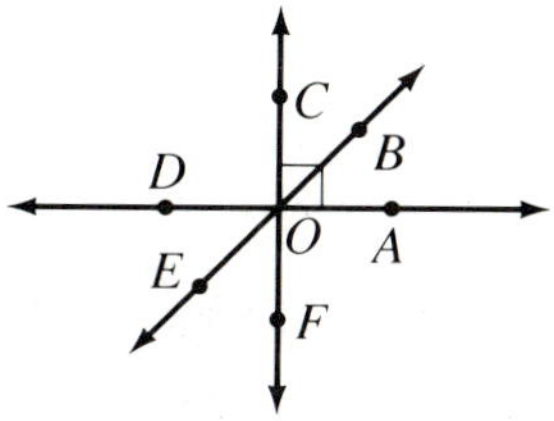

2. In the figure for Exercise 1, name two pairs of
   (a) complementary angles
   (b) supplementary angles
3. What is the measure of an angle that has the same measure as its supplement?
4. Find the measure of the complement of each of the following angles:
   (a) 25°
   (b) 60°
   (c) 50°
   (d) $42\frac{1}{2}°$
   (e) $x°$ $(0 < x < 90)$
5. Find the measure of the supplement of each of the following angles:
   (a) 30°
   (b) 120°
   (c) 45°
   (d) 155°
   (e) $y°$ $(0 < y < 90)$
6. Refer to the following figures and tell which of the following pairs of angles are adjacent. If they are not, give a reason.
   (a) $\angle ABC$ and $\angle DBC$
   (b) $\angle ABD$ and $\angle ABC$
   (c) $\angle QPV$ and $\angle PVU$
   (d) $\angle ROS$ and $\angle SOT$
   (e) $\angle ROT$ and $\angle TOS$

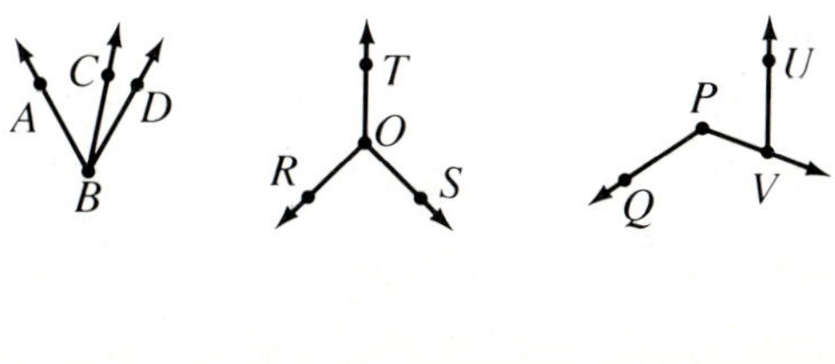

7. Refer to the accompanying figure and tell whether each of the following is acute, right, obtuse, or straight.
   (a) $\angle AOB$ **(b)** $\angle DOG$
   (c) $\angle BOF$ **(d)** $\angle COF$
   (e) $\angle COG$ **(f)** $\angle EOF$
   (g) $\angle AOG$ **(h)** $\angle GOF$
   (i) $\angle FOB$

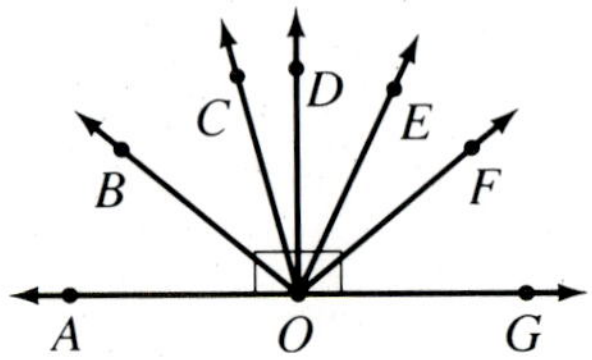

8. In the following figure, name
   (a) two sets of complementary angles
   (b) two sets of supplementary angles

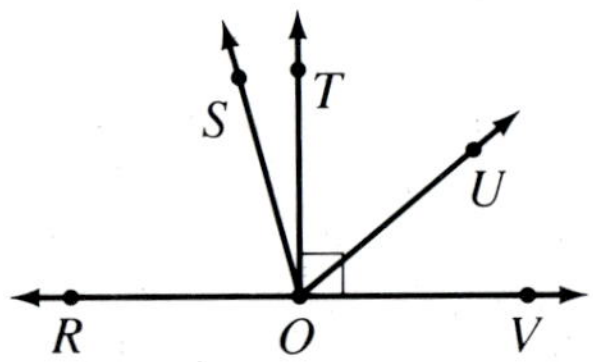

9. Find the measure of an angle if it is $\frac{3}{4}$ of the measure of its complement.
10. Find the measure of the angle formed by the hands of a clock at
    (a) 1 o'clock **(b)** 4 o'clock **(c)** 5 o'clock
11. The measure of an angle is 30 more than that of its supplement. What is the measure of each angle?
12. If two angles are supplementary and the measure of one is 5 more than twice that of the other, what is the measure of each angle?
13. Two angles are supplementary and the difference of their measures is 90. What is the measure of each angle?
14. The measure of the supplement of a certain angle is 25 more than twice that of its complement. What is the measure of the angle?

Congruent Angles

The definition of congruent angles is similar to that of congruent segments.

Definition 3-8 Two **angles** are **congruent** if and only if their measures are equal. Therefore, $\angle A \cong \angle B$ if and only if $m\angle A = m\angle B$.

Examples

1. If $m\angle ABC = 28$ and $m\angle RST = 28$, then $\angle ABC \cong \angle RST$ by Def. 3-8. (The reader should state the entire definition.)
2. In the following figure $\overleftrightarrow{AB}$, $\overleftrightarrow{CD}$, and $\overleftrightarrow{GF}$ intersect at O, $\angle AOC$ and $\angle BOC$ are right angles, $m\angle a = 30$, and $m\angle c = 60$.

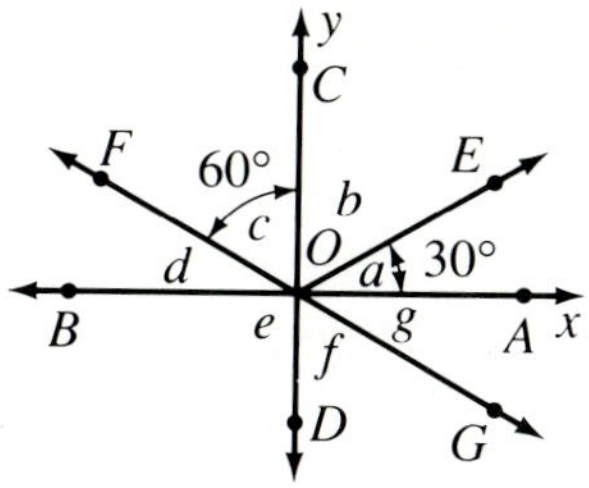

Complete the following:

(a) $m\angle b =$ _______ **(b)** $m\angle d =$ _______

(c) $\angle a$ _______ $\angle d$ **(d)** $\angle b$ _______ $\angle c$

(e) $\angle a$ and $\angle b$ are _______ **(f)** $\angle a$ and $\angle c$ are _______

Solution (a) We are given that *BOC* is a right angle; hence by definition its measure is 90. Since $m\angle a = 30$, $m\angle b = 90 - 30 = 60$. (b) *AOC* is a right angle and $m\angle c = 60$, so $m\angle d = 90 - 60 = 30$. (c) and (d) Since $m\angle a = m\angle d = 30$, $\angle a \cong \angle d$ and $m\angle b = m\angle c = 60$, $\angle b \cong \angle c$. (e) $m\angle a = 30$ and $m\angle b = 60$, so $\angle a$ and $\angle b$ are complementary by definition of complementary angles (Def. 3-7). (f) $\angle a$ and $\angle c$ are complementary since $m\angle a = 30$ and $m\angle c = 60$.

3. In the following figure we have $\overleftrightarrow{AB}$ and $\overleftrightarrow{CD}$ intersecting at *O*. Show that (a) $\angle a$ and $\angle d$ are supplementary; (b) $\angle a$ and $\angle b$ are supplementary; (c) $\angle b \cong \angle d$.

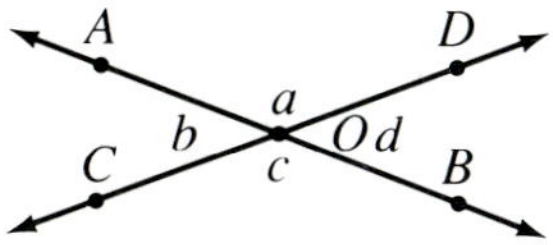

Solution (a) $\angle AOB$ and $\angle COD$ are straight angles, so $m\angle AOB = m\angle COD = 180$ by the definition of angle measure (Def. 3-3(a)). By Postulate 3-3 $m\angle a + m\angle d = m\angle AOB$. Thus $\angle a$ and $\angle d$ are supplementary, according to Defs. 3-4 and 3-6.
(b) As in part (a), $m\angle a + m\angle b = m\angle COD = 180$, so $\angle a$ and $\angle b$ are supplementary. (c) From Post. 3-3, we know that $m\angle d = m\angle AOB - m\angle a = 180 - m\angle a$, and $m\angle b = m\angle COD - m\angle a = 180 - m\angle a$. Thus $m\angle b = m\angle d$ by the Transitive Axiom for Equality of real numbers, and $\angle b \cong \angle d$ by the definition of congruent angles (Def. 3-8). ◁

Exercise Set 3.5

Draw the figure for each problem.

1. If $m\angle ABC = 30$, $m\angle CBD = 60$, and $\angle RST$ is complementary to $\angle CBD$, what is the measure of $\angle RST$?
2. If $m\angle ABC = 135$, $m\angle CBP$ is 45, and $\angle QPT$ is supplementary to $\angle ABC$, what is $m\angle QPT$?

3. Given $\overleftrightarrow{AB}$ intersects $\overleftrightarrow{CD}$ at O.

(a) $m\angle a + m\angle b = 180$. Why?
(b) $m\angle b + m\angle c =$ ________. Why?
(c) $m\angle c + m\angle d =$ ________. Why?
(d) $m\angle d + m\angle a =$ ________. Why?
(e) $m\angle a + m\angle b = m\angle b + m\angle c$. Why?
(f) $m\angle b + m\angle c$ ________ $m\angle c + m\angle d$. Why?

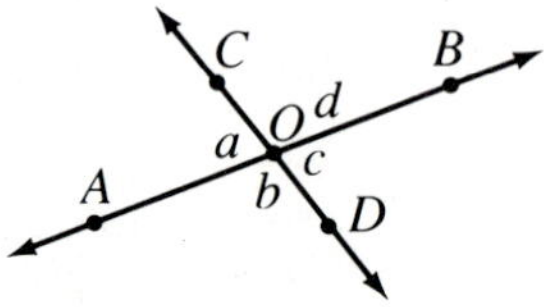

Since congruence for angles has been defined as equality of their measure, we have another equivalence relation, which can be stated as a theorem:

Theorem 3-1 For any angles A, B and C
(a) $\angle A \cong \angle A$. (Congruence of angles is reflexive.)
(b) If $\angle A \cong \angle B$, then $\angle B \cong \angle A$. (Congruence of angles is symmetric.)
(c) If $\angle A \cong \angle B$ and $\angle B \cong \angle C$, then $\angle A \cong \angle C$. (Congruence of angles is transitive.)

The next set of theorems is needed for ease of development in our future work.

Theorem 3-2 If P is in the interior of $\angle ABC$ and Q is in the interior of $\angle RST$, so that $\angle ABP \cong \angle RSQ$ and $\angle PBC \cong \angle QST$, then $\angle ABC \cong \angle RST$.

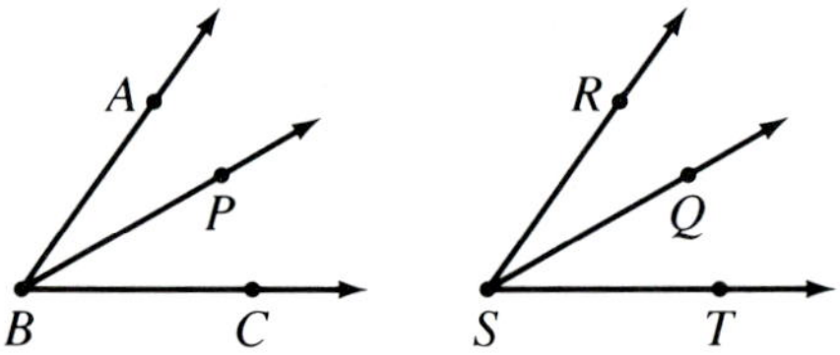

Given: $\angle ABP \cong \angle RSQ$ and $\angle PBC \cong \angle QST$
Prove: $\angle ABC \cong \angle RST$

Proof

Statements	*Reasons*
1. $\angle ABP \cong \angle RSQ$ and $\angle PBC \cong \angle QST$	1. Given
2. $m\angle ABP = m\angle RSQ$ and $m\angle PBC = m\angle QST$	2. If two angles are congruent, their measures are equal (Def. 3-8)
3. $m\angle ABP + m\angle PBC = m\angle RSQ + m\angle QST$	3. Addition Law for Equals

Statements	*Reasons*
4. $m\angle ABP + m\angle PBC = m\angle ABC$ and $m\angle RSQ + m\angle QST = m\angle RST$	4. If P is in the interior of $\angle ABC$, then $m\angle ABP + m\angle PBC = m\angle ABC$ (Post. 3-3)
5. $m\angle ABC = m\angle RST$	5. Substitution
6. $\angle ABC \cong \angle RST$	6. If the measures of two angles are equal, the angles are congruent (Def. 3-8)

The proofs of the theorems below can be done in a manner similar to that of Theorem 3-2 and are left as exercises for the reader.

Theorem 3-3(a) If P is in the interior of $\angle ABC$ and Q is in the interior of $\angle RST$, so that $\angle ABC \cong \angle RST$ and $\angle ABP \cong \angle RSQ$, then $\angle PBC \cong \angle QST$. (Refer to the figure with Theorem 3-2.)

Theorem 3-3(b) If ABC and RST are straight angles, P is a point not on ABC and Q is a point not on RST, and $\angle ABP \cong \angle RSQ$, then $\angle PBC \cong \angle QST$.

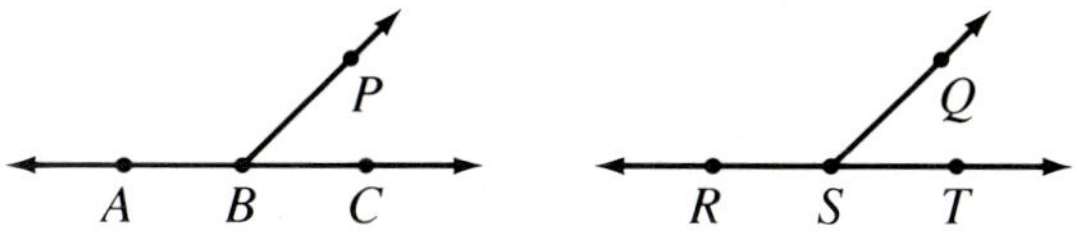

Exercise Set 3.6

1. In the following figure, $\overleftrightarrow{TV}$ and $\overleftrightarrow{RS}$ intersect at B.
   (a) $\angle TBR \cong \angle TBR$. Why?
   (b) $\angle SBT \cong \angle VBR$. Why?

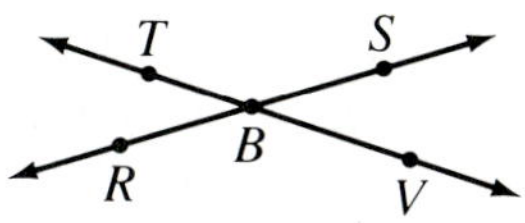

2. In the following figure, $\angle ABP \cong \angle PBC$, $\angle BAP \cong \angle PAC$, and $\angle BAP \cong \angle ABP$.
   (a) $\angle PAC \cong \angle PBC$. Why?
   (b) $\angle CAB \cong \angle CBA$. Why?

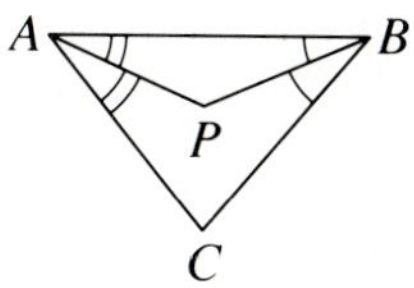

3. Prove Theorem 3-1(c). *Hint:* Refer to the proof of Theorem 2-1(b) in Sec. 2-4.
4. **(a)** Prove Theorem 3-3(a).
   (b) Prove Theorem 3-3(b).

5. In the following figure, $\angle GMU \cong \angle SMC$. Prove that $\angle GMS \cong \angle UMC$.

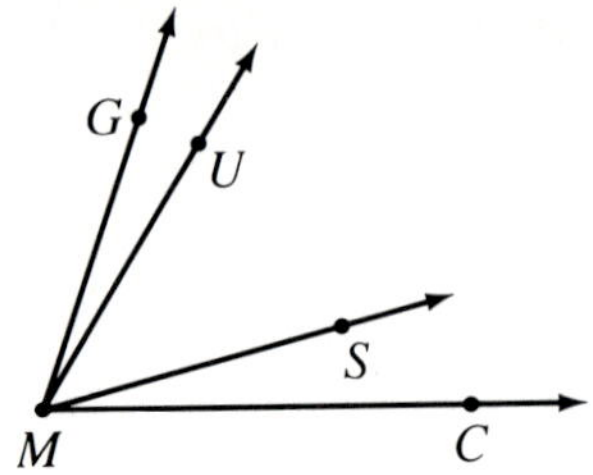

Vertical Angles

In Example 3 on page 51 and in the preceding Exercise 1, on page 53 we found that if two straight lines intersect, the angles formed by the opposite rays are congruent.

Definition 3-9 Two nonadjacent angles formed by two intersecting lines are called **vertical angles.**

In Fig. 3-15 angles a and c are vertical angles, and angles b and d are also vertical angles. Note that vertical angles are *pairs* of angles, so that $\angle a$ alone is not a vertical angle, and $\angle a$ and $\angle b$ are not vertical angles with respect to one another.

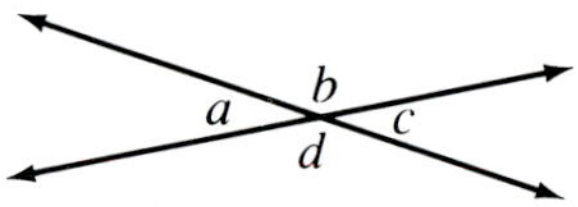

Figure 3-15

Theorem 3-4 Vertical angles are congruent.

(We have demonstrated the proof of this theorem in Example 3 on page 51, although at the time the angles were not identified as vertical angles.)

Exercise Set 3-7

1. In the following figure, name two pairs of:
   (a) adjacent angles
   (b) supplementary angles
   (c) vertical angles

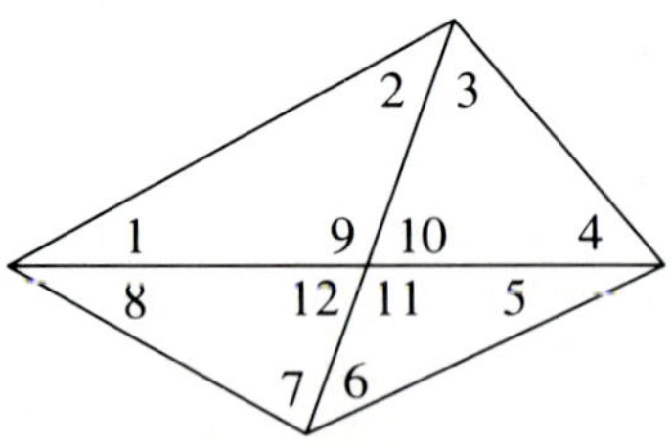

2. In the figure for Exercise 1, if $\angle 9$ has measure 110, find
   (a) $m\angle 10$ **(b)** $m\angle 11$
   (c) $m\angle 12$
3. If in the following figure $\angle c$ and $\angle b$ are complementary and $\angle a$ and $\angle b$ are supplementary, find the measure of each angle.

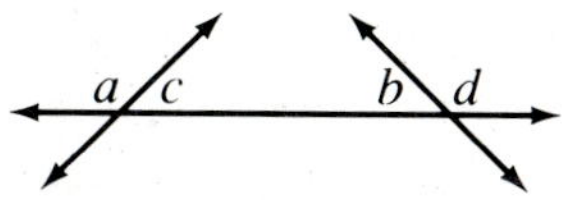

4. In the following figure, $m\angle a = 30$ and $\angle a \cong \angle g$. Find
   (a) $m\angle b$ **(b)** $m\angle c$
   (c) $m\angle d$ **(d)** $m\angle e$
   (e) $m\angle f$ **(f)** $m\angle h$

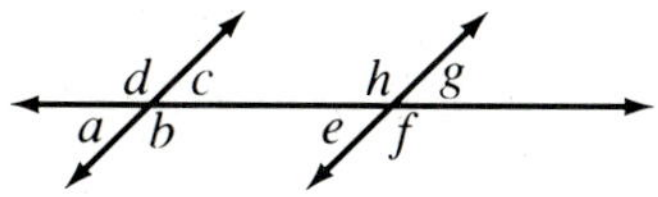

5. If, in the figure for Exercise 4, $\angle a \cong \angle e$, prove that $\angle c \cong \angle g$.

3-4 PROVING THEOREMS

Hypothesis and Conclusion

A statement in the form "If *P*, then *Q*" is said to have *hypothesis P* and *conclusion Q*. This means that the statements contained in *P* are assumed to be true or "given," whereas those in *Q* are to be proved. Not all theorems are stated in the "if *P*, then *Q*" form, but it is usually possible to change them so that hypothesis and conclusion are easily identified. Since "If *Q*, then *P*" is the converse of "If *P*, then *Q*" and we have established that an implication and its converse are not always equivalent, we must be careful not to interchange hypothesis and conclusion. For example, consider the theorem "All right angles are congruent." This can be changed to "If $\angle A$ and $\angle B$ are right angles, then they are congruent." It would not be correct to say "If $\angle A$ and $\angle B$ are congruent, then they are right angles." It is unlikely that we would make this particular mistake, but we must be alert to the possibility.

Example

If $\overleftrightarrow{AB}$ intersects $\overleftrightarrow{CD}$, the vertical angles formed are congruent.

Hypothesis: $\overleftrightarrow{AB}$ intersects $\overleftrightarrow{CD}$.

Conclusion: The vertical angles formed are congruent. ◁

Formal Proofs

The theorems that follow are essential, so be sure to work through the proofs.

Although no specific rules for proving theorems can be given, the following may prove useful as a model in the development of proofs.

1. Read the problem (theorem) carefully, being certain that you understand the meaning of all the terms used.
2. Determine the hypothesis and conclusion, then draw an appropriate figure in terms of the given information and label it.
3. State the hypothesis and conclusion in terms of the figure.
4. When writing the proof, consider the definitions, postulates, and theorems that might aid you or give you information on proving the theorems. Then write an orderly sequence of statements, with supporting reasons, that lead from the hypothesis to the conclusion.

Theorem 3-5 If the exterior sides of two adjacent angles form a straight angle, the angles are supplementary.

Hypothesis: $\angle ABP$ is adjacent to $\angle PBC$ and $\angle ABC$ is a straight angle
Conclusion: $\angle ABP$ is the supplement of $\angle PBC$.

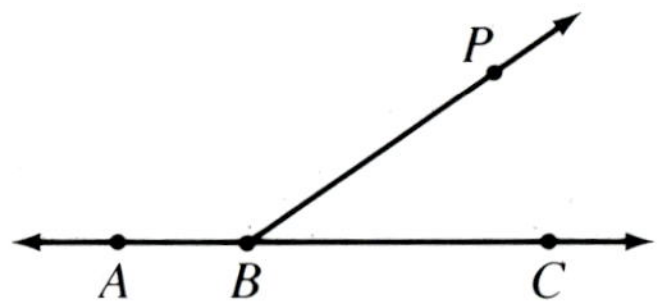

Proof

Statements	*Reasons*
1. $\angle ABP$ is adjacent to $\angle PBC$ and $\angle ABC$ is a straight angle	1. Given
2. $m\angle ABP + m\angle PBC = m\angle ABC$	2. Angle Addition Postulate (Postulate 3-3)
3. $m\angle ABC = 180$	3. Definition of measure of straight angle (Def. 3-4(d))
4. $m\angle ABP + m\angle PBC = 180$	4. Transitive Axiom of Equality for real numbers
5. $\therefore \angle ABP$ is the supplement of $\angle PBC$	5. Two angles are supplementary if the sum of their measures is 180 (Def. 3-6)

Theorem 3-6 All right angles are congruent.
Hypothesis: $\angle A$ and $\angle B$ are right angles
Conclusion: $\angle A \cong \angle B$

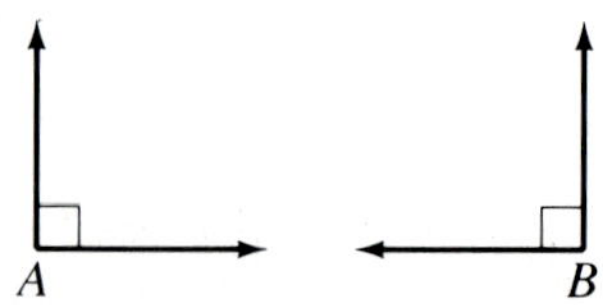

Proof

Statements	*Reasons*
1. $\angle A$ and $\angle B$ are right angles	1. Given
2. $m\angle A = 90$ and $m\angle B = 90$	2. The measure of a right angle is 90 (Def. 3-4(b))
3. $m\angle A = m\angle B$	3. Transitive Axiom of Equality for real numbers
4. $\therefore \angle A \cong \angle B$	4. If two angles have equal measures, then they are congruent (Def. 3-8)

A theorem which follows from a previous theorem with little additional proof is called a *corollary*. You will be asked to prove the following two corollaries for Theorem 3-6 in the exercises.

Corollary 3-6(a) An angle congruent to a right angle is a right angle.

Corollary 3-6(b) If a pair of angles are equal in measure and supplementary, they are right angles.

Theorem 3-7 Complements of the same angle are congruent.
Hypothesis: $\angle a$ is the complement of $\angle b$ and $\angle b$ is the complement of $\angle c$
Conclusion: $\angle a \cong \angle c$

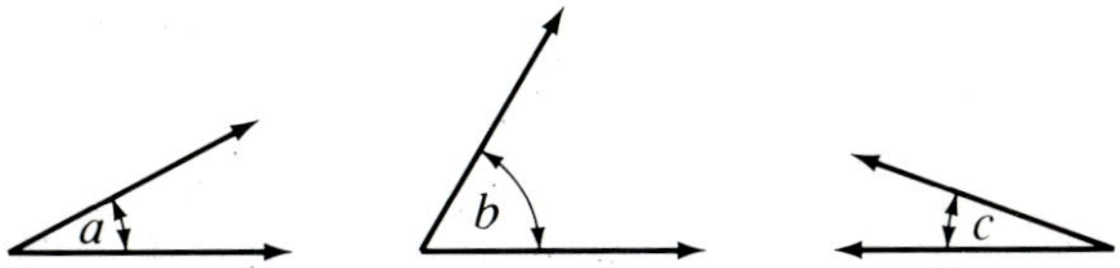

Proof

Statements	*Reasons*
1. $\angle a$ is the complement of $\angle b$; $\angle b$ is the complement of $\angle c$	1. Given
2. $m\angle a + m\angle b = 90$; $m\angle b + m\angle c = 90$	2. If two angles are complementary the sum of their measures is 90 (Def. 3-7)
3. $m\angle a + m\angle b = m\angle b + m\angle c$	3. Substitution Axiom of Equality for real numbers
4. $m\angle a = m\angle c$	4. Cancellation Law for Equals
5. $\therefore \angle a \cong \angle c$	5. Definition of congruent angles (Def. 3-8)

Theorem 3-8 Complements of congruent angles are congruent. (The proof is left to the reader.)

Theorem 3-9 Supplements of the same angle are congruent. (The proof is left to the reader.)

Theorem 3-10 Supplements of congruent angles are congruent.
Hypothesis: $\angle a$ and $\angle b$, $\angle c$ and $\angle d$ are supplementary; $\angle a \cong \angle c$
Conclusion: $\angle b \cong \angle d$

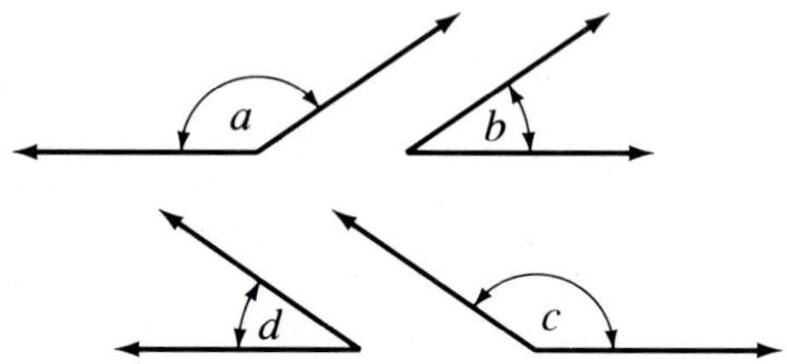

Proof

Statements	*Reasons*
1. $\angle a$ and $\angle b$, $\angle c$ and $\angle d$ are supplementary; $\angle a \cong \angle c$	1. Given
2. $m\angle a + m\angle b = 180$, $m\angle c + m\angle d = 180$	2. Definition of supplementary angles (Def. 3-6)
3. $m\angle a + m\angle b = m\angle c + m\angle d$	3. Transitive Axiom of Equality for real numbers
4. $m\angle a = m\angle c$	4. Definition of congruent angles (Def. 3-8)
5. $m\angle b = m\angle d$	5. Cancellation Law for Equals
6. $\therefore \angle b \cong \angle d$	6. Definition of congruent angles (Def. 3-8)

In writing our proofs in a step-by-step manner we sometimes lose sight of what we are doing and leave out an important part of the proof. Let us recall the Law of Syllogism and the Law of Detachment, which were discussed in Section 1-3. The Law of Syllogism states $[(P \rightarrow Q) \text{ and } (Q \rightarrow R)] \rightarrow (P \rightarrow R)$ and the Law of Detachment says that $[(P \rightarrow Q) \text{ and } P] \rightarrow Q$.

In proving Theorem 3-10 (supplements of congruent angles are congruent), we could make use of a diagram such as Fig. 3-16.

Here we see a sequence of implications, the conclusion of which is true because of the statement in the hypothesis. You will be asked to draw such a diagram for some of the proofs in the exercises which follow.

Solutions for Exercises

It is often necessary to present reasons for statements in the solution to an exercise or problem that is not general or significant enough to be considered as a theorem. Following are some examples of proofs of this type.

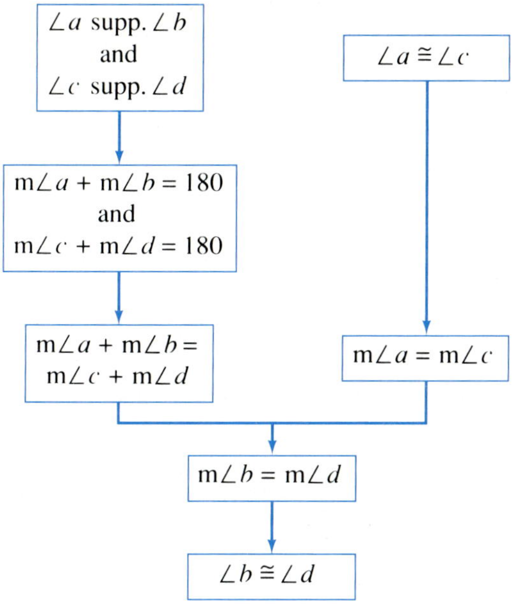

Figure 3-16

Example

In the illustration below, $\angle ABC$ and $\angle DEF$ are right angles and $\angle 1 \cong \angle 3$. Prove that $\angle 2 \cong \angle 4$.

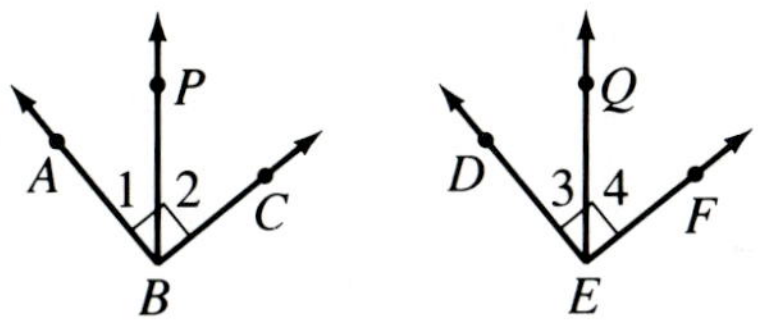

Proof

Statements	*Reasons*
1. $\angle ABC$ and $\angle DEF$ are right angles; $\angle 1 \cong \angle 3$	1. Given
2. $\angle ABC \cong \angle DEF$	2. All right angles are congruent (Theorem 3-6)
3. $\therefore \angle 2 \cong \angle 4$	3. If P is in the interior of $\angle ABC$ and Q is in the interior of $\angle RST$, so that $\angle ABP \cong \angle RSQ$, then $\angle PBC \cong \angle QST$. (Theorem 3-3(a))

Exercise Set 3.8

1. Identify the hypothesis and conclusion of each of the following.
   (a) If $a = b$, then $b = a$.
   (b) If $a + c = b + c$, then $a = b$.

(c) Straight angles are congruent.
(d) If two angles are supplements of the same angle, then they are congruent.

2. Write each of the following in "if P, then Q" form.
(a) Complements of the same angle are congruent.
(b) The sum of the measures of supplementary angles is 180°.

3. In the following figure, $\angle 1 \cong \angle 3$ and AB is a line. Prove $\angle 2 \cong \angle 4$.

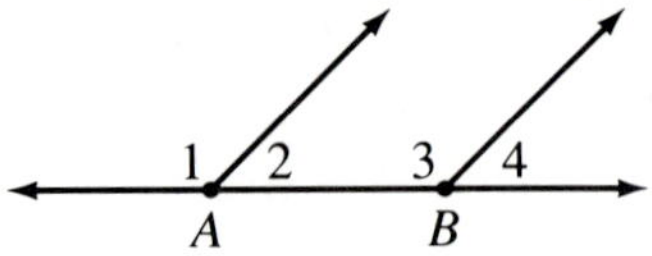

4. Prove Corollary 3-6(a).

5. **(a)** Prove Corollary 3-6(b).
(b) Using the model following Theorem 3-10, diagram the proof of Corollary 3-6(b).

6. **(a)** Prove Theorem 3-8.
(b) Prove Theorem 3-9.

7. If, in the following figure, $\angle DPC$ is a right angle and P is on $\overleftrightarrow{AB}$, prove that $\angle APD$ is the complement of $\angle BPC$.

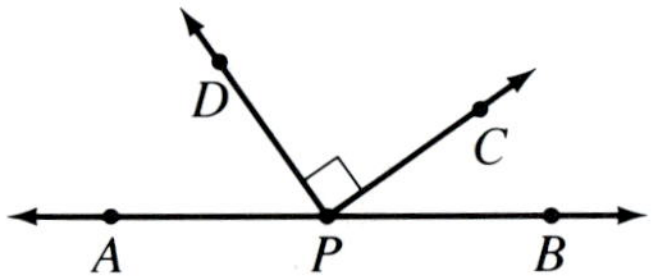

8. In the plane $\mathscr{E}$, $\overleftrightarrow{AB}$, $\overleftrightarrow{CD}$, $\overleftrightarrow{EF}$, $\overleftrightarrow{GH}$, and $\overleftrightarrow{KM}$ intersect at O so that $\angle COB$, $\angle COA$, $\angle AOD$, and $\angle DOB$ are right angles.
Prove: $m\angle a + m\angle b + m\angle c + m\angle j = m\angle k$.

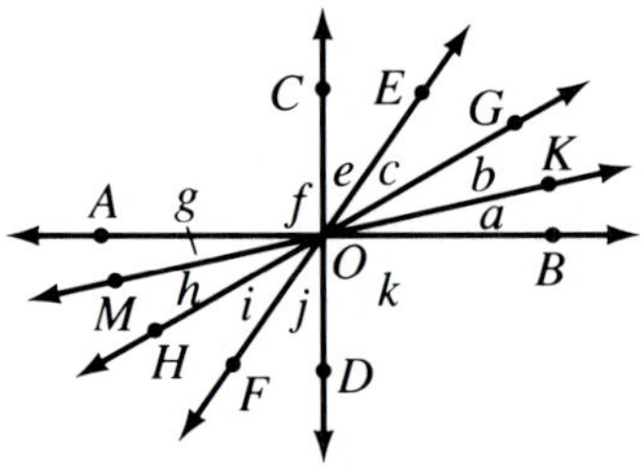

9. In the accompanying figure, $\angle a$ and $\angle b$ are supplementary angles on $\overleftrightarrow{AB}$.

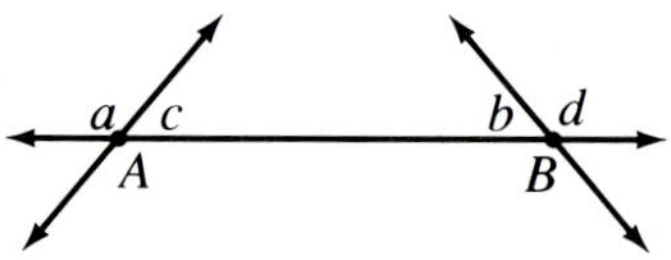

Prove:
(a) $\angle c \cong \angle b$
(b) $\angle c$ is supplementary to $\angle d$
(c) $\angle a \cong \angle d$
(d) Using the model following Theorem 3-10, diagram the proof of parts (a), (b), and (c).

10. Supply reasons for each statement in the following proof:
Given: ℓ_1 and ℓ_2 intersect line ℓ_3 so that $\angle 1 \cong \angle 4$
Prove: $\angle 3$ is supplementary to $\angle 4$; $\angle 1 \cong \angle 5$

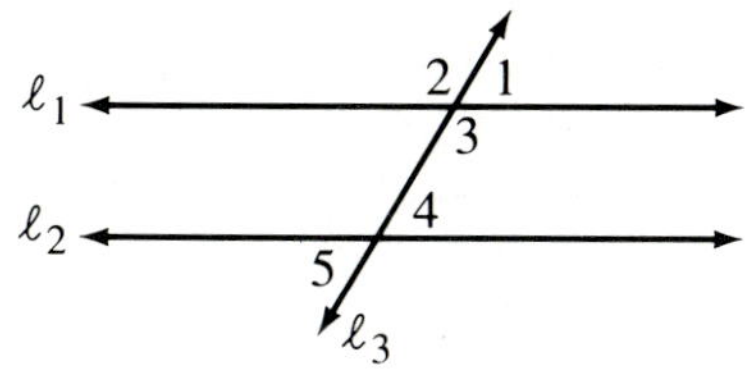

Proof

Statements	*Reasons*
1. $\angle 1$ and $\angle 3$ are adjacent	1.
2. $\angle 1$ and $\angle 3$ are supplementary	2.
3. $m\angle 1 + m\angle 3 = 180$	3.
4. $\angle 1 \cong \angle 4$	4.
5. $m\angle 1 = m\angle 4$	5.
6. $m\angle 4 + m\angle 3 = 180$	6.
7. $\therefore \angle 3$ and $\angle 4$ are supplementary	7.
8. $\angle 4$ and $\angle 5$ are vertical angles	8.
9. $\angle 4 \cong \angle 5$	9.
10. $\therefore \angle 1 \cong \angle 5$	10.

11. **(a)** Prove that if $\angle x$ and $\angle y$ are complements and $\angle x \cong \angle z$, then $\angle y$ is complementary to $\angle z$.
(b) Using the model following Theorem 3-10, diagram the proof of part (a).

12. Prove that if $\angle x$ and $\angle y$ are supplements and $\angle x \cong \angle z$, then $\angle y$ and $\angle z$ are supplements.

3-5 PERPENDICULAR LINES; THE COORDINATE SYSTEM

From the definition of supplementary angles, we see that if two lines intersect so that the measures of the adjacent angles formed are equal, the adjacent angles are right angles. In Fig. 3-17, $a + a = 180$, so $a = 90$ and $\angle MOS$ is a right angle.

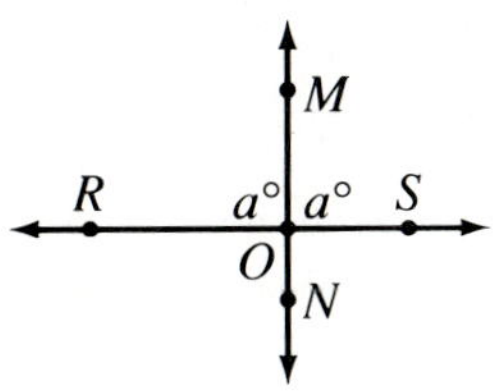

Figure 3-17

Definition 3-10 Two intersecting sets, each of which is either a segment, ray, half-line or line, are said to be **perpendicular** if and only if they determine at least one right angle.

Some of the possibilities for perpendicularity are illustrated in Fig. 3-18. The symbol $\perp$ is used to indicate that two sets are perpendicular.

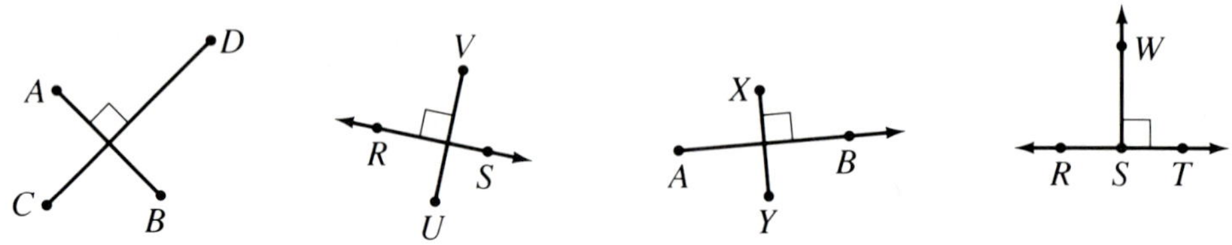

Figure 3-18

From Fig. 3-18 we can write

$\overline{AB} \perp \overline{CD}$: segment AB is perpendicular to segment CD
$\overleftrightarrow{RS} \perp \overline{UV}$: line RS and segment UV are perpendicular
$\overline{XY} \perp \overrightarrow{AB}$: segment XY and ray AB are perpendicular
$\overleftrightarrow{RT} \perp \overrightarrow{SW}$: line RT is perpendicular to ray SW

Theorem 3-11 If one of the four angles formed by the intersection of two lines is a right angle, the other three are right angles. (The proof is left to the student.)

We now have a series of statements with which we can prove angles to be right angles (Theorem 3-6, Corollaries 3-6(a) and 3-6(b), and Theorem 3-11). Since these are all closely related, we will put them all under the title "Right angle theorems" and then designate the particular one to which we are referring.

Exercise Set 3.9

1. In the following figure, $\overleftrightarrow{AD}$ and $\overleftrightarrow{BC}$ intersect at B and $\angle ABC \cong \angle CBD$.
   (a) How is $\overline{AB}$ related to $\overline{BC}$?
   (b) How is $\overline{BC}$ related to $\overline{BD}$?
   (c) What kind of angle is $\angle ABC$? $\angle CBD$?

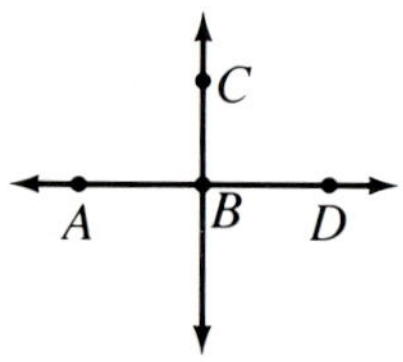

2. In a given plane, how many perpendiculars can be drawn to a given line which contain a given point on the line?
3. Given $\overrightarrow{BA}$ and $\overrightarrow{BC}$ are opposite rays, $m\angle CBT = 40$, and $m\angle RBT = 90$.
   (a) Name a pair of complementary angles.
   (b) Name a pair of perpendicular figures.
   (c) Name a pair of supplementary angles.

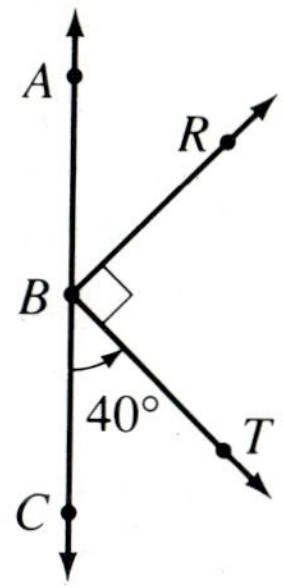

4. Refer to the figure for Exercise 3 and tell whether each of the following angles is acute, right, obtuse or straight.
   (a) $\angle RBA$ **(b)** $\angle ABT$
   (c) $\angle RBT$ **(d)** $\angle TBC$
   (e) $\angle RBC$ **(f)** $\angle ABC$
5. If two angles of the same measure are complementary, what is the measure of each angle?
6. Prove Theorem 3-11.

3-6 CONSTRUCTION OF ANGLES

The construction that follows is given here since we have been discussing angles in this chapter. The construction can be proved to be valid, but we need to wait until the next chapter in order to have the theorems necessary for the proof. The procedure should be intuitively clear, however.

Construction 2: Construct an angle congruent to a given angle.

To duplicate an angle MON at a point P on ray PQ

1. Draw an arc of a circle with center at O and with arbitrary radius r_1, intersecting $\overrightarrow{OM}$ at D and $\overrightarrow{ON}$ at E.
2. Draw another arc at P with the same radius r_1, intersecting $\overrightarrow{PQ}$ at G.
3. Set the compass points at D and E so that $r_2 = \text{m}\overline{DE}$; then draw an arc with radius r_2 and center G that intersects the second arc at R.
4. Draw $\overrightarrow{PR}$, and $\angle RPQ \cong \angle MON$.

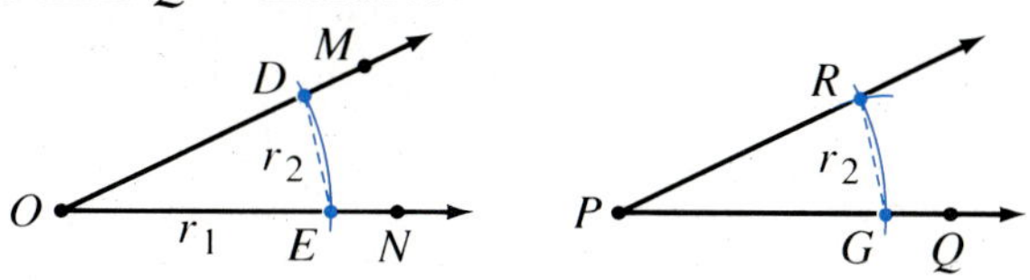

Exercise Set 3.10

When drawing the following constructions, plan to use at least half a page for each problem. It would probably be a good idea to work each problem out on scratch paper first and then try it again for your homework. Be sure to label your results and show all construction lines.

1. Construct an angle on your paper that is congruent to angle MON in Construction 2.
2. Draw an obtuse angle ABC. Construct an angle PDQ that is congruent to angle ABC.
3. Draw an acute angle RST with $\overrightarrow{ST}$ horizontal. At point R construct an angle MRS that is congruent to $\angle RST$. How would you describe the way in which lines ST and MR are related?
4. Draw an acute angle FGH with $\overrightarrow{GH}$ horizontal and point H to the right of G. At point H construct an angle GHD congruent to angle FGH with point D to the left of H. Label the point of intersection of $\overrightarrow{GF}$ and $\overrightarrow{HD}$ as E. How do the measures of $\overline{GE}$ and $\overline{HE}$ compare?

RENÉ DESCARTES (1596–1650) The great French philosopher revolutionized geometry with the invention of the concept of a coordinate system, which enabled mathematicians to interpret algebraic equations geometrically through the use of curves. This system is known today as the Cartesian coordinate system, or rectangular coordinate system. Descartes also developed the first systematic classification of curves. (*New York Public Library Picture Collection*)

Coordinate System in a Plane

In Chapter 2 we developed a coordinate system for a line, based upon the Ruler Postulate (Post. 2-4), which states that the points on a line can be associated with the set of real numbers so that

(a) to every point there corresponds a unique real number, and
(b) to every real number there corresponds one and only one point.

We also recall, from Def. 2-6, that the number corresponding to each point on a line is called the coordinate of the point and the correspondence is called a *coordinate system*. Thus on the horizontal line OX shown in Fig. 3-19, every number locates a point and every point is determined by its coordinate x. Let us call this line the *x-axis*. A line OY, perpendicular to $\overleftrightarrow{OX}$ at 0, may be designated

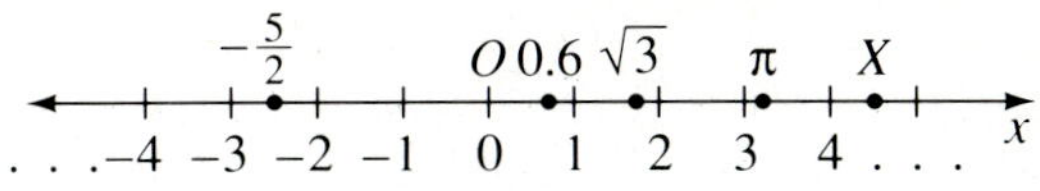

Figure 3-19

as the *y-axis*, with the positive direction upward, the negative downward (Fig. 3-20). According to Post. 2-4 and Def. 2-6 there is a coordinate system on the y-axis, and Post. 2-8 tells us that the x-axis and y-axis determine a plane. Now we can establish a method for naming and locating each of the points in the plane by means of an "ordered pair" of real numbers. In Fig. 3-21 we see the ordered pair $(a, 0)$ associated with A, $(b, 0)$ associated with B, $(0, c)$ associated with C, and $(0, d)$ associated with D. Note that the ordered pairs of the form $(x, 0)$ are on the x-axis and that the ordered pairs of the form $(0, y)$ are on the y-axis. The point at the intersection of the two axes is called the *origin* and is designated $(0, 0)$. Note that this association is a *one-to-one correspondence*, that is, to each point on an axis there corresponds a unique ordered pair and for each ordered pair of the form $(x, 0)$ or $(0, y)$ there is exactly one point. The ordered pairs $(x, 0)$ and $(0, y)$ are *coordinates* of the points on the axes.

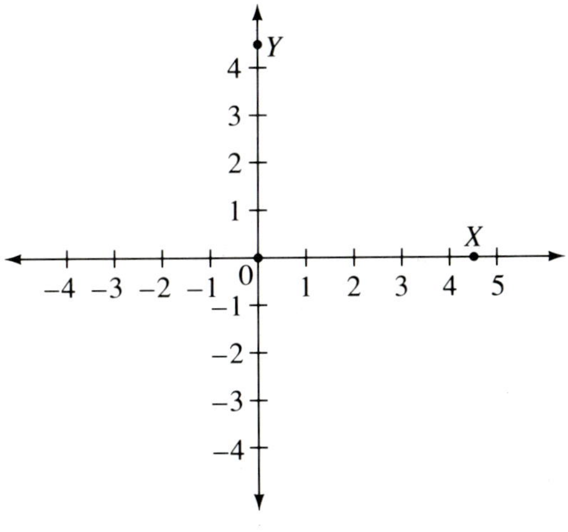

Figure 3-20

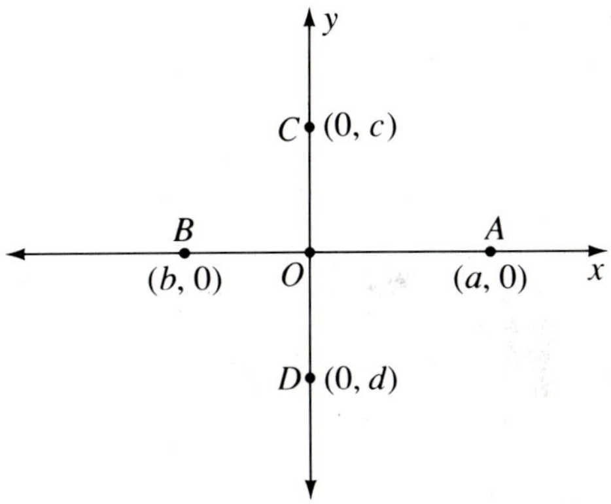

Figure 3-21

Recall that a line separates the plane into two half-planes. Lines OX and OY separate the plane into four parts known as *quadrants*, which are identified by numbers as shown in Fig. 3-22. A point P not the origin will be in one of the four quadrants, on line OX, or on line OY.

Definition 3-11 If P is a point in one of the quadrants determined by the x-axis $\overleftrightarrow{OX}$ and the y-axis $\overleftrightarrow{OY}$, there is a point Q with coordinates $(x, 0)$ such that $\overleftrightarrow{PQ} \perp \overleftrightarrow{OX}$ and there is a point R with coordinates $(0, y)$ such that $\overleftrightarrow{PR} \perp \overleftrightarrow{OY}$. Point Q is the **projection of P on $\overleftrightarrow{OX}$** and point R is the **projection of P on $\overleftrightarrow{OY}$** (Fig. 3-23).

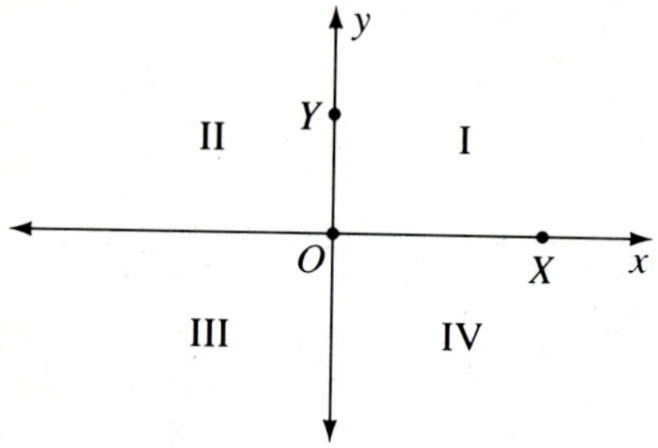

Figure 3-22

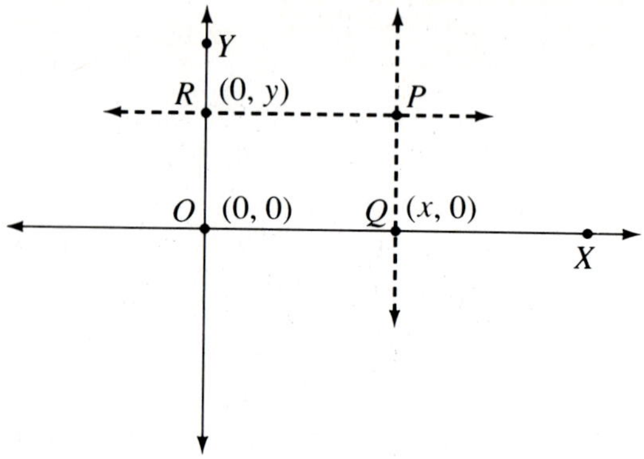

Figure 3-23

Example

Refer to the following figure. List the quadrants in which points A through G are found:

Solution A is in I, B none (on line OY), C in II, D in III, E in IV, F none (on $\overleftrightarrow{OX}$), G none (on $\overleftrightarrow{OX}$), and H none (on $\overleftrightarrow{OY}$). ◁

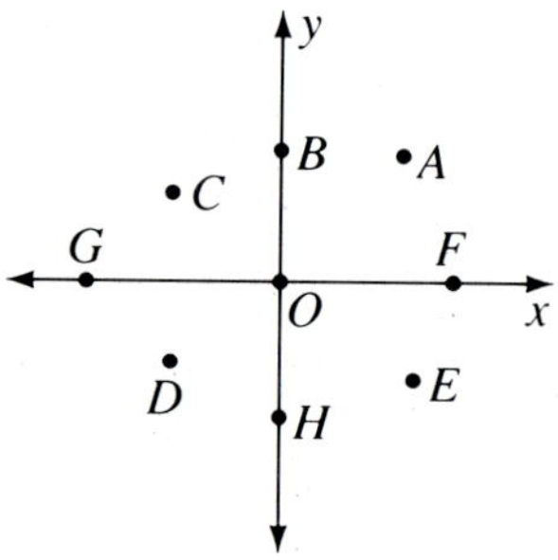

Exercise Set 3.11

1. In the illustration, how many projections from points on segment MN will fall on $\overleftrightarrow{OX}$? How many will fall on $\overleftrightarrow{OY}$?

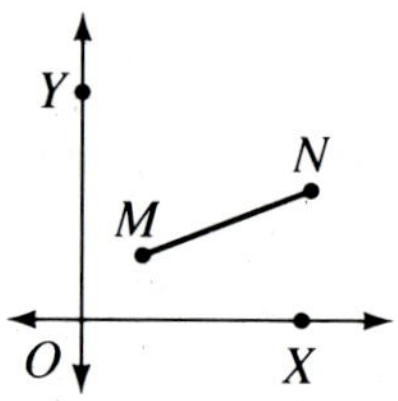

2. In the illustration below, how many projections from points on line AB will fall on $\overleftrightarrow{OX}$?

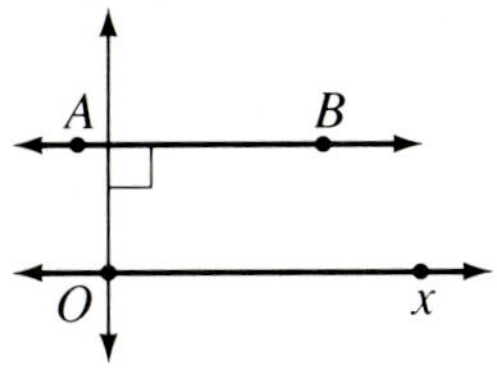

3. How many lines may be drawn which are perpendicular to $\overleftrightarrow{OX}$ in the plane OXY?
4. In what quadrant does P lie if the coordinates of its projections are:
   (a) (5, 0) and (0, -3) **(b)** (-2, 0) and (0, -4)
   (c) (-4, 0) and (0, 1) **(d)** (3, 0) and (0, 6)
   (e) ($5\frac{1}{2}$, 0) and (0, -2)
5. Locate each point described in Exercise 4, using a scale as indicated in the following illustration.

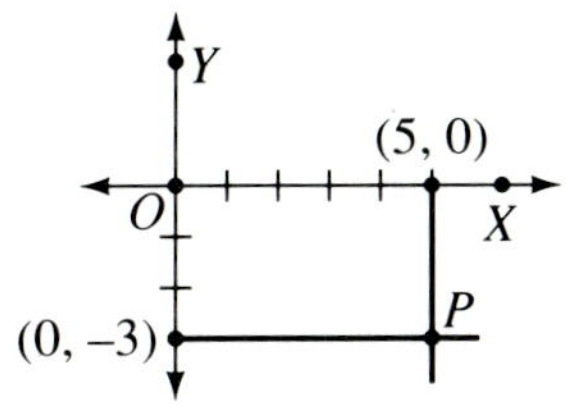

Coordinates of Any Point

Through any point in $\overleftrightarrow{OX}$ we can draw a perpendicular line, as indicated in Fig. 3-24. With each of these points is associated an ordered pair of numbers (x, 0).

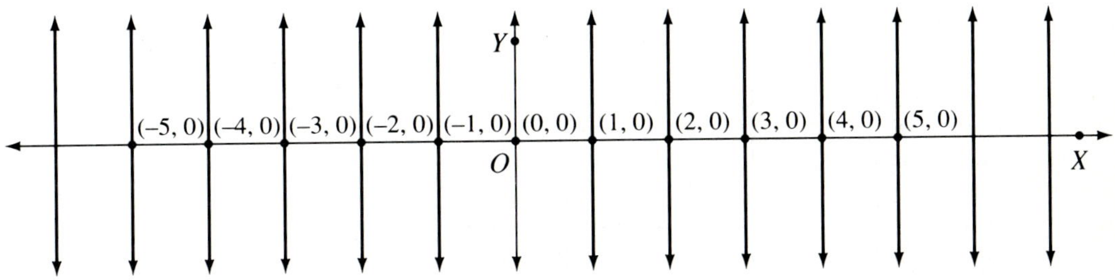

Figure 3-24

In like fashion we can draw a perpendicular line through any point in $\overleftrightarrow{OY}$ (Fig. 3-25); each point of intersection has coordinates (0, y). Combining the two sets would establish a grid that resembles Fig. 3-26.

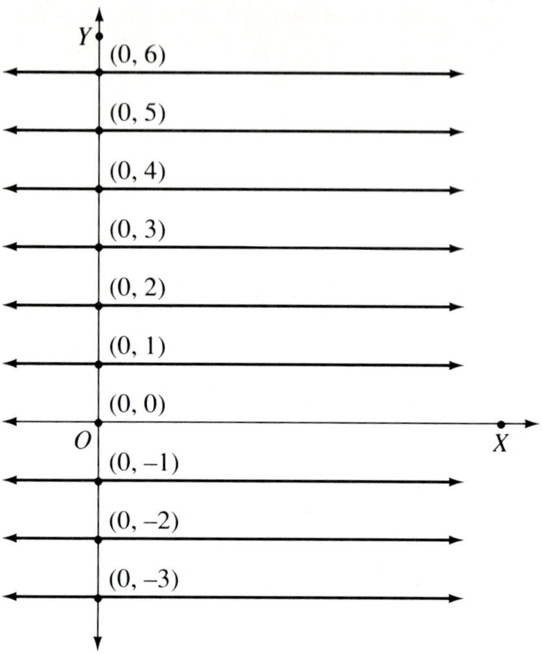

Figure 3-25

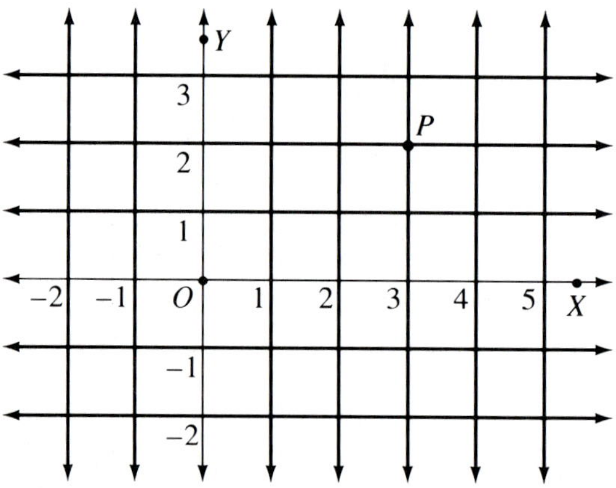

Figure 3-26

According to the Ruler Postulate (Post. 2-4) it is possible to establish a number line along any horizontal or vertical line, using the point of intersection with the axis for the zero point. On the line perpendicular to $\overleftrightarrow{OX}$ at $(x_1, 0)$ in Fig. 3-27, for example, we see the points labeled $(x_1, 1)$, $(x_1, 2)$, $(x_1, 3)$, etc.

Similarly we could select a point $(0, y_1)$ of $\overleftrightarrow{OY}$ and show the set of points on the perpendicular through the point which we could designate $(1, y_1)$, $(2, y_1)$, $(3, y_1)$, and so forth.

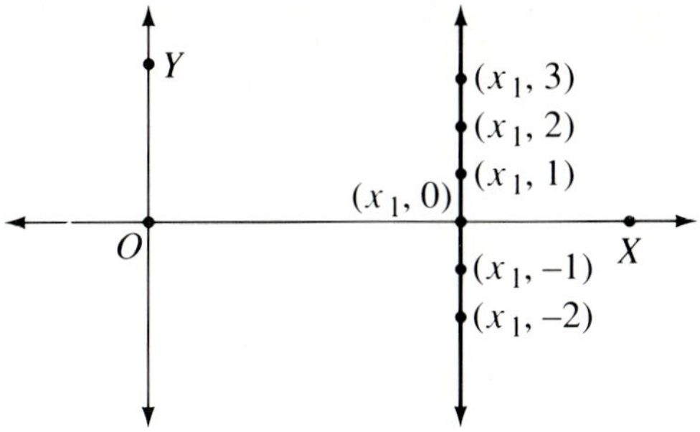

Figure 3-27

Since each P not on $\overleftrightarrow{OX}$ or $\overleftrightarrow{OY}$ has projections with coordinates $(x, 0)$ and $(0, y)$, we will say that the coordinates of each such point are the ordered pairs of real numbers (x, y).

Definition 3-12 The ordered pair of real numbers (x, y) are the **coordinates of any point *P*** with respect to a coordinate system on $\overleftrightarrow{OX}$ and $\overleftrightarrow{OY}$. The number x is the x-coordinate or **abscissa** and y is the y-coordinate or **ordinate** of the point. The symbol $P(x, y)$ will denote "the point P with coordinates (x, y)."

Example

Refer to the following coordinate system.

(a) Give the x and y projections of the points P, T, and S.

(b) Give the quadrants of the points T, U, and R.

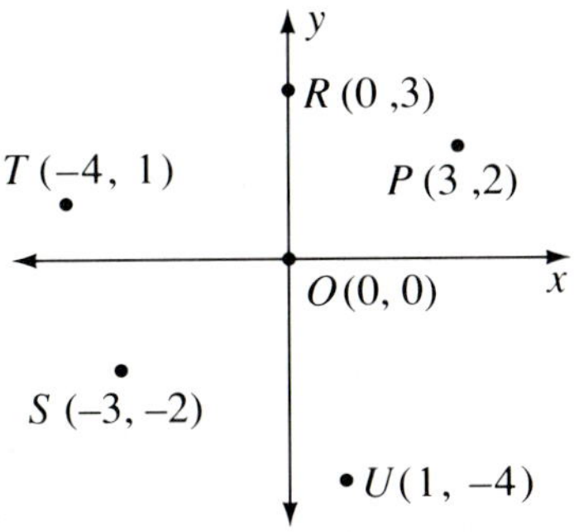

Solution

(a) For P, the x projection is $(3, 0)$; the y projection is $(0, 2)$.
For T, the x projection is $(-4, 0)$; the y projection is $(0, 1)$.
For S, the x projection is $(-3, 0)$; the y projection is $(0, -2)$.

(b) T is in II, U is in IV, and R is in none (it is on $\overleftrightarrow{OY}$). ◁

Exercise Set 3.12

1. Given the following figure, list the ordered pairs of real numbers that locate each point A through L.

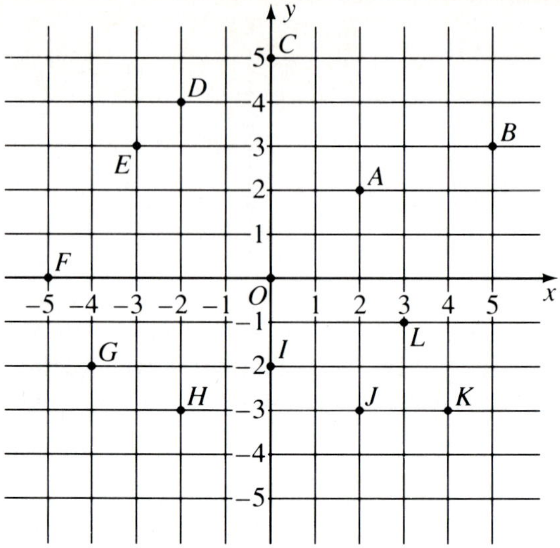

2. What is the ordinate of point (3, −4)?
3. What is the abscissa of the point (5, 2)?
4. Draw a pair of axes. Place equally spaced coordinates on the axes and plot each of the following points.
   (a) (1, 2) **(b)** (1, −2)
   (c) (3, −5) **(d)** (−2, −3)
   (e) (−5, 4) **(f)** (0, 2)
   (g) (−3, 0) **(h)** (5, −5)
5. What are the coordinates of the origin?
6. Which two points are closer together? (Locate each set of points on coordinate axes and compare.)
   (a) (4, 1) and (1, 4) or (4, 1) and (4, 0)
   (b) (3, 2) and (2, 3) or (3, 2) and (3, 0)
   (c) (7, 4) and (4, 7) or (4, 7) and (7, 0)
7. Name (give coordinates for) the point which is the *x* projection of each of the following points:
   (a) (3, 1) **(b)** (2, 4)
   (c) (3, 0) **(d)** (−7, 1)
   (e) (7, −1) **(f)** (4, −5)
8. Name the *y* projection of each point given in Problem 7.
9. Rearrange the following list of points so that their projections on the *x*-axis will fall in order from left to right:

   (5, 3), (−7, 1), (2, 6), (4, −5), (6, 6)

10. **(a)** Plot the points given in Exercise 9 on coordinate axes.
   (b) Interchange their *x*- and *y*-coordinates and plot the resulting points.
   (c) Did interchanging the coordinates determine any new points?
   (d) If you interchange the coordinates of the origin will you get a new point?

(e) List three points for which interchanging the coordinates will not result in a new point.

11. In which quadrant is each point of Exercise 4?

12. If a and b are both positive numbers, in which quadrant does each of the following points lie?

(a) (a, b) **(b)** $(-a, b)$
(c) $(a, -b)$ **(d)** $(-a, -b)$
(e) (b, a) **(f)** $(+b, -a)$
(g) $(-b, a)$ **(h)** $(-b, -a)$

13. If $\overleftrightarrow{OX}$ separates the plane into an upper half-plane $\mathcal{H}_1$ and a lower half plane $\mathcal{H}_2$, whereas $\overleftrightarrow{OY}$ divides the plane into a right half-plane $\mathcal{H}_3$ and a left half-plane $\mathcal{H}_4$, quadrant I is the intersection of $\mathcal{H}_1$ and $\mathcal{H}_3$.

(a) What quadrant is $\mathcal{H}_1 \cap \mathcal{H}_4$?
(b) Quadrant III is the intersection of which two half-planes?
(c) Illustrate the intersection of $\mathcal{H}_2$ and $\mathcal{H}_4$ by drawing horizontal lines in $\mathcal{H}_2$ and vertical lines in $\mathcal{H}_4$.

14. Computer-controlled machinery is being used extensively in industry today. The computer translates a pattern into points and determines their x- and y-coordinates. It then directs the cutting head to a first point where it begins to cut a piece of metal or plastic, then moves to the next point, etc., until the last point is reached. (It is possible to have the depth of the cut controlled by using a third coordinate.) Plot the following set of points consecutively and connect them with a smooth curved line.
(20.0, 0.0), (19.9, 1.7), (19.5, 3.4), (19.0, 5.1), (18.2, 6.6), (17.3, 8.1), (16.2, 9.3), (14.9, 10.4), (13.5, 11.4), (12.1, 12.1), (10.6, 12.6), (9.0, 12.9), (7.5, 13.0), (6.0, 12.9), (4.6, 12.6), (3.3, 12.2), (2.0, 11.6), (0.9, 10.8), (0.0, 10.0), (−0.8, 9.1), (−1.4, 8.1), (−1.9, 7.2), (−2.3, 6.2), (−2.4, 5.2), (−2.5, 4.3), (−2.4, 3.5), (−2.3, 2.7), (−2.1, 2.1), (−1.8, 1.5), (−1.5, 1.0), (−1.2, 0.7), (−0.8, 0.4), (−0.6, 0.2), (0.0, 0.0), (−0.6, −0.2), (−0.8, −0.4), (−1.2, −0.7), (−1.5, −1.0), (−1.8, 1.5), (−2.1, −2.1), (−2.3, −2.7), (−2.4, −3.5), (−2.5, −4.3), (−2.4, −5.2), (−2.3, −6.2), (−1.9, −7.2), (−1.4, −8.1), (−0.8, −9.1), (0.0, −10.0), (0.9, −10.8), (2.0, −11.6), (3.3, −12.2), (4.6, −12.6), (6.0, −12.9), (7.5, −13.0), (9.0, −12.9), (10.6, −12.6), (12.1, −12.1), (13.5, −11.4), (14.9, −10.4), (16.2, −9.3), (17.3, −8.1), (18.2, −6.6), (19.0, −5.1), (19.5, −3.4), (19.9, −1.7), (20.0, 0.0)

3-7 DISTANCE BETWEEN TWO POINTS

The idea of distance between two points is of some importance to most of us as we go through life. We are concerned with the distance from home to school, from home to work, or perhaps from San Francisco to New York.

We recall from the Distance Postulate (Post. 2-6) that for each pair of points A and B on the number line there is a unique number d called the distance from A to B. If A has coordinates $(x_1, 0)$ and B has coordinates $(x_2, 0)$, then $d = \text{m}\overline{AB} = |x_2 - x_1|$. For a pair of points C and D on the y-axis we see that $\text{m}\overline{CD} = |y_2 - y_1|$. See Fig. 3-28.

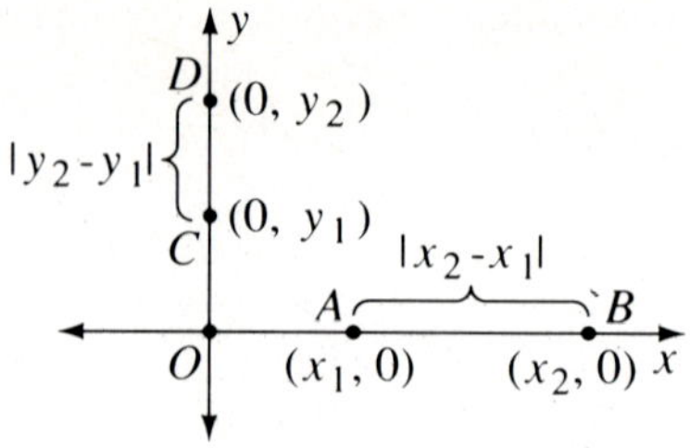

Figure 3-28

Example

Given points $A(2, 0)$, $B(5, 0)$, $C(-3, 0)$, $D(0, 3)$, and $E(0, 5)$. Find (a) m$\overline{AB}$, (b) m$\overline{BC}$, (c) m$\overline{CA}$, (d) m$\overline{DE}$, (e) m$\overline{AC}$.

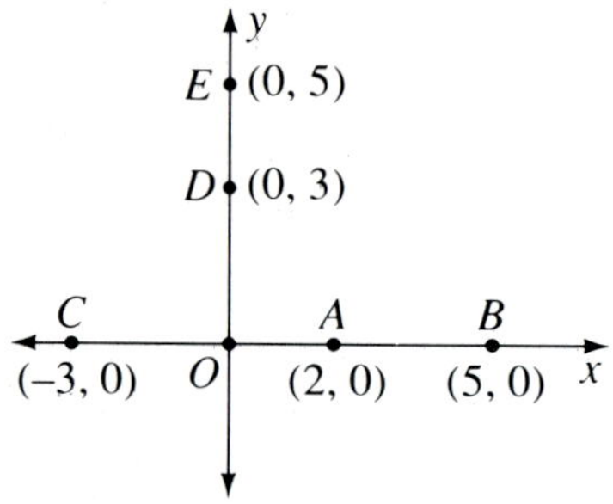

Solution

(a) m$\overline{AB} = |5 - 2| = 3$
(b) m$\overline{BC} = |(-3) - 5| = |-8| = 8$
(c) m$\overline{CA} = |2 - (-3)| = 5$
(d) m$\overline{DE} = |5 - 3| = 2$
(e) m$\overline{AC} = |(-3) - 2| = |-5| = 5$ ◁

Now consider a point P, not on an axis, which has projections $A(x_1, 0)$ and $B(0, y_1)$ on the x and y axes. The distance from origin O to A is $|x_1 - 0| = |x_1|$, and the distance from O to B is $|y_1 - 0| = |y_1|$.

We call $|x_1|$ the *x-distance* of P from the origin and $|y_1|$ the *y-distance* of P from the origin. In Fig. 3-29(b) the projection of the set of points of $\overline{RS}$ on the

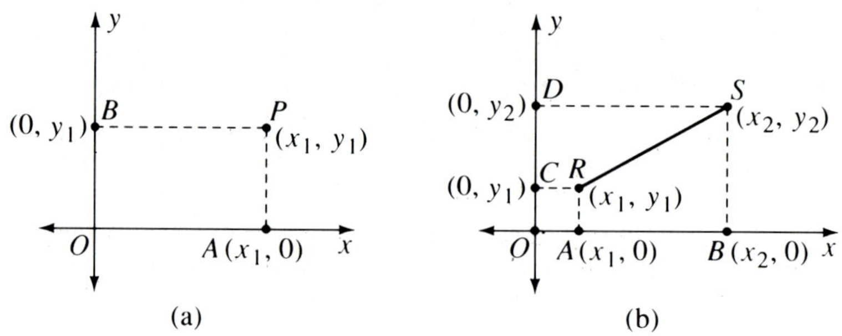

Figure 3-29

x-axis is the segment AB, which has length $|x_2 - x_1|$, and the projection of $\overline{RS}$ on the y-axis is the segment CD, for which the length is $|y_2 - y_1|$. In this case we will call $|x_2 - x_1|$ the x-distance from R to S, that is, it is the distance of the x projection, and $|y_2 - y_1|$ will be the y-distance from R to S. Using these two lengths we can determine the length of $\overline{RS}$ by means of the following theorem (proof of which we defer until Chapter 6).

Theorem (*Distance Formula*) The distance from $R(x_1, y_1)$ to $S(x_2, y_2)$ is the real number $d(\overline{RS}) = \sqrt{(x_2 - x_1)^2 + (y_2 - y_1)^2}$.

Examples

1. The distance from $A(3, 1)$ to $B(4, 3)$ is given by $d(\overline{AB}) = \sqrt{(4 - 3)^2 + (3 - 1)^2} = \sqrt{1 + 4} = \sqrt{5}$.

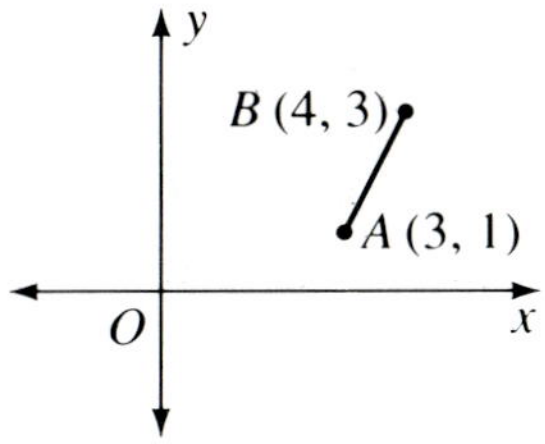

2. Find the distance from $G(5, -3)$ to $H(8, 1)$.

$$d(\overline{GH}) = \sqrt{(8 - 5)^2 + [1 - (-3)]^2} = \sqrt{3^2 + 4^2} = \sqrt{25} = 5$$

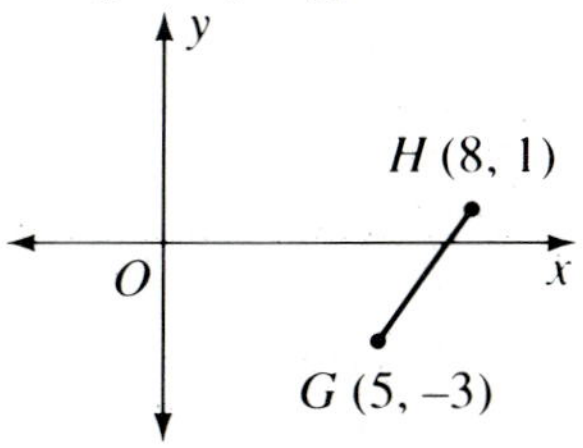

Since the distances $|x_2 - x_1|$ and $|y_2 - y_1|$ are *squared* in the distance formula, it makes no difference whether $(x_2 - x_1)$ and $(y_2 - y_1)$ are positive or negative and the absolute value signs are not used.

The Distance Formula may be used for the distance between any two points on the plane. If R and S are on a horizontal line,

$$d(\overline{RS}) = \sqrt{(x_2 - x_1)^2 + (y_1 - y_1)^2} = \sqrt{(x_2 - x_1)^2} = |x_2 - x_1|.$$

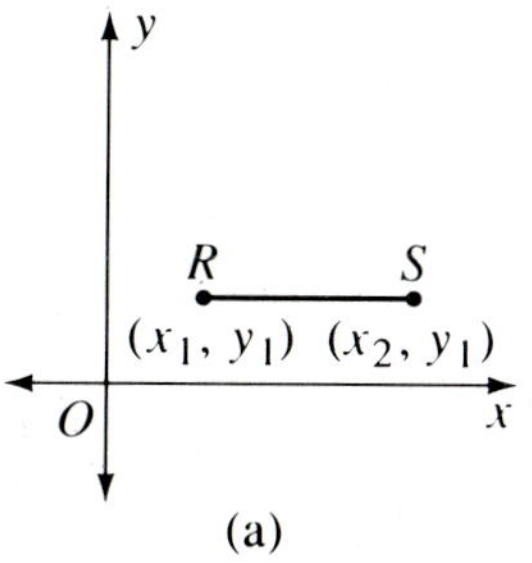

(a)

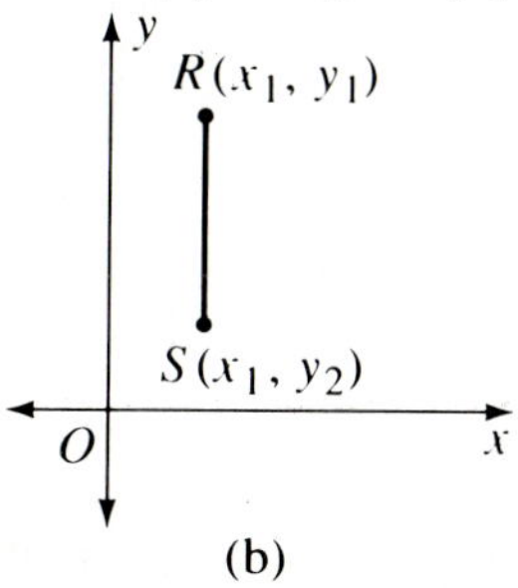

(b)

If the points are on a vertical line, $x_2 = x_1$, or $x_2 - x_1 = 0$, so $d(\overline{RS}) = \sqrt{(y_2 - y_1)^2} = |y_2 - y_1|$. Note that the radical sign indicates a positive number, so we use the absolute value signs again. ◁

Exercise Set 3.13

1. On the x- and y-axis find the lengths of the projections of the segments with endpoints whose coordinates are given below.
   (a) (0, 0) and (3, 4)
   (b) (0, 0) and (5, 12)
   (c) (0, 0) and (−4, 3)
   (d) (−1, 3) and (3, 0)
   (e) (2, 3) and (−3, −9)
   (f) (0, 0) and ($\frac{3}{4}$, 1)
   (g) (−2, 3) and (−1, 4)
   (h) (2, −3) and (1, −4)
   (i) (3, 4) and (5, 2)
   (j) (−4, −6) and (−1, −2)
   (k) (−4, 2) and (4, −4)
   (l) (0, 2) and (4, 2)
   (m) (−3, 2) and (−3, 5)
   (n) (−3, 3) and (2, −2)
2. Draw the segments determined by the points whose coordinates are given in Exercise 1, using a separate pair of axes for each segment.
3. Using the Distance Formula, find the distance between the points whose coordinates are given in Exercise 1.

3-8 SLOPE OF A LINE

In working the last set of exercises perhaps you noticed that some of the segments slanted "slowly" upward and to the right, some slanted "steeply" upward and to the right, while others slanted downward to the right. The line through $A(2, 4)$ and $B(5, 6)$ in Fig. 3-30 rises 2 units while it moves to the right 3 units. To be more precise with respect to the steepness of the line, we can define a numerical measure of this quality.

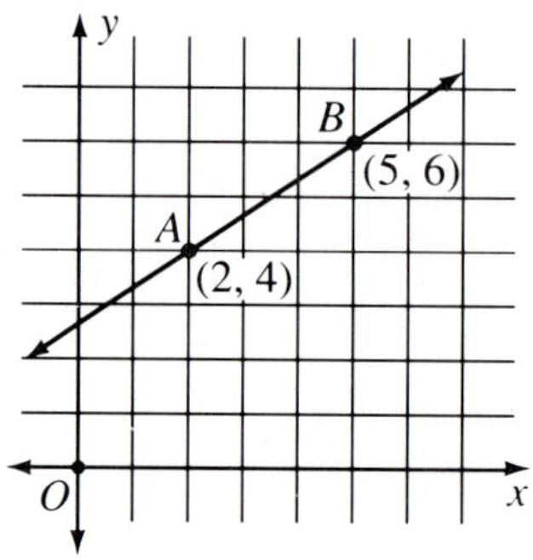

Figure 3-30

Definition 3-13 The **slope** of a nonvertical line passing through $A(x_1, y_1)$ and $B(x_2, y_2)$ is the real number m, $m = (y_2 - y_1)/(x_2 = x_1)$. (The slope of a segment or ray is the slope of the line containing it).

The slope of line AB in Fig. 3-30 is

$$m = \frac{6 - 4}{5 - 2} = \frac{2}{3}$$

If we are concerned with the slope of more than one line in a discussion we will denote the slope of the line through A and B by $m_{\overleftrightarrow{AB}}$ and the slope of the line through P and Q by $m_{\overleftrightarrow{PQ}}$.

Vertical lines are excluded because if $A(x_1, y_1)$ and $B(x_2, y_2)$ are on the same vertical line, $x_2 = x_1$ (see Fig. 3-31(a)), so

$$m = \frac{y_2 - y_1}{x_1 - x_1} = \frac{y_2 - y_1}{0}$$

which is not a real number (division by zero is not defined). Hence, if a line has no slope, it is vertical. On the other hand if the line is horizontal, $y_2 = y_1$ so

$$m = \frac{y_1 - y_1}{x_2 - x_1} = \frac{0}{x_2 - x_1} = 0$$

hence the slope of a *horizontal* line is zero (Fig. 3-31(b)).

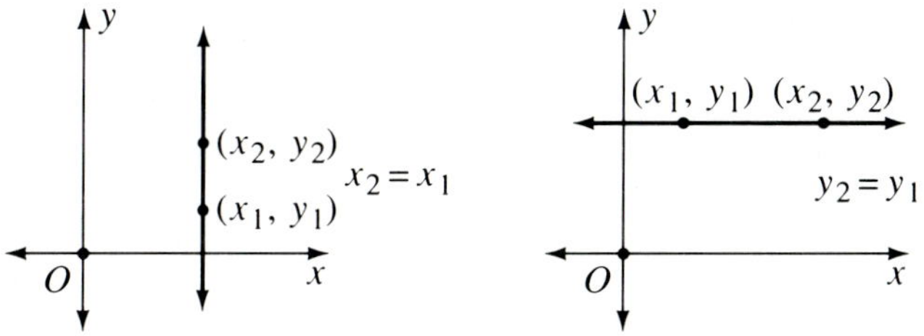

(a) (b) **Figure 3-31**

Examples

1. Plot the following points and determine the slope of the line passing through them.

(a) $A(3, 5)$ and $B(2, 4)$.

$$m_{\overleftrightarrow{AB}} = \frac{4 - 5}{2 - 3} = \frac{-1}{-1} = 1$$

(b) $P(0, 2)$ and $Q(7, 0)$.

$$m_{\overleftrightarrow{PQ}} = \frac{0 - (+2)}{7 - 0} = \frac{-2}{7}$$

(c) $R(2, 5)$ and $S(-3, 5)$.

$$m_{\overleftrightarrow{RS}} = \frac{5 - 5}{-3 - 2} = \frac{0}{-5} = 0$$

(d) $F(7, 1)$ and $G(7, -2)$.

$m_{\overleftrightarrow{FG}} = \dfrac{-2 - 1}{7 - 7} = \dfrac{-3}{0}$; thus there is no slope and the line is vertical.

(e) $U(-3, 3)$ and $V(4, -7)$.

$$m_{\overleftrightarrow{UV}} = \frac{-7 - 3}{4 - (-3)} = \frac{-10}{7} = -\frac{10}{7}$$

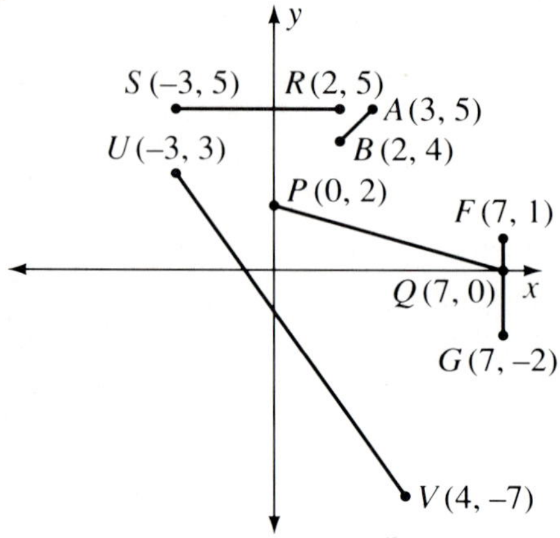

2. In the following figure we see that the segment slopes upward as we travel from left to right. Here $m_{\overleftrightarrow{P_1P_2}} > 0$, since the numerator and denominator will both be of the same sign; i.e., $x_2 - x_1 > 0$ and $y_2 - y_1 > 0$.

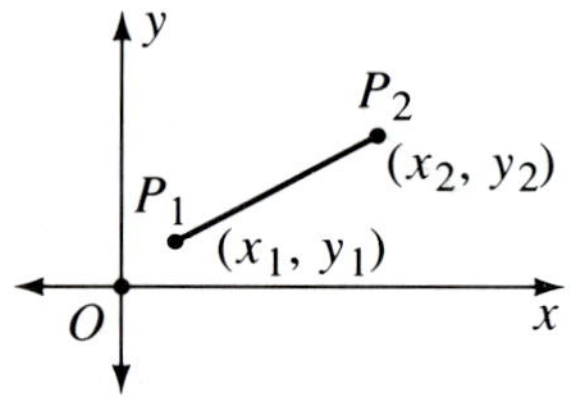

3. If the segment slopes downward from left to right, we find that $m_{\overleftrightarrow{P_1P_2}} < 0$, since the numerator and denominator will be of opposite signs; i.e., $x_2 - x_1 > 0$, but $y_2 - y_1 < 0$.

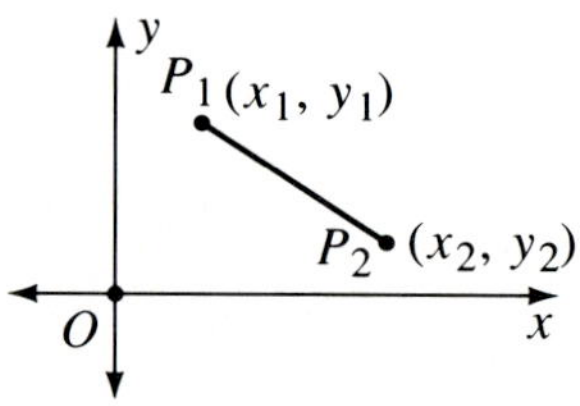

Exercise Set 3.14

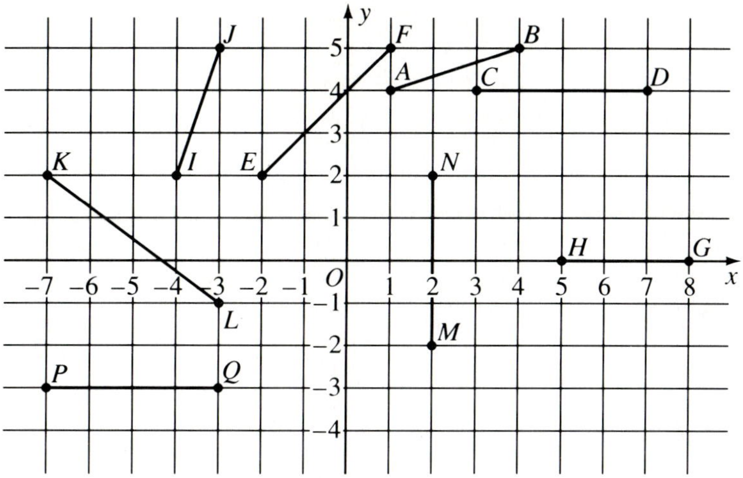

1. Refer to the accompanying figure and give the coordinates of the points $A, B, C, D, E, F, G, H, I, J, K, L, M, N, P, Q$.
2. Give the slope of each segment in the figure.
3. State whether each of the following pairs of points determines a horizontal or a vertical line.
   (a) (3, 4) and (3, 7) **(b)** (2, 5) and (7, 5)
   (c) (−3, 2) and (3, 2) **(d)** (−7, 4) and (−7, −4)
   (e) (a, b) and (c, b) **(f)** (a, b) and (a, c)
4. Find the slope of the line that contains the pairs of points with the following coordinates:
   (a) (3, 5) and (0, 0) **(b)** (4, 8) and (8, 6)
   (c) (3, 3) and (−4, 1) **(d)** (3, 0) and (0, 4)
   (e) $(\frac{1}{2}, 2)$ and $(\frac{3}{4}, \frac{2}{3})$ **(f)** (25, 7) and (−3, −7)
   (g) $(2\sqrt{3}, \sqrt{3})$ and $(\sqrt{75}, \sqrt{12})$ **(h)** $(2r, s)$ and $(5r, -3s)$
5. Find a value for b in each of the following so that the line determined by each pair of points will be horizontal.
   (a) $(3, 7)$ and $(5, b)$ **(b)** $(-4, b)$ and $(2, -5)$
   (c) (x_1, y_1) and (x_2, b)
6. Find a value for a in each of the following so that the line determined by each pair of points will be vertical.
   (a) $(2, 4)$ and $(a, -3)$ **(b)** $(a, -3)$ and $(4, 5)$
   (c) (x_1, y_1) and (a, y_2)

7. If a line has slope 2 and passes through the point (3, 4), find the y-coordinate of another point on the line with an abscissa of 4.
8. Find the distance between each pair of points given in Problem 4.

In Problems 9–12 draw each pair of lines determined by the points on separate sets of axes, then determine the slope of each line.

9. $(-1, -1)$ and $(1, 3)$; $(0, 2)$ and $(2, 1)$
10. $(1, -5)$ and $(-1, 1)$; $(3, 3)$ and $(-3, 1)$
11. $(0, -2)$ and $(3, 0)$; $(-2, 4)$ and $(2, -2)$
12. $(0, 0)$ and $(5, 3)$; $(0, -2)$ and $(-3, 3)$
13. **(a)** Find the product of the slopes of each pair of lines in Problems 9–12.
    (b) What seems to be the geometric relationship between the pairs of lines in Problems 9–12?

3-9 MIDPOINT OF A SEGMENT

Let C be the midpoint of a segment AB on the x-axis, as shown in Fig. 3-32. If $x_1 < x_2$, we can express the coordinate of C in terms of x_1 and x_2 in the following way: C is the midpoint so $m\overline{AC} = m\overline{CB}$, also $m\overline{AC} = |x - x_1| = x - x_1$, and $m\overline{CB} = |x_2 - x| = x_2 - x$. By substitution we obtain $x - x_1 = x_2 - x$, so $2x = x_1 + x_2$ and $x = (x_1 + x_2)/2$. (Since interchanging x_2 and x_1 as labels for the coordinates of A and B does not affect their numerical values, it really makes no difference in the value of x if $x_2 < x_1$.)

If y is the coordinate for the midpoint of two points D and E on the y-axis, with y_1 and y_2 their coordinates, we can demonstrate in a similar manner that $y = (y_1 + y_2)/2$.

If the segment is not contained in either axis, we can still readily find the midpoint. In Fig. 3-33 the projections of A, M, and B on the x-axis have coordinates x_1, x, and x_2, respectively. The projection has midpoint $x = (x_1 + x_2)/2$. Likewise the projection of $\overline{AB}$ on the y-axis has midpoint $y = (y_1 + y_2)/2$.

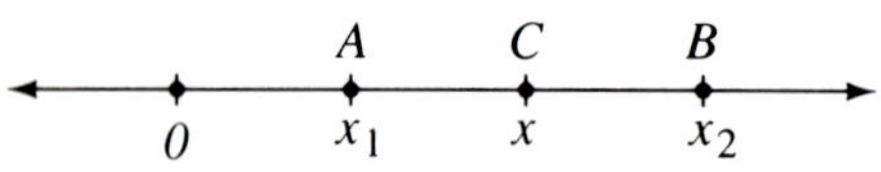

Figure 3-32

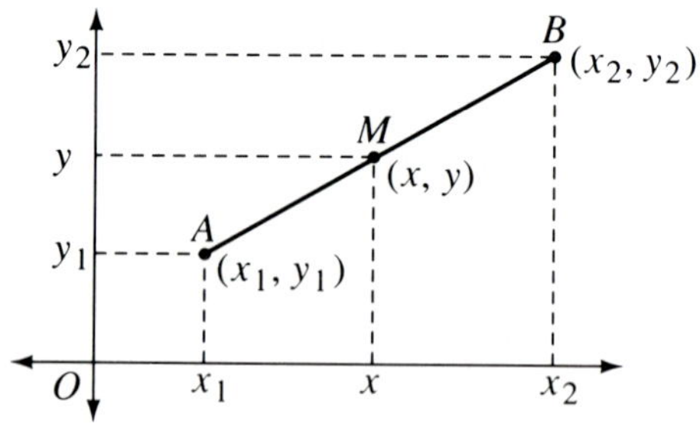

Figure 3-33

Definition 3-14 If $A(x_1, y_1)$ and $B(x_2, y_2)$ are endpoints of a segment AB, the coordinates of M, the **midpoint** of $\overline{AB}$, are $((x_1 + x_2)/2, (y_1 + y_2)/2)$.

Example

Find the midpoint of the segment joining $P(3, 5)$ and $Q(13, 15)$.

Solution

$$\left(\frac{3 + 13}{2}, \frac{5 + 15}{2}\right) = (8, 10)$$

◁

Exercise Set 3.15

1. Find the coordinates of the midpoint of each of the following pairs of points:
   (a) (6, 0) and (0, 8) **(b)** (4, 3) and (8, 15)
   (c) (0, 0) and (0, 10) **(d)** (7, 5) and (6, 8)
   (e) $(\frac{1}{2}, \frac{3}{4})$ and $(\frac{1}{8}, \frac{2}{5})$ **(f)** (a, b) and (c, d)
2. **(a)** Find the slope of each segment in Exercise 1.
   (b) Find the distance between each pair of points in Exercise 1.
3. Using the midpoint formula, find the coordinates of one end of a segment if the coordinates of its midpoint are (4, 3) and the other end is at (−2, 2).
4. Given $A(7, r)$, $M(s, 2)$ and $B(2, 5)$, find r and s so that M is the midpoint of $\overline{AB}$.
5. Verify Def. 3-14 by using the Distance Formula to show that $d(\overline{AM}) = d(\overline{MB})$ given $A(3, -4)$ and $B(5, 10)$.

REVIEW EXERCISES

1. If an angle has measure a such that $90 < a < 180$, it is called a(n) ________ angle.
2. The point S of an angle RST is the ________ of the angle.
3. The sum of the measures of two ________ angles is 90.
4. If P is in the interior of an angle ABC, then $m\angle ABC =$ ________ + ________.
5. If $\angle A$ and $\angle B$ are supplementary and $m\angle A = 65$, then $m\angle B =$ ________.
6. If $m\angle A = m\angle B$, then $\angle A$ and $\angle B$ are ________.
7. If $\angle R \cong \angle S$ and $\angle S \cong \angle T$, then ________ $\cong$ ________ according to ________ (theorem, postulate or definition).
8. When two lines intersect as in the following figure, $\angle 1$ and $\angle 3$ are ________ angles.

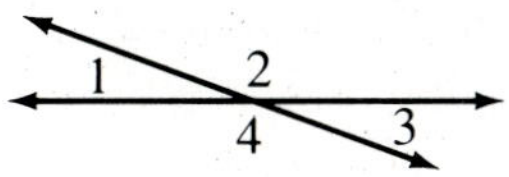

9. If $\angle A \cong \angle B$ and $\angle A$ and $\angle B$ are supplements, then $\angle A$ and $\angle B$ are ________ angles.
10. If two rays meet so that a right angle is determined, the rays are ________.
11. The distance between points with coordinates (0, 6) and (0, −6) is ________.
12. The projection of a point P with coordinates (3, −5) onto the x-axis is the point ________.

13. In the following figure, which pairs of angles listed are not adjacent angles?
(a) *AOE* and *EOD* **(b)** *AOD* and *DOB*
(c) *AOE* and *DOC* **(d)** *AOC* and *BOC*
(e) *AOD* and *EOC*

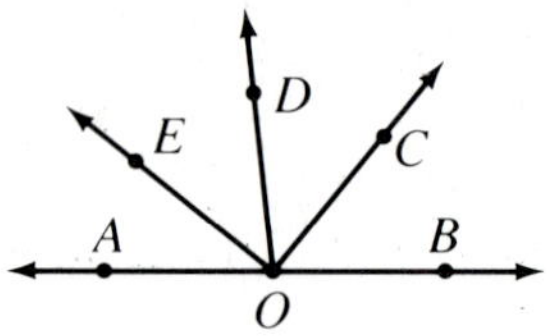

14. If the line through (5, 6) and (a, -3) is vertical, then a must equal ______.

15. If (7, 0) is the x projection of a point P and (0, -3) is its y projection, the coordinates of P are ______.

16. If the measure of an angle is 20 more than that of its complement, what is the measure of each angle?

17. If two lines intersect so that the angles formed have measures as indicated below, find the values of x and y.

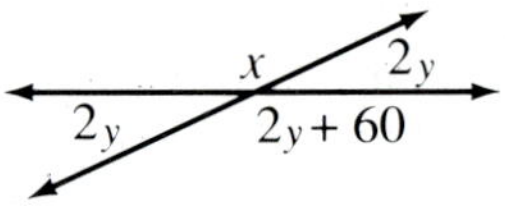

18. If (2, 7) is an endpoint of a segment and (4, 5) is its midpoint, find the coordinates of the other endpoint.

19. Given $P(3, -5)$ and $Q(7, -3)$, determine
(a) the midpoint of $\overline{PQ}$
(b) the slope of $\overline{PQ}$
(c) the length of $\overline{PQ}$.

20. *Given:* $\overleftrightarrow{AR}$, $\overleftrightarrow{AD}$ and $\overleftrightarrow{KT}$ intersecting at B, and $\angle RAB \cong \angle KBA$
Prove: $\angle RAB \cong \angle DBT$

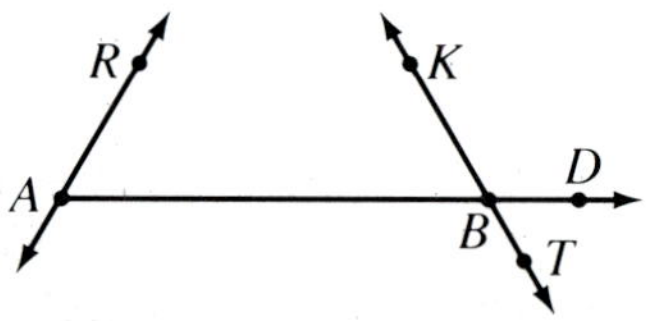

21. If the following points describe one endpoint of a segment and the origin is the midpoint, find the other endpoint for each.
(a) (2, 4) **(b)** (4, -2)
(c) (5, 0) **(d)** (-3, 5)
(e) (-3, -4) **(f)** (0, 9)

22. If O is the origin and $A(-5, 0)$, $B(-6, 5)$, $C(-1, 6)$, $D(3, 5)$, $E(6, 1)$, $F(7, -4)$, $G(-4, -4)$, $H(-5, -2)$, find the measure of each of the following with a protractor:
(a) $\angle AOB$ **(b)** $\angle AOC$
(c) $\angle AOD$ **(d)** $\angle AOE$
(e) $\angle AOF$ **(f)** $\angle AOG$
(g) $\angle AOH$

23. Construct an angle congruent to $\angle AOD$ in Exercise 22(c). Measure the angle with your protractor and compare with your measurement in Exercise 22(c).

Cumulative Review—Chapters 2 and 3

Complete each of the following statements:

1. When we refer to a line, we are referring to a _______ line.
2. A _______ figure is a nonempty set of points.
3. To indicate that T lies between R and S, we write _______.
4. A _______ is formed by two distinct points and the set of all points between them: _______ $= \{X | A - X - B\}$.
5. A figure formed by a segment AB and the set of points beyond A or B, but not both, is called a _______.
6. If a number can be written in the form p/q, where p and q are integers and $q \neq 0$, the number is said to be a _______ _______.
7. The number corresponding to each point on a line is called the _______ of that point.
8. The number $\sqrt{a}$, $a > 0$, is always a _______ number.
9. The absolute value of x, if x is a negative number, is _______.
10. If $\overline{AB} \cong \overline{CD}$ and $\overline{CD} \cong \overline{AB}$, the congruence of the segments is said to be _______.
11. An angle is the union of two _______ that have the same endpoint.
12. Two segments with a common endpoint determine a(n) _______.
13. An angle separates a plane into three distinct sets: (1) _______, (2) _______, (3) _______.
14. A degree is $\frac{1}{180}$ of a _______ angle.
15. To every angle there corresponds a real number a such that _______.
16. If two acute angles are congruent, their complements are _______.
17. If two intersecting lines determine a right angle, they are said to be _______.
18. A point $(3, -4)$ is in quadrant _______.

Problems

1. How many lines can contain one given point?
2. In what way does a ray differ from a half line?
3. **(a)** Draw a line and locate the following points on it. $A(3)$, $B(4)$, $C(-1)$, $D(-3)$, $E(0)$
   (b) Find $m\overline{AC}$, $m\overline{BD}$, $m\overline{AD}$, $m\overline{BE}$, $m\overline{CA}$
4. Given three points A, B, and C, when is C a midpoint of $\overline{AB}$?
5. If we are given the set of points A, B, C, and D on a line, place the appropriate symbol for line, ray or segment that will make the following paragraph correct.
   AB contains neither the point C or D and AB contains both points C and D. C belongs to AB while D does not. (Illustrate your answer.)
6. Consider the sets of points listed below. Identify each as a line, ray, half line, segment, or point.
   (a) $\{x | x > 2\}$
   (b) $\{x | -2 \leq x \leq 5\}$
   (c) $\{x | \, |x| \leq 2\}$
   (d) $\{x | \, |x| \geq 0\}$

(e) $\{x|x = 4\}$ **(f)** $\{x|x < 4\}$
(g) $\{x|2 \leq x \leq 15\}$ **(h)** $\{x|x \geq 0\}$

7. Simplify each of the following:
(a) $|-11 + 5|$ **(b)** $|8 - 3|$
(c) $|-5 - 7|$ **(d)** $|4 + 9|$
(e) $|-5| + |4|$ **(f)** $|3| - |-7|$
(g) $|7| - |14|$ **(h)** $|-4| - |-7|$
(i) $|2x + 8 - (3x + 8)|$ **(j)** $|-4x + 5x|$

8. Which of the following sets of points are convex sets? If they are not, tell why.
(a) a line
(b) a segment
(c) a half line
(d) two points
(e) a ray with one point removed
(f) a segment with an endpoint removed
(g) a plane with a point removed

9. (a) Is the set of points of a half plane a convex set? Why?
(b) Is the union of two half planes a convex set?

10. Consider a set of points A, B, C, D, and E with $\overleftrightarrow{AB}$ intersecting $\overleftrightarrow{CD}$ at E. Consider the half-plane with edge $\overleftrightarrow{AB}$ and containing the point C, and the half-plane with edge $\overleftrightarrow{CD}$ containing the point A.
(a) Draw the figure, shading the half-planes.
(b) Is the intersection of the two half-planes a convex set? Why?

11. In the following figure, $A-B-C$.
(a) $m\angle EBF + m\angle EBD = m\angle$________
(b) $m\angle DBF + m\angle CBF = m\angle$________
(c) $m\angle DBE + m\angle EBF + m\angle FBC = m\angle$________
(d) Using a protractor, verify parts (a), (b), and (c).

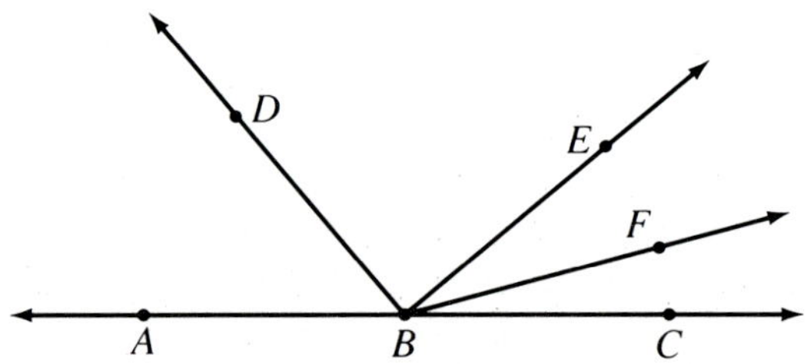

12. Determine the measure of the supplement of each of the following angles.
(a) $55°$ **(b)** $105°$
(c) $27.8°$ **(d)** $a°$
(e) $(180 - a)°$ **(f)** $(90 - a)°$

13. If the measure of the complement of $\angle A$, when increased by 20, is half the measure of its supplement, what is the measure of $\angle A$?

14. Consider the following figure.

(a) Is $\angle ABC$ supplementary to $\angle CBE$?

(b) Is $m\angle GAE = m\angle GAB$?

(c) Is $m\angle FCD = m\angle BCD$?

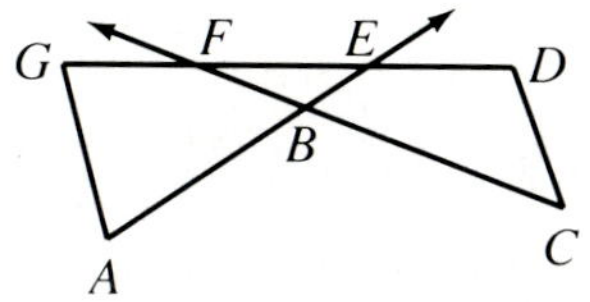

15. Consider the set of points $A(4, 4)$, $B(-4, 4)$, $C(-3, 2)$, $D(3, -3)$ and $E(6, 2)$.

(a) Locate the points on a coordinate system.

(b) Determine $d(\overline{AB})$, $d(\overline{AC})$, $d(\overline{EC})$.

(c) Determine the coordinates of the midpoints of $\overline{AB}$, $\overline{BC}$, $\overline{CD}$.

(d) Find the slopes of the lines containing $\overline{AB}$, $\overline{AD}$, $\overline{AE}$.

16. An angle ABC has measure 58. An angle DBE has measure 28. If the intersection of the interior of the angles ABC and DBC is empty and ABE is a line, draw the figure and find the measure of

(a) $\angle CBD$

(b) $\angle ABD$

(c) $\angle CBE$

4

Triangles, Congruence and Proofs

Mathematical proofs, like diamonds, are hard as well as clear, and will be touched with nothing but strict reasoning.

JOHN LOCKE (1632–1704) English philosopher.

The triangle is one of the basic geometric figures. Its applications are endless since it has mathematical properties which are relatively easy to analyze and physical properties of considerable importance, one of which is its rigidity. All sorts of strengthening and bracing arrangements for construction of everything from the simplest furniture to the most complex bridges or buildings depend upon triangles. Navigation, surveying, and design in our technological age depend heavily upon the triangle.

4-1 TRIANGLES

Definition 4-1 For any three noncollinear points A, B, and C, the union of the segments AB, BC, and CA ($\overline{AB} \cup \overline{BC} \cup \overline{CA}$) is called a **triangle.** It is denoted by $\triangle ABC$, has **vertices** A, B, and C, and **sides** $\overline{AB}$, $\overline{BC}$, and $\overline{CA}$.

In Fig. 4-1 $\overline{AB}$ is the **side opposite** angle C, $\overline{AC}$ is the side opposite angle B, and $\overline{BC}$ is the side opposite angle A.

The sides of the triangle determine three angles, $\angle ABC$, $\angle CAB$, and $\angle BCA$. A triangle determines a unique plane, since the vertices are three noncollinear points.

Figure 4-2 suggests a method for determining the interior and exterior of the $\triangle ABC$. A point such as D which is interior to all three angles is in the interior of the triangle. Points E, F, and G are not interior to *all* three angles, and hence are exterior to the triangle.

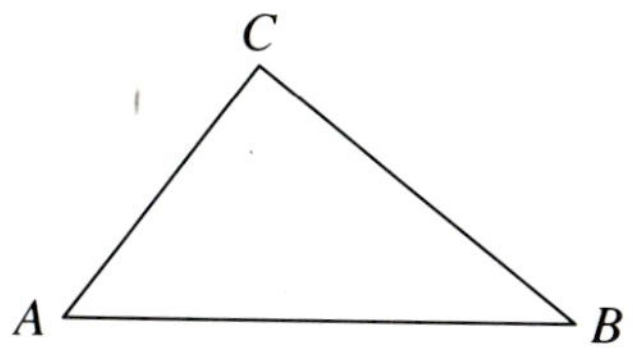

Figure 4-1

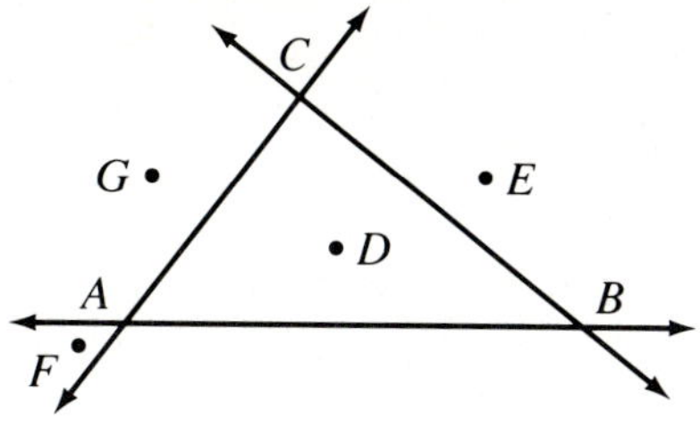

Figure 4-2

Definition 4-2 The **interior of a triangle** is that set of points interior to each angle of the triangle. The **exterior of a triangle** is the set of points not on the triangle or in its interior.

Thus we see that a triangle separates a plane in a manner much like an angle separates the plane, i.e., into three disjoint sets: the triangle, its interior, and its exterior.

Identification of Triangles and Parts of Triangles

Triangles are classified with respect to their angles and sides, and some properties depend upon the relationships so identified. Our next definitions will help to distinguish among various types.

Definition 4-3 (*Triangles Classified According to Angles*) A triangle is called an **acute triangle** if all of its angles are acute; it is an **obtuse triangle** if one angle is obtuse. A **right triangle** is a triangle with a right angle; the sides of the triangle which are part of the right angle are the **legs** of the triangle and the third side (opposite the right angle) is the **hypotenuse** (Fig. 4-3).

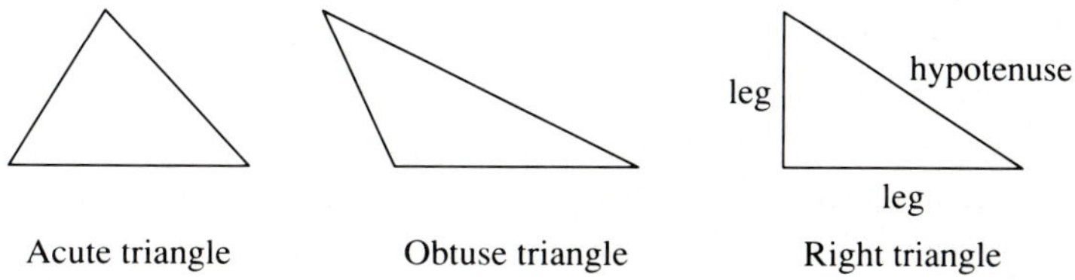

Figure 4-3

Definition 4-4 (*Triangles Classified According to Sides*) A **scalene** triangle is a triangle with no two sides congruent. An **isosceles** triangle is a triangle with at least two sides congruent. The angle included between the congruent sides of an isosceles triangle is the **vertex angle,** the side opposite the vertex angle is the **base,** and the angles which include the base are the **base angles.** The congruent sides of an isosceles triangle are called its **legs.** An **equilateral triangle** is a triangle with all three sides congruent (Fig. 4-4).

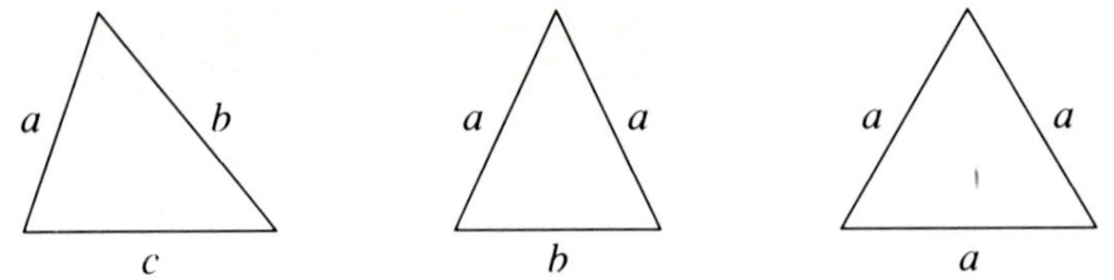

Scalene triangle $a \neq b,\ a \neq c,\ b \neq c$ | Isosceles triangle | Equilateral triangle

Figure 4-4

Example

Consider a triangle in the coordinate plane with vertices $A(3, -1)$, $B(3, 1)$ and $C(-1, 1)$.

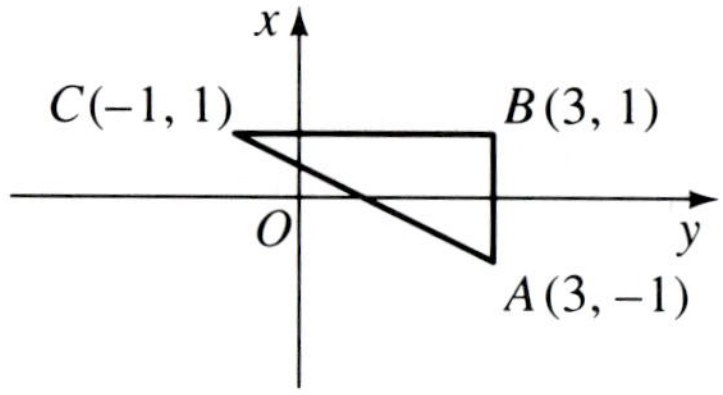

Using the distance formula we find $m\overline{AB} = \sqrt{(3-3)^2 + [1-(-1)]^2} = \sqrt{4} = 2$, $m\overline{BC} = \sqrt{[3-(-1)]^2 + (1-1)^2} = \sqrt{16} = 4$, and $m\overline{CA} = \sqrt{(-1-3)^2 + [1-(-1)]^2} = \sqrt{20} = 2\sqrt{5}$. Since no two sides have the same length, the triangle is a scalene triangle. (It is also a right triangle, since $\overline{BC}$ is horizontal and $\overline{AB}$ is vertical.) ◁

Construction of Isosceles and Equilateral Triangles

Construction 3: Construct an isosceles triangle.
To construct triangle ABC with $\overline{AB} \cong \overline{AC}$:

1. Draw a horizontal line BC. (The length of segment BC is arbitrary.)
2. Set compass points for $m\overline{AB}$ and with B as a center draw an arc above $\overleftrightarrow{BC}$.
3. Retaining this compass setting, draw an arc with C as center that intersects the arc drawn in Step 2. Call the intersection point A.
4. Draw $\overrightarrow{BA}$ and $\overrightarrow{CA}$; then triangle ABC has $\overline{AB} \cong \overline{AC}$.

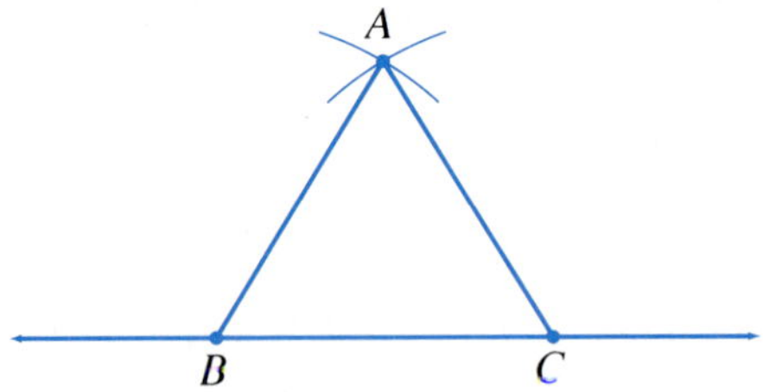

(*Comment:* In Step 2 $m\overline{AB}$ must be greater than $\frac{1}{2}m\overline{BC}$. What happens to the construction if $m\overline{AB} \leq \frac{1}{2}m\overline{BC}$?)

Construction 4: Construct an equilateral triangle.
(This construction is illustrated next.)

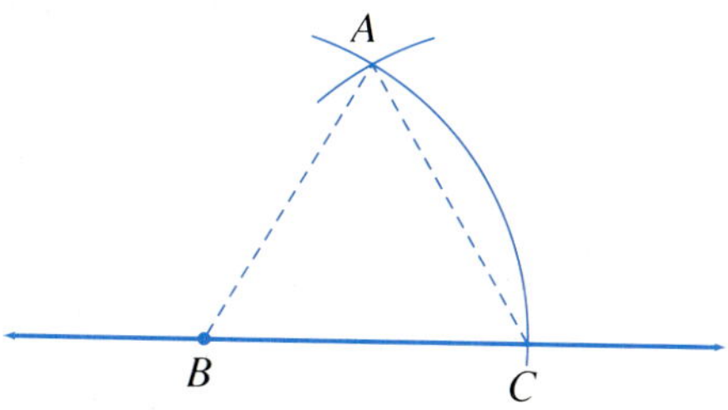

Exercise Set 4.1

1. Could a point in the interior of an angle of a triangle also be in the exterior of the triangle? (Explain.)
2. Could an angle of a triangle be a straight angle?
3. **(a)** How are the interior of a triangle and an angle alike?
   (b) How do they differ?
4. **(a)** In the following figure, name four triangles with $\overline{DE}$ as a side.
   (b) Name all the triangles that have point A as one vertex.
   (c) Name all the triangles that have point F as one vertex.

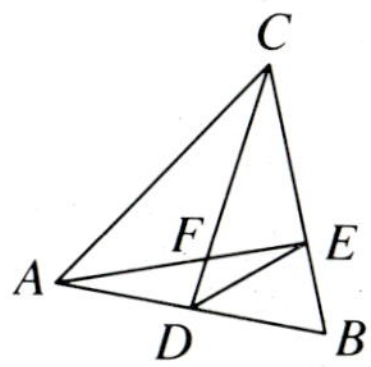

5. Draw a segment RS and points A and B on opposite sides of $\overline{RS}$ so that $\overline{AB}$ intersects $\overline{RS}$ at G. Draw $\overline{AR}$, $\overline{BR}$, $\overline{BS}$, and $\overline{AS}$.
   (a) Name three triangles having $\overline{BS}$ as a side.
   (b) Name three triangles having $\overline{AS}$ as a side.
   (c) Name four different triangles having G as a vertex. Name at least four different triangles which have A as a vertex.
6. Classify the following triangles if the numbers indicate the measure of their *sides*.
   (a) 2, 2, 3 **(b)** 4, 5, 6
   (c) 6, 6, 6 **(d)** 5, 4, 5
7. Classify the following triangles if the numbers indicate the measures of their *angles*.
   (a) 15, 75, 90 **(b)** 110, 30, 40
   (c) 45, 55, 80
8. Draw a representative triangle of the type named, indicating lengths of sides and measures of angles as appropriate.
   (a) scalene **(b)** right
   (c) equilateral **(d)** obtuse
   (e) isosceles **(f)** isosceles right
   (g) isosceles obtuse **(h)** equilateral obtuse

9. **(a)** Construct an isosceles triangle with $\overline{AB} \cong \overline{AC}$. Measure the angles with your protractor. How do the measures of $\angle B$ and $\angle C$ compare?
   (b) Draw another isosceles triangle RST with $\overline{RS} \cong \overline{RT}$ and $m\angle R$ greater than $m\angle A$ of part (a). Measure the angles with your protractor. How do the measures of $\angle S$ and $\angle T$ compare?
10. Construct an equilateral triangle XYZ. Measure the angles with your protractor. How do the measures of the angles compare?
11. **(a)** Construct a triangle OPQ with $\angle P \cong \angle Q$. Measure the sides to the nearest sixteenth of an inch. How do the measures of OP and OQ compare?
    (b) Construct another triangle FGH with $\angle G \cong \angle H$ but with $m\angle G$ less than $m\angle P$ of part (a). Measure the sides and compare them.
12. If the vertices of a triangle have coordinates $A(0, 0)$, $B(4, 0)$, $C(2, 4)$, find whether the following points lie in the interior or the exterior of the triangle (draw the figure).
    (a) (1, 1) **(b)** (1, 2)
    (c) (1, 3) **(d)** (2, 1)
    (e) (3, 4)
13. In Exercise 12, state which of the given points are in the interior of
    (a) $\angle A$ **(b)** $\angle B$
    (c) $\angle C$
14. Locate the vertices of each of the following, draw the triangles, use the distance formula to determine the lengths of the sides, and classify with respect to sides and angles.
    (a) $A(0, 3)$, $B(5, 0)$, $C(0, 0)$
    (b) $A(-1, 2)$, $B(2, 5)$, $C(2, -1)$
    (c) $A(0, 0)$, $B(0, 6)$, $C(4.8, 2.4)$
    (d) $A(-2, 4)$, $B(5, -1)$, $C(-2, -1)$
    (e) $A(1, 0)$, $B(3, 0)$, $C(7, 5)$
15. Prove that a right triangle cannot be equilateral. (*Hint:* Let the vertices of the triangle be $(0, a)$, $(b, 0)$, and $(0, 0)$. There are two cases, i.e., $a = b$ and $a \neq b$.)

4-2 CONGRUENT TRIANGLES

Since triangles are composed of line segments and angles, and since we have established requirements for congruent segments and congruent angles, it is from these congruences that we establish the congruence for triangles.

Definition 4-5 Two **triangles** ABC and DEF are said to be **congruent** ($\triangle ABC \cong \triangle DEF$) if and only if $\overline{AB} \cong \overline{DE}$, $\overline{BC} \cong \overline{EF}$, $\overline{CA} \cong \overline{FD}$, $\angle A \cong \angle D$, $\angle B \cong \angle E$ and $\angle C \cong \angle F$. In a manner of speaking we can say that geometric figures are congruent if they have the same size and shape. Note that in Fig. 4-5 the sides and angles are marked to indicate the congruent parts.

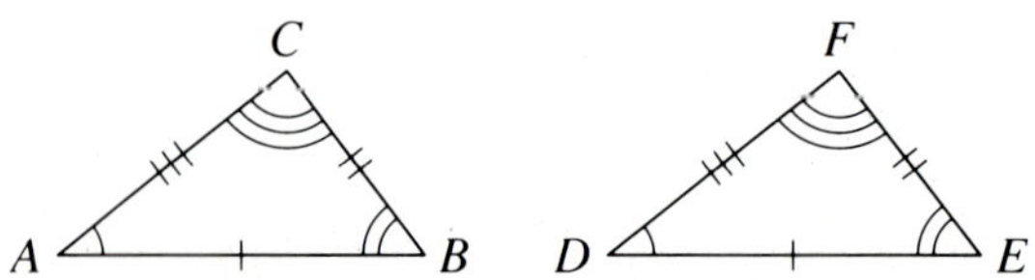

Figure 4-5

One of the characteristics of triangle congruence is *correspondence* of the congruent parts; that is, if the sides and angles of two triangles are corresponding congruent sets, then the triangles are congruent. For two congruent triangles ABC and DEF we can symbolize this correspondence in the following way:

$$\angle A \leftrightarrow \angle D \quad \text{and} \quad \overline{AB} \leftrightarrow \overline{DE}$$
$$\angle B \leftrightarrow \angle E \qquad \overline{BC} \leftrightarrow \overline{EF}$$
$$\angle C \leftrightarrow \angle F \qquad \overline{CA} \leftrightarrow \overline{FD}$$

or, $\triangle ABC \leftrightarrow \triangle DEF$, which implies both sets of correspondences above. As a result of this correspondence, we may state a definition analogous to Def. 4-5.

Definition 4-5(a) Corresponding parts of congruent triangles are congruent. (CPCTC)

The correspondence implied by the definition is very important and the congruence relation must be written so as to obtain the correct correspondence. If from Fig. 4-5 we had written $\triangle ABC \cong \triangle EFD$, this would have implied that $\angle A \cong \angle E$, $\angle B \cong \angle F$, $\angle C \cong \angle D$, and $\overline{AB} \cong \overline{EF}$, $\overline{BC} \cong \overline{FD}$, and $\overline{AC} \cong \overline{ED}$, which is not true.

Another consequence of the definition is that a triangle is congruent to itself, so we may say that triangle congruence is reflexive.

Theorem 4-1(a) $\triangle ABC \cong \triangle ABC$. (Triangle congruence is reflexive.)

Given $\triangle ABC$, we want to prove that $\triangle ABC \cong \triangle ABC$.

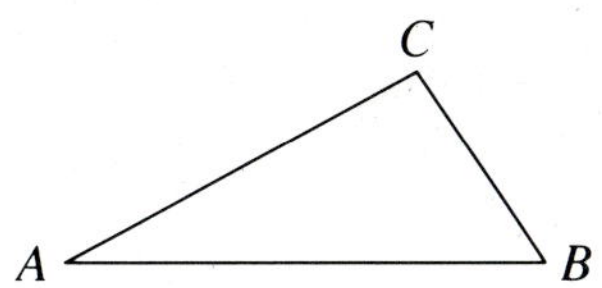

Proof

Statements	*Reasons*
1. In $\triangle ABC$, $\angle A \cong \angle A$, $\angle B \cong \angle B$, $\angle C \cong \angle C$	1. Congruence of angles is reflexive (Theorem 3-1(a))
2. In $\triangle ABC$, $\overline{AB} \cong \overline{AB}$, $\overline{BC} \cong \overline{BC}$, and $\overline{CA} \cong \overline{CA}$	2. Congruence of segments is reflexive (Theorem 2-1(a))
3. $\triangle ABC \cong \triangle ABC$	3. Two triangles are congruent if and only if their corresponding angles and corresponding sides are congruent (Def. 4-5)

We can also prove that triangle congruence is symmetric and transitive, and we state this to complete the equivalence relation.

Theorem 4-1(b) If $\triangle ABC \cong \triangle DEF$, then $\triangle DEF \cong \triangle ABC$. (Triangle congruence is symmetric.)

Theorem 4-1(c) If $\triangle ABC \cong \triangle DEF$ and $\triangle DEF \cong \triangle GHI$, then $\triangle ABC \cong \triangle GHI$. (Triangle congruence is transitive.)

In practice it is not actually necessary to prove *all* of the corresponding parts congruent in order to establish that two triangles are congruent. Suppose you were to draw two congruent angles A and D, then locate points B and C on the sides of angle A, points E and F on the sides of angle D so that $\overline{AB} \cong \overline{DE}$ and $\overline{AC} \cong \overline{DF}$, and finally draw $\overline{BC}$ and $\overline{EF}$. The resulting triangles would look somewhat like Fig. 4-6. The triangles *appear* to be congruent; as a matter of fact you would find by measuring that $\angle C \cong \angle F$, $\angle B \cong \angle E$, and $\overline{BC} \cong \overline{EF}$. The sides $\overline{AB}$ and $\overline{AC}$ of $\triangle ABC$ are said to *include* $\angle A$, and sides $\overline{DE}$ and $\overline{DF}$ of $\triangle DEF$ include $\angle D$. From our drawing we may gather that if two triangles have two sides and their included angles congruent, the triangles are congruent, and we state this as a postulate.

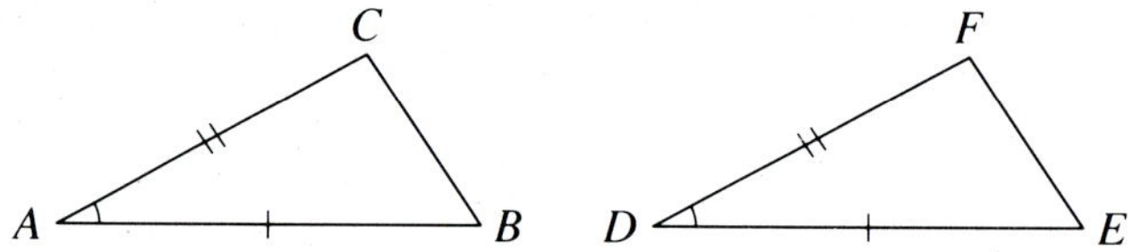

Figure 4-6

Postulate 4-1 (SAS) $\triangle ABC \cong \triangle DEF$ if $\overline{AB} \cong \overline{DE}$, $\overline{AC} \cong \overline{DF}$, and $\angle A \cong \angle D$.

We call this the SAS, or "side-angle-side," postulate, since the two sides and included angles are given as congruent.

We again emphasize here that the correspondence of parts is an important requirement for the congruence, because if we had drawn the triangles described above so that $\overline{AB} \cong \overline{DE}$, $\angle A \cong \angle D$, and $\overline{BC} \cong \overline{EF}$ as in Fig. 4-7 (the angles are not *included* between the given sides), then the triangles are not necessarily congruent. We see that although the given parts are congruent as marked, $\overline{AC} \not\cong \overline{DF}$ ($\overline{AC}$ is not congruent to $\overline{DF}$), $\angle B \not\cong \angle E$, and $\angle C \not\cong \angle F$.

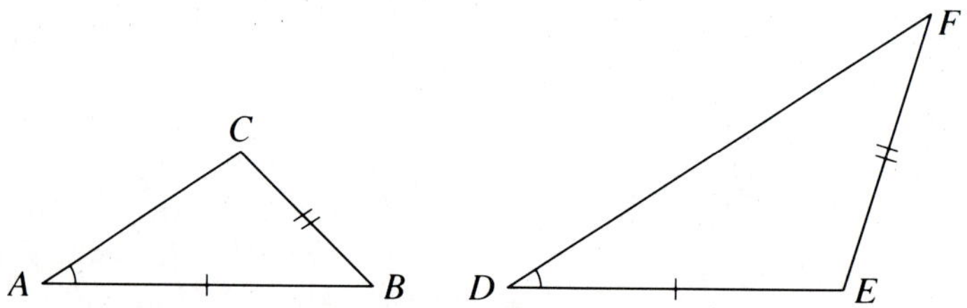

Figure 4-7

We can now begin to prove various relationships among triangles and their parts.

Examples

1. *Given:* $\overline{AB}$ intersects $\overline{CD}$ at E, $\overline{EC} \cong \overline{ED}$ and $\overline{AE} \cong \overline{BE}$
Prove: (a) $\overline{AC} \cong \overline{BD}$
(b) $\angle A \cong \angle B$

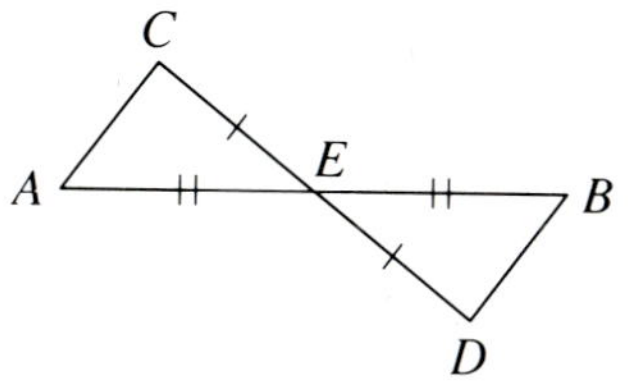

Proof

Statements	*Reasons*
1. $\overline{EC} \cong \overline{ED}$ and $\overline{AE} \cong \overline{BE}$	1. Given
2. $\angle AEC \cong \angle BED$	2. Vertical angles are congruent (Theorem 3-4)
3. $\triangle AEC \cong \triangle BED$	3. SAS (Post. 4-1)
4. $\overline{AC} \cong \overline{BD}$ and $\angle A \cong \angle B$	4. Corresponding parts of congruent triangles are congruent (Def. 4-5(a))

2. *Given:* $\overline{AC} \cong \overline{BC}$, $\angle ACD \cong \angle BCD$
Prove: $\triangle ACD \cong \triangle BCD$

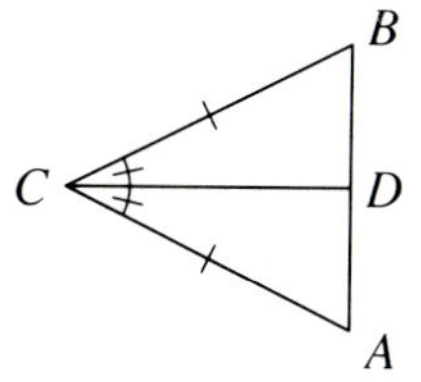

Proof

Statements	*Reasons*
1. $\overline{AC} \cong \overline{BC}$ and $\angle ACD \cong \angle BCD$	1. Given
2. $\overline{CD} \cong \overline{CD}$	2. Congruence of segments is reflexive (Theorem 2-1(a))
3. $\triangle ACD \cong \triangle BCD$	3. SAS (Post. 4-1)

3. Which of the following pairs of triangles are congruent as a consequence of the SAS postulate or definition of congruent triangles?

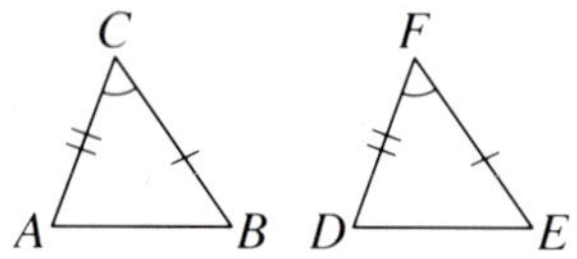

(a)

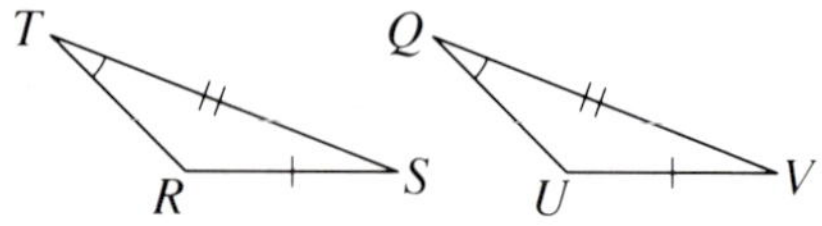

(b)

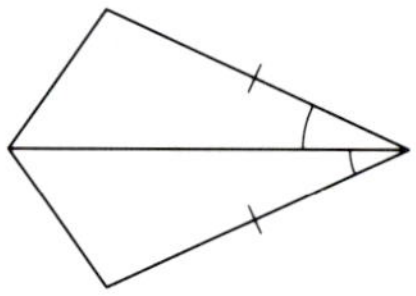

(c)

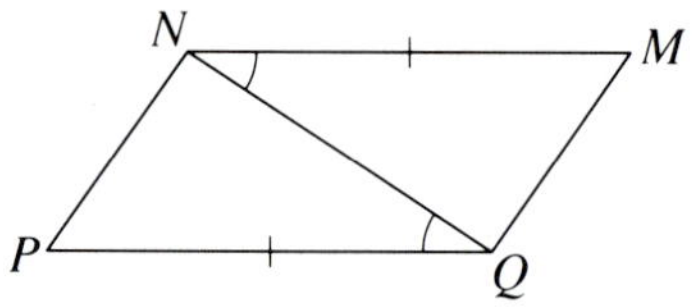

(d)

Solution In (a) the triangles are congruent according to the SAS postulate; the triangles in (b) cannot be assumed to be congruent because the congruent angles are not included between the congruent sides; and in (c) and (d) the pairs are congruent since the common sides are congruent. ◁

Exercise Set 4.2

1. Given the correspondence $\triangle RST \leftrightarrow \triangle QVW$, list the correspondence between the vertices. Is this the same correspondence as $\triangle TSR \leftrightarrow \triangle WVQ$?
2. Write the correspondence $A \leftrightarrow R$, $B \leftrightarrow S$ and $C \leftrightarrow T$ as one statement.
3. If $\triangle LMN \cong \triangle PQR$, write the congruency of the corresponding parts.
4. Given that $\triangle ABC \cong \triangle BAC$ and $\triangle ABC \cong \triangle ACB$. List the correspondences that are a congruence between the pairs of congruent triangles. What kind of a triangle is ABC?
5. Using a protractor and a ruler, draw $\angle RST$ and $\angle XYZ$ measuring 40° each, with $m\overline{SR} = m\overline{XY} = 3$ in. and $m\overline{ST} = m\overline{YZ} = 4$ in. With the ruler find $m\overline{RT}$ and $m\overline{XZ}$. Are they equal? With the protractor find $m\angle STR$ and $m\angle YZX$. Are they equal? With the protractor find $m\angle TRS$ and $m\angle ZXY$. Are they equal? Are $\triangle RST$ and $\triangle XYZ$ congruent? Should they be congruent?
6. In the second example on page 91, are A, D, and B necessarily collinear? Could the triangles be congruent if they are not?
7. Given $\triangle ABC$ with $\overline{CD} \perp \overline{AB}$ and D the bisector of $\overline{AB}$. Prove that $\overline{AC} \cong \overline{BC}$.

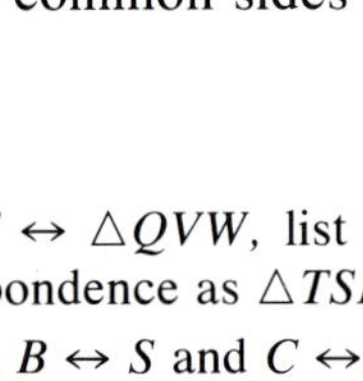

8. Given R, S, T, U are collinear with $\overline{RS} \cong \overline{TU}$, $\angle R \cong \angle U$, and $\overline{RM} \cong \overline{UN}$. Prove that $\triangle RMT \cong \triangle UNS$.

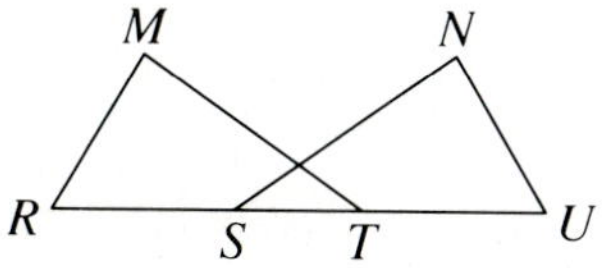

9. Prove that if the two legs of one right triangle are congruent to the two legs of another right triangle, the two triangles are congruent.

10. Draw triangles ABC and DEF with $m\angle A = m\angle D = 25$, $m\angle B = m\angle E = 50$, and $m\overline{AB} = m\overline{DE} = 5$ in. Measure to find

(a) $m\angle C$ **(b)** $m\angle F$
(c) $m\overline{AC}$ **(d)** $m\overline{DF}$
(e) $m\overline{BC}$ **(f)** $m\overline{EF}$.

Does it appear that $\triangle ABC \cong \triangle DEF$?

11. Given $\overline{AB} \perp \overline{BC}$, $\overline{DC} \perp \overline{BC}$, and $\overline{AB} \cong \overline{DC}$.
Prove $\overline{AC} \cong \overline{BD}$.

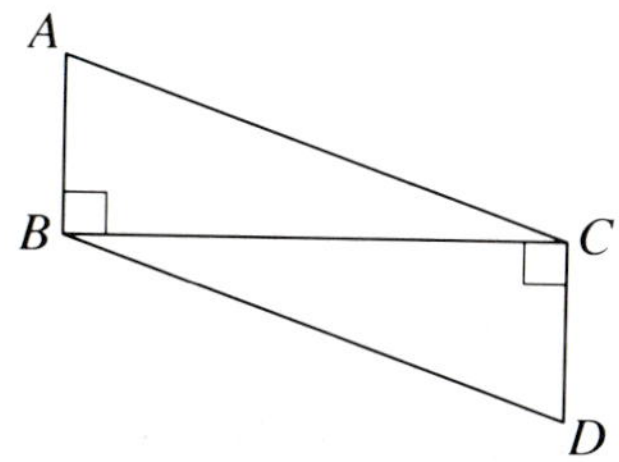

12. Given A-O-C and $\overline{AO} \cong \overline{CO}$, $\angle AOD \cong \angle COB$ and $\overline{BO} \cong \overline{DO}$.
Prove $\triangle AOB \cong \triangle COD$.

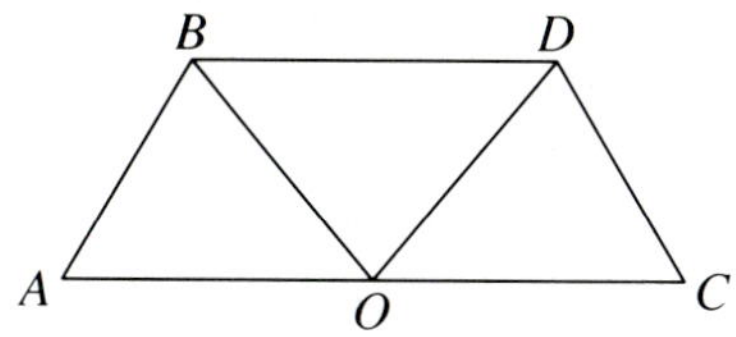

4-3 THE ASA AND SSS POSTULATES

There are two additional postulates that are useful in proving triangles congruent.

Postulate 4-2 (ASA) Given triangles ABC and DEF, if $\angle A \cong \angle D$, $\overline{AB} \cong \overline{DE}$ and $\angle B \cong \angle E$, then $\triangle ABC \cong \triangle DEF$ (Fig. 4-8).

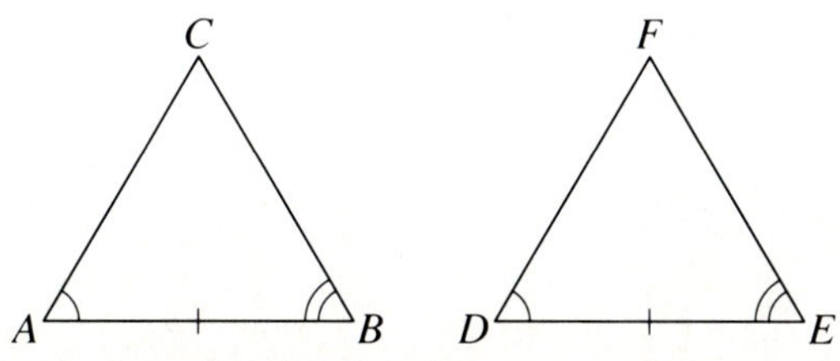

Figure 4-8

This "angle-side-angle" postulate states that if two angles and the included side of one triangle are congruent to two angles and the included side of another, the triangles are congruent.

Postulate 4-3 (SSS) Given triangles ABC and DEF, if $\overline{AB} \cong \overline{DE}$, $\overline{BC} \cong \overline{EF}$, and $\overline{AC} \cong \overline{DF}$, then $\triangle ABC \cong \triangle DEF$ (Fig. 4-9).

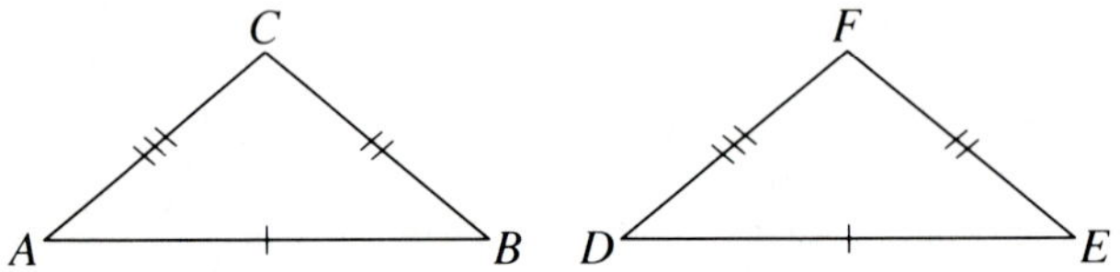

Figure 4-9

The "side-side-side" postulate states that if three sides of one triangle are congruent to three sides of another, the triangles are congruent.

Examples

1. Given the congruences indicated, determine which pairs of triangles are congruent and state the postulate that justifies your answer.

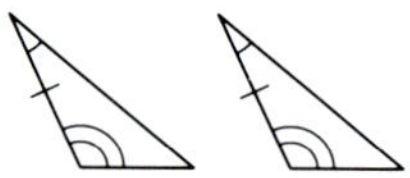

(a)

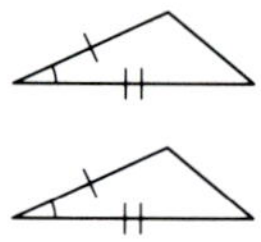

(b)

(c)

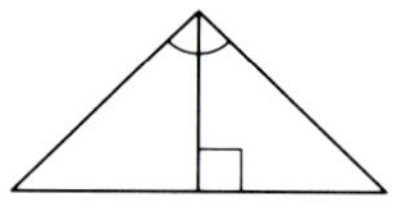

(d)

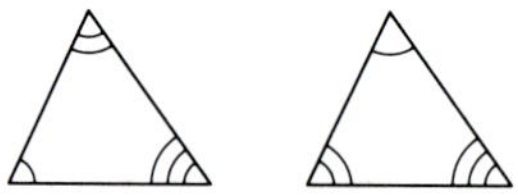

(e)

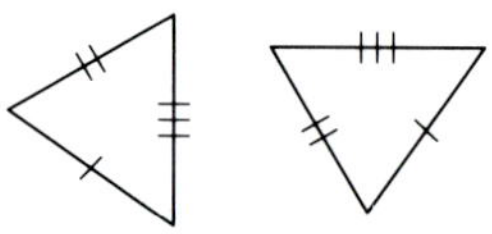

(f)

Solution We see that (a), (c), and (d) are congruent by ASA (Post. 4-2), the pair in (b) are congruent by SAS (Post. 4-1), and the pair in (f) by SSS (Post. 4-3). The triangles in (e) are not necessarily congruent since no conditions were placed on the sides. Even though the angles may be congruent, we are not assured of the congruency of the segments.

2. *Given:* $\overline{AB} \cong \overline{CB}$ and $\overline{AD} \cong \overline{CD}$
Prove: $\angle A \cong \angle C$

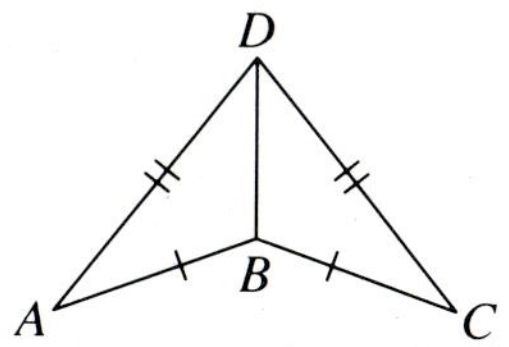

Proof

Statements	Reasons
1. $\overline{AB} \cong \overline{CB}$ and $\overline{AD} \cong \overline{CD}$	1. Given
2. $\overline{BD} \cong \overline{BD}$	2. Congruence of segments is reflexive (Theorem 2-1(a))
3. $\triangle ABD \cong \triangle CBD$	3. SSS (Post. 4-3)
4. $\angle A \cong \angle C$	4. Corresponding parts of congruent triangles are congruent (CPCTC) (Def. 4-5(a))

3. *Given:* $\triangle ABC$ and $\triangle BAD$ with $\angle DAB \cong \angle CBA$ and $\angle CAB \cong \angle DBA$
Prove: (a) $\triangle ABC \cong \triangle BAD$
(b) $\angle C \cong \angle D$
(c) $\angle CAE \cong \angle DBE$
(d) $\triangle AEC \cong \triangle BED$

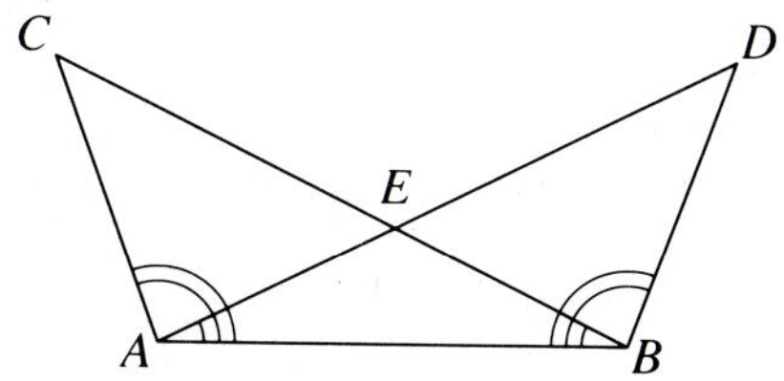

Proof

Statements	Reasons
1. $\angle DAB \cong \angle CBA$, $\angle CAB \cong \angle DBA$	1. Given
2. $\overline{AB} \cong \overline{AB}$	2. Congruence of segments is reflexive (Theorem 2-1(a))
3. $\triangle ABC \cong \triangle BAD$	3. ASA (Post. 4-2)
4. $\angle C \cong \angle D$ and $\overline{AC} \cong \overline{BD}$	4. CPCTC (Def. 4-5(a))
5. *C-E-B* so *E* is in the interior of $\angle CAB$, and *A-E-D* so *E* is in the interior of $\angle ABD$	5. Definition of interior of an angle (Def. 3-2(b))

Statements	*Reasons*
6. $\angle CAE \cong \angle DBE$	6. If P is in the interior of $\angle ABC$ and Q is in the interior of $\angle RST$, so that $\angle ABP \cong \angle RSQ$, then $\angle PBC \cong \angle QST$ (Theorem 3-3(a))
7. $\triangle AEC \cong \triangle BED$	7. ASA (Post. 4-2) ◁

In the preceding example we have a figure that contains two *overlapping* triangles. In this particular figure it is relatively easy to distinguish the separate triangles, but when we write $\triangle ABC \cong \triangle BAD$, we establish the correspondence $ABC \leftrightarrow BAD$, with segments and angles corresponding as follows:

$$\overline{AB} \leftrightarrow \overline{BA} \quad \text{and} \quad \angle ABC \leftrightarrow \angle BAD$$
$$\overline{BC} \leftrightarrow \overline{AD} \qquad \angle BCA \leftrightarrow \angle ADB$$
$$\overline{CA} \leftrightarrow \overline{DB} \qquad \angle CAB \leftrightarrow \angle DBA$$

From these we obtain the congruences required to establish the congruence of the other triangles.

Example

Given: C-D-A and C-F-B, $\overline{CA} \cong \overline{CB}$ and $\overline{AD} \cong \overline{BF}$
Prove: $\overline{AF} \cong \overline{BD}$

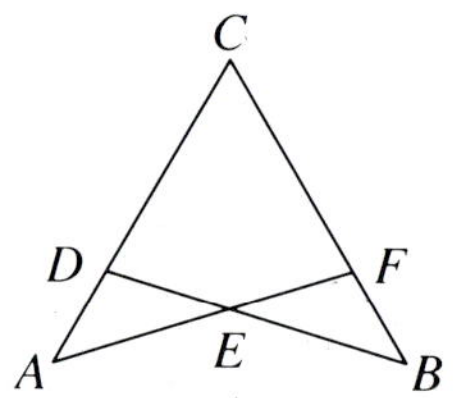

Proof

Statements	*Reasons*
1. $\overline{CA} \cong \overline{CB}$ and $\overline{AD} \cong \overline{BF}$	1. Given
2. $\angle C \cong \angle C$	2. Congruence of angles is reflexive (Theorem 3-1(a))
3. $\overline{CD} \cong \overline{CF}$	3. If A-B-C and R-S-T such that $\overline{AC} \cong \overline{RT}$ and $\overline{AB} \cong \overline{RS}$, then $\overline{BC} \cong \overline{ST}$ (Theorem 2-3)
4. $\triangle CDB \cong \triangle CFA$	4. SAS (Post. 4-1)
5. $\overline{AF} \cong \overline{BD}$	5. CPCTC ◁

An Application of ASA

Airplanes usually carry one or more radio receivers, which pick up directional signals from special transmitters called ''VORs'' (for VHF Omni Range). These VORs located at airports and waypoints throughout the country. The first purpose of this equipment is to help a pilot navigate from point to point—the receiver

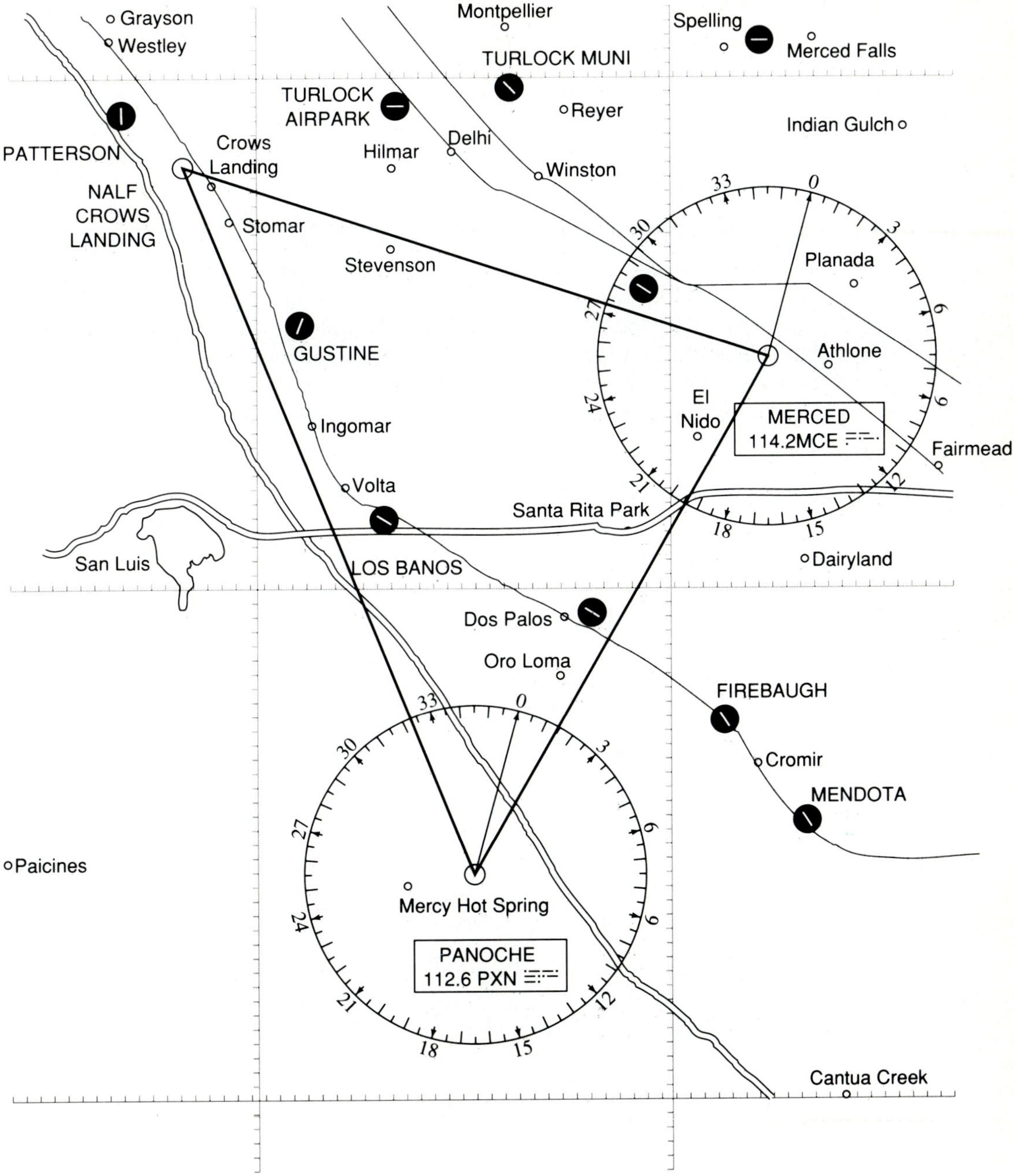
Grayson
Westley
Montpellier
Spelling
Merced Falls
TURLOCK MUNI
TURLOCK AIRPARK
Reyer
Indian Gulch
PATTERSON
Crows Landing
Hilmar
Delhi
Winston
NALF CROWS LANDING
Stomar
Stevenson
Planada
GUSTINE
Athlone
El Nido
MERCED
114.2MCE
Ingomar
Fairmead
Volta
Santa Rita Park
San Luis
LOS BANOS
Dairyland
Dos Palos
Oro Loma
FIREBAUGH
Cromir
MENDOTA
Paicines
Mercy Hot Spring
PANOCHE
112.6 PXN
Cantua Creek

Figure 4-10

aboard the plane can be tuned to the frequency of the destination VOR station, and a dial is then turned to indicate the bearing to the station. The second purpose is to locate the position of the airplane when it is in unfamiliar territory or when poor visibility conditions are encountered. The receiver in the plane can find the bearings from VOR station A and VOR station B; by drawing lines along these bearings, the pilot or copilot can determine the aircraft's present position on an aeronautical chart.

In Fig. 4-10, an airplane that has bearing 325° from the Panoche VOR and 270° from the Merced VOR will be nearly overhead the Navy Auxiliary Landing Field at Crows Landing. Since these navigational aids are in operation in all types of weather, pilots can determine where they are even if they are unable to see the ground. The application of the ASA Postulate is that the distance between the VOR stations at Panoche and Merced is fixed and the angles of bearing to the airplane are known, so that the triangle drawn on the chart will locate only one point, the third vertex of the triangle, which is the position of the plane.

Exercise Set 4.3

1. Tell which of the following pairs of triangles are congruent, and give the postulate or theorem that justifies your answer.

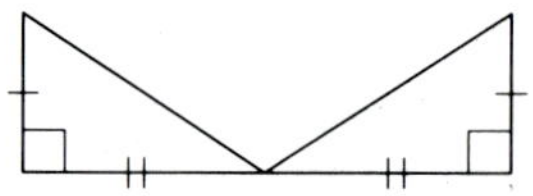

(a)

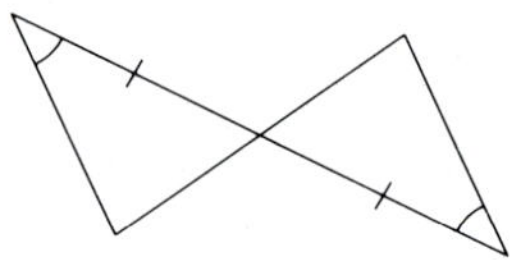

(b)

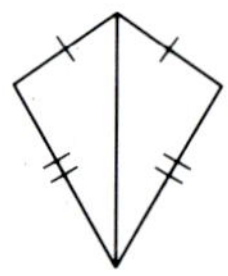

(c)

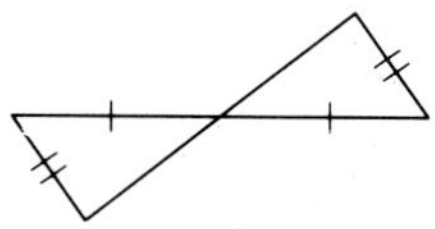

(d)

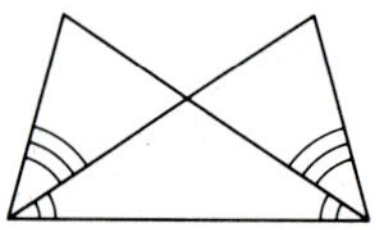

(e)

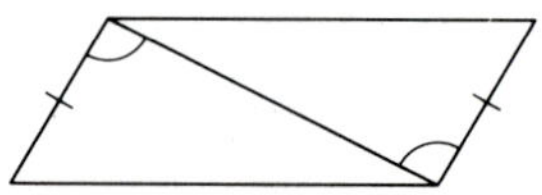

(f)

2. *Given:* $\overline{AB} \cong \overline{CD}$ and $\overline{AD} \cong \overline{CB}$
Prove: $\triangle ABC \cong \triangle CDA$

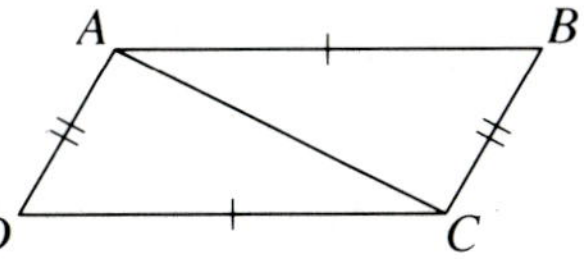

3. *Given:* $\overline{RT} \perp \overline{RS}$, $\overline{SU} \perp \overline{SR}$, $\overline{RT} \cong \overline{SU}$, and $\overline{TS}$ and $\overline{UR}$ intersect at E.

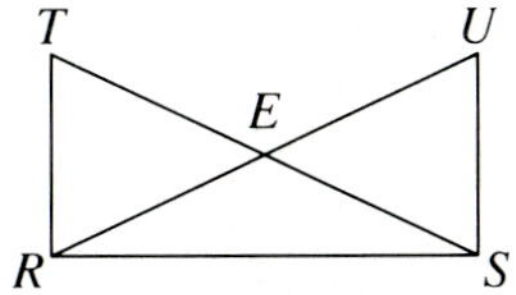

Prove: **(a)** $\angle URS \cong \angle TSR$ **(b)** $\angle T \cong \angle U$
(c) $\angle TRU \cong \angle UST$ **(d)** $\overline{RE} \cong \overline{SE}$

4. *Given:* $\overline{CD}$ and $\overline{AB}$ intersect at E, with E the bisector of $\overline{CD}$, $\angle C \cong \angle D$
Prove: $\angle B \cong \angle A$

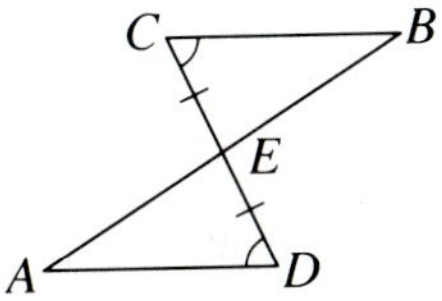

5. Prove that the segment joining the vertex of an isosceles triangle to the midpoint of the base separates it into two congruent triangles.

6. *Given:* A, B, C, and D are collinear with $\overline{AB} \cong \overline{DC}$, $\angle A \cong \angle D$, and $\angle x \cong \angle y$
Prove: **(a)** $\triangle ABE \cong \triangle DCE$
(b) $\triangle ADE$ is isosceles

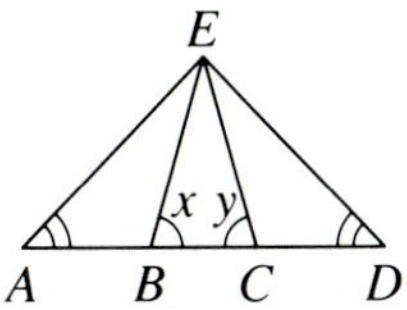

7. *Given:* $\overline{AB} \cong \overline{AC}$, $B \in \overline{AD}$, $C \in \overline{AF}$, $\overline{BD} \cong \overline{CF}$
Prove: **(a)** $\triangle ADC \cong \triangle AFB$
(b) $\triangle BCD \cong \triangle CBF$
(c) $\angle ABC \cong \angle ACB$

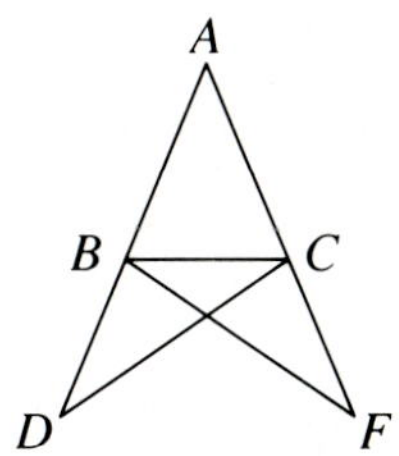

8. *Given:* $\overline{RS}$ intersects $\overline{TU}$ so that $\angle TVS$ is a right angle and $\overline{TV} \cong \overline{UV}$
Prove: **(a)** $\angle TSV \cong \angle USV$
(b) $\triangle RVT \cong \triangle RVU$
(c) $\triangle STR \cong \triangle SUR$

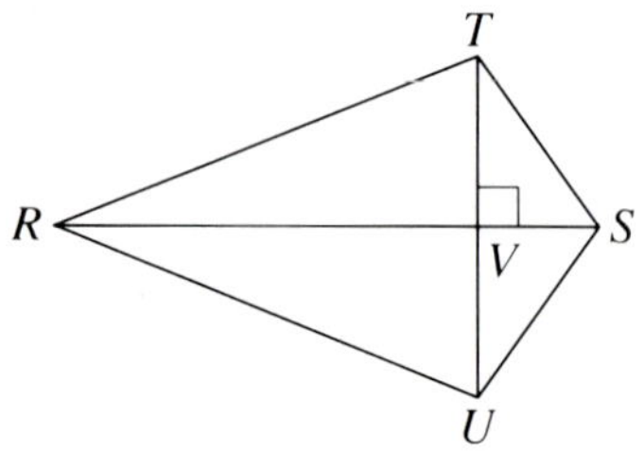

9. *Given:* U the midpoint of $\overline{SZ}$, $\angle S \cong \angle Z$, $\overline{SE} \cong \overline{ZI}$, and $\angle SEI \cong \angle ZIE$
Prove: **(a)** $\overline{EU} \cong \overline{IU}$
(b) $\angle IEU \cong \angle EIU$

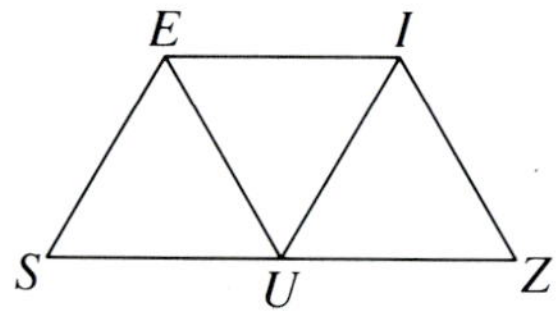

10. Prove that the segments joining the midpoints of the sides of an equilateral triangle form four congruent equilateral triangles.
11. Locate the points $A(0, 0)$, $B(6, 0)$, $C(2, 3)$, and $D(-4, 3)$ on coordinate axes. Prove that triangle ABC is congruent to triangle CDA.
12. Prove that Construction 2 (construct an angle congruent to a given angle) on page 63 is valid.
13. In order to illustrate the principle of VOR navigation described previously, do the following: (a) Assume that VOR station A is located at the origin of a rectangular coordinate system and that the point $(0, 5)$ is north of A, $(5, 0)$ is east of A, $(0, -5)$ is south, and so on. (b) Assume that VOR station B is located at a point with coordinates $(12, 2)$. (c) You are the pilot of an airplane that is on a bearing of 025° from A and 326° from B. What are the coordinates of your position? (Bearings were discussed in Section 3-2, page 46.)

4-4 MEDIANS AND ANGLE BISECTORS OF A TRIANGLE; CONSTRUCTION OF TRIANGLES

It would seem obvious that for any angle ABC, there is a half-line BD that "bisects" the angle. We will define angle bisector and then prove that there exists exactly one for a given angle, since this is an essential fact for some of the later proofs.

Definition 4-6 If D is in the interior of $\angle ABC$ and $\angle ABD \cong \angle DBC$, then $\overrightarrow{BD}$ is the **bisector** of $\angle ABC$ (Fig. 4-11).

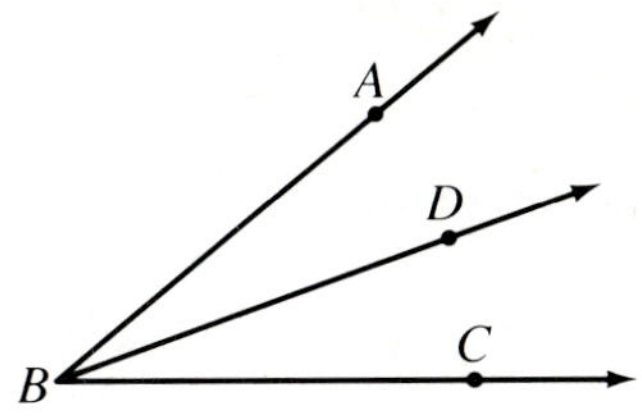

Figure 4-11

Theorem 4-2 Every angle has exactly one bisector.

Let $\angle B$ be the angle. Prove:

a. There is a bisector of $\angle B$.

b. There is only one bisector of $\angle B$.

a. There is one bisector of $\angle B$.

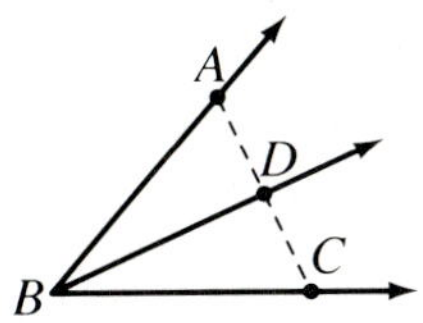

Proof

Statements	*Reasons*
1. Choose A and C on the sides of $\angle B$ so that $\overline{BA} \cong \overline{BC}$	1. Segment Construction (Theorem 2-4)
2. Let D be the midpoint of $\overline{AC}$	2. Every segment has exactly one midpoint (Theorem 2-5)
3. $\overline{AD} \cong \overline{CD}$	3. Midpoint divides a segment into congruent segments (Theorem 2-6)
4. $\overline{BD} \cong \overline{BD}$	4. Congruence of segments is reflexive (Theorem 2-1(a))
5. $\triangle ABD \cong \triangle CBD$	5. SSS (Post. 4-3)
6. $\angle ABD \cong \angle CBD$	6. CPCTC
7. $\overrightarrow{BD}$ bisects $\angle ABC$	7. Definition of angle bisector (Def. 4-6)

b. There is only one bisector of $\angle B$.

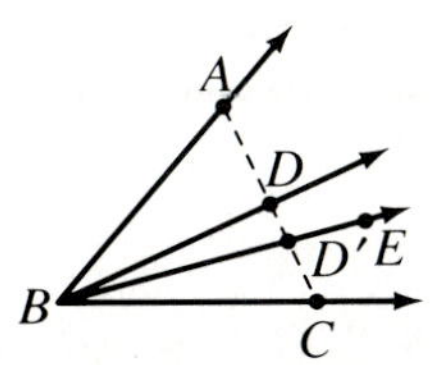

Proof

Statements	*Reasons*
1. Choose A and C on the sides of $\angle B$ so that $\overline{AB} \cong \overline{BC}$	1. Segment Construction (Theorem 2-4)
2. Let $\overrightarrow{BE}$ be a bisector of $\angle B$	2. Part a of proof
3. $\angle ABE \cong \angle CBE$	3. Definition of angle bisector (Def. 4-6)
4. Let D' be the intersection of $\overrightarrow{BE}$ and $\overline{AC}$	4. Definition of interior of an angle (Def. 3-2(b)) and two different lines intersect in at most one point (Theorem 2-8)
5. $\overline{BD'} \cong \overline{BD'}$	5. Congruence of segments is reflexive (Theorem 2-1(a))
6. $\triangle ABD' \cong \triangle CBD'$	6. SAS (Post. 4-1)
7. $\therefore \overline{AD'} \cong \overline{CD'}$	7. CPCTC
8. D' is the midpoint of $\overline{AC}$	8. If A-D'-C such that $\overline{AD'} \cong \overline{CD'}$, then D' is the midpoint of $\overline{AC}$ (Theorem 2-6)
9. Point D of Part a and D' are the same point	9. Every segment has exactly one midpoint (Theorem 2-5)
10. $\therefore \overrightarrow{BD} = \overrightarrow{BD'} = \overrightarrow{BE}$	10. For any two distinct points there is exactly one line containing them (Post. 2-1(a))

Corollary 4-2 The angles formed by bisectors of congruent angles are congruent.

The following definitions will also be useful in determining triangle relationships.

Definition 4-7 A **median** of a triangle is a segment whose endpoints are a vertex of one of its angles and the midpoint of the opposite side.

In Fig. 4-12, $\overline{CD}$ is the median to side $\overline{AB}$ and $\overline{AE}$ is the median to side $\overline{BC}$.

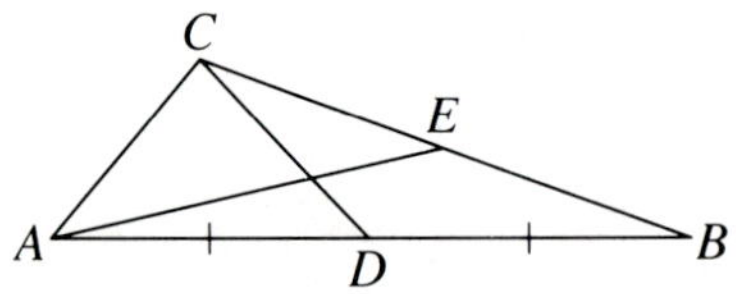

Figure 4-12

Definition 4-8 If A is a vertex of triangle ABC and $\overrightarrow{AE}$, the bisector of $\angle A$, intersects $\overline{BC}$ at D, then $\overline{AD}$ is an **angle bisector** of $\triangle ABC$ (Fig. 4-13).

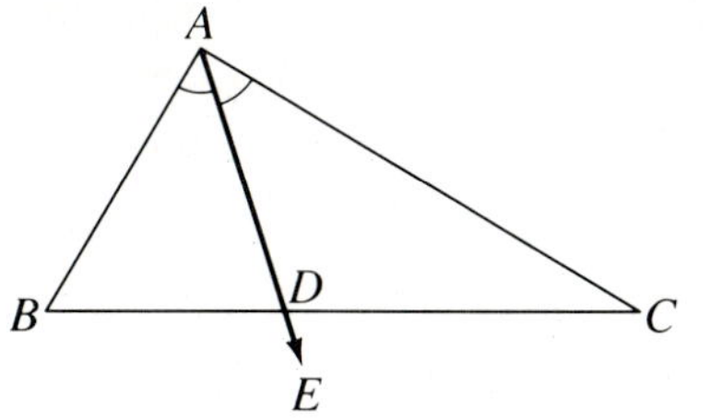

Figure 4-13

Examples

1. In $\triangle GHI$, how many angle bisectors and how many medians can you draw?

Solution A triangle determines three angles; therefore, it will have three angle bisectors. Each of the three segments that form the triangle has a midpoint; hence there will be three segments that can be drawn from opposite vertices to the midpoints.

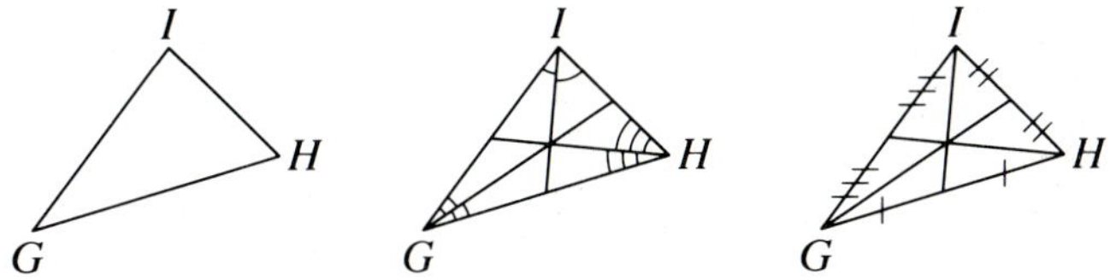

2. In the following figure, $\overline{TU}$ is a median of the triangle and $\overline{TV}$ is the bisector of $\angle RTS$.

(a) How are $\overline{RU}$ and $\overline{US}$ related?

(b) How are $\angle RTV$ and $\angle STV$ related?

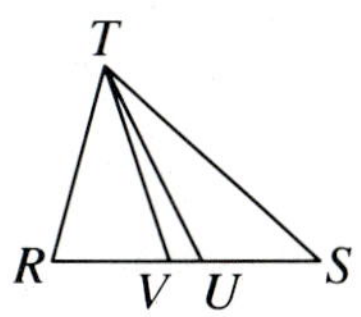

Solution (a) $\overline{RU} \cong \overline{US}$, because the median joins a vertex and the midpoint of the opposite side. (b) $\angle RTV \cong \angle STV$ since an angle bisector divides the angle into two congruent angles.

3. If $\triangle LMN$ is an isosceles triangle with base angles L and N, and $\overline{MQ}$ is a median, show

(a) $\angle 1 \cong \angle 2$

(b) $\angle 3 \cong \angle 4$

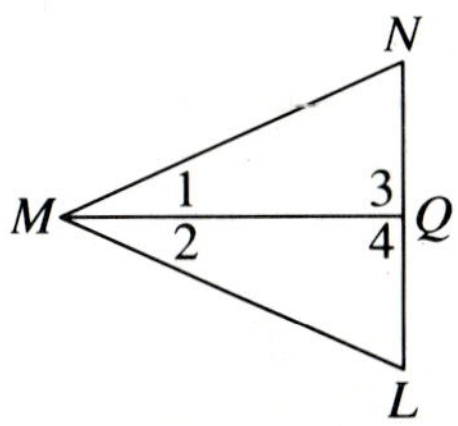

Solution $\overline{LM} \cong \overline{NM}$ since the triangle is isosceles, and $\angle N$ and $\angle L$ are base angles. $\overline{LQ} \cong \overline{NQ}$ by Def. 4-7, $\overline{MQ} \cong \overline{MQ}$, thus $\triangle LQM \cong \triangle NQM$ by SSS. Hence $\angle 1 \cong \angle 2$ and $\angle 3 \cong \angle 4$ since CPCTC.
(What properties does $\overline{MQ}$ have with respect to the triangle other than being a median?) ◁

Theorem 4-3 The midpoint of the hypotenuse of a right triangle is equidistant from the vertices.

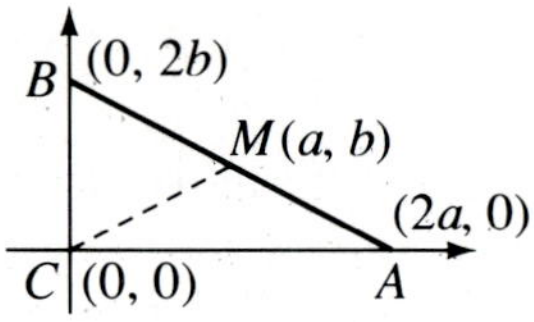

Proof We can place a coordinate system on the legs of the right triangle with the right angle at the origin. Let $A(2a, 0)$, $B(0, 2b)$ and $C(0, 0)$ be the coordinates of the vertices. By the midpoint formula

$$M = \left(\frac{2a + 0}{2}, \frac{0 + 2b}{2}\right) = (a, b)$$

Applying the distance formula

$$\begin{aligned} m\overline{AM} &= \sqrt{(2a - a)^2 + (0 - b)^2} = \sqrt{a^2 + b^2} \\ m\overline{BM} &= \sqrt{(0 - a)^2 + (2b - b)^2} = \sqrt{a^2 + b^2} \\ m\overline{CM} &= \sqrt{(0 - a)^2 + (0 - b)^2} = \sqrt{a^2 + b^2} \end{aligned}$$

Therefore, M is equidistant from A, B, and C.

Theorem 4-4 If two medians of a triangle are congruent, the triangle is isosceles. (The proof of this theorem is left as an exercise.)

Construction of Triangles and Angle Bisectors

Construction 5: Construct a triangle congruent to a given triangle.
To construct $\triangle A'B'C' \cong ABC$:

1. Construct $\overline{A'B'} \cong \overline{AB}$ on line ℓ.
2. Draw an arc with center at A' and $r_1 = m\overline{AC}$ and another arc with center at B' and $r_2 = m\overline{BC}$ that intersect above $\overline{A'B'}$ at C'. $\triangle A'B'C'$ is the required triangle.

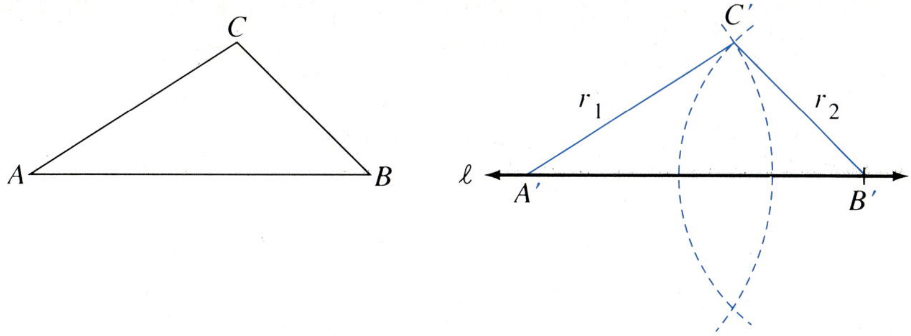

Proof $\overline{A'B'} \cong \overline{AB}$, $\overline{A'C'} \cong \overline{AC}$, and $\overline{B'C'} \cong \overline{BC}$ by construction. $\triangle A'B'C' \cong \triangle ABC$ by SSS.

Construction 6: Construct the bisector of a given angle. To bisect angle ABC:

1. Draw an arc with center at B that intersects $\overrightarrow{BA}$ at S and $\overrightarrow{BC}$ at T.
2. Using S and T as centers and $r > \frac{1}{2}\text{m}\overline{ST}$, draw arcs that intersect at P. Then $\overrightarrow{BP}$ is the bisector of $\angle ABC$.

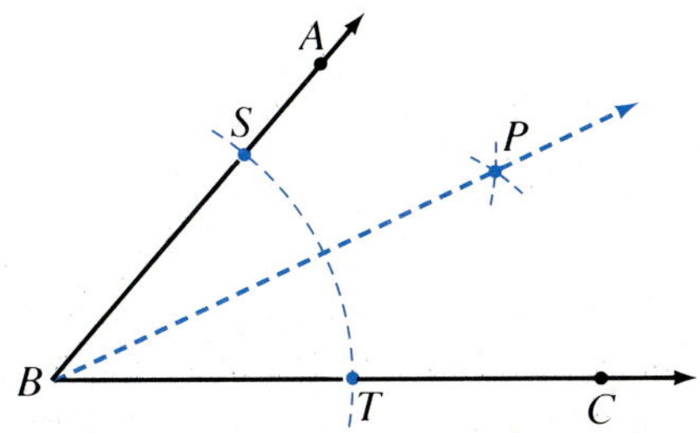

Proof $\overline{BS} \cong \overline{BT}$ and $\overline{SP} \cong \overline{TP}$ by construction. $\overline{PB} \cong \overline{PB}$, since segment congruence is reflexive, and so $\triangle BSP \cong \triangle BYP$ by SSS (Post. 4-3) and $\angle PBS \cong \angle PBT$ by CPCTC.

Exercise Set 4.4

1. In the triangle ABC below $\overline{AD}$ bisects $\angle BAC$ and $\overline{AC} \cong \overline{AB}$.
   Prove: $\triangle ABD \cong \triangle ACD$

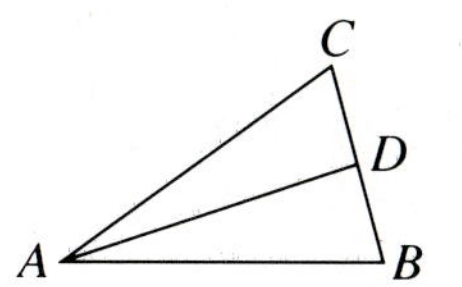

2. In the triangle RST $\overline{RP}$ bisects $\angle TRS$, $\overline{SQ}$ bisects $\angle RST$, and $\angle TRS \cong \angle TSR$.
Prove: $\triangle RSQ \cong \triangle SRP$

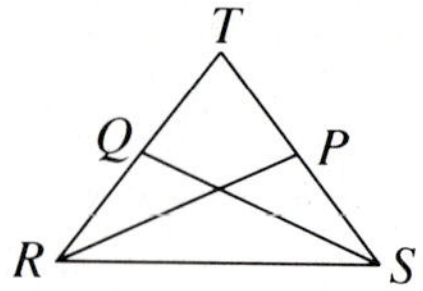

3. If $\overline{DG}$ is a median in the following figure and $\angle FGD$ is a right angle, prove
(a) $\overline{DG}$ bisects $\angle FDE$
(b) $\triangle DEF$ is isosceles

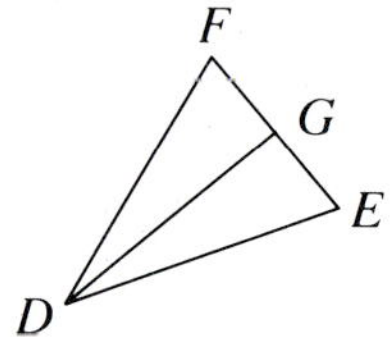

4. Draw triangles with sides and angles as given. Use protractor and ruler to determine the measures of the other sides and angles.

	$\angle A$	$\overline{BC}$	$\angle B$	$\overline{AC}$	$\angle C$	$\overline{AB}$
a	40°		40°			3 in.
b	40°			$2\frac{1}{2}$ in.		3 in.
c	40°		95°			3 in.
d	40°			3 in.		3 in.
e		3 in.		3 in.		4 in.
f		$2\frac{1}{2}$ in.	60°		60°	
g		3 in.		4 in.		5 in.

5. Prove Corollary 4-2(a).
6. Prove Theorem 4-4 by means of a coordinate proof.
7. If $\triangle RST$ has vertices $R(0, 0)$, $S(13, 1)$, $T(5, 5)$, and $\triangle XYZ$ has vertices $X(5, -10)$, $y(18, -9)$, and $Z(10, -5)$, show that the triangles are congruent.
*8. Suppose $\triangle UVP$ has vertices with coordinates $U(0, 0)$, $V(3, -3)$, $P(5, 5)$, and $\triangle MNQ$ has vertices with coordinates $M(8, 3)$, $N(11, 1)$ and $Q(13, 8)$.
(a) What kind of triangles are formed?
(b) Are the triangles congruent? Why or why not?
9. Draw a right triangle RST with right angle at S. Measure $\overline{RT}$ and locate the midpoint, U. Place the points of your compass at U and draw a circle with radius $m\overline{UR}$. Does

your circle pass through the points S and T? Why *should* the circle pass through points S and T?

10. Draw a scalene triangle ABC. Construct the bisector of each of the three angles. How do angle bisectors $\overline{AP}$, $\overline{BQ}$, and $\overline{CR}$ meet?

Isosceles and Equilateral Triangles

By now you have probably deduced that the isosceles triangle has some unique properties. We shall formalize some of these now, first by proving that the base angles are congruent.

Theorem 4-5 (*Isosceles Triangle Theorem*) If two sides of a triangle are congruent, the angles opposite these sides are congruent.

Given: $\triangle ABC$ with $\overline{AC} \cong \overline{BC}$

Prove: $\angle A \cong \angle B$

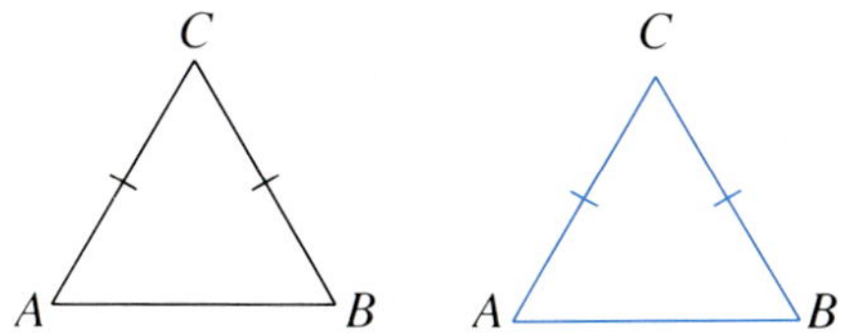

Proof

Statements	*Reasons*
1. $\overline{AC} \cong \overline{BC}$	1. Given
2. $\overline{BC} \cong \overline{AC}$	2. Congruence of segments is symmetric (Theorem 2-1(b))
3. $\angle C \cong \angle C$	3. Congruence of angles is reflexive (Theorem 3-1(a))
4. $\triangle ABC \cong \triangle BAC$	4. SAS (Post. 4-1)
5. $\angle A \cong \angle B$	5. CPCTC

Note that this proof depends upon the correspondence between $\triangle ABC$ and itself, i.e.,

$$\triangle ABC \leftrightarrow \triangle BAC$$

Under this correspondence we have $\overline{AC} \leftrightarrow \overline{BC}$, $\overline{BC} \leftrightarrow \overline{AC}$, and $\angle C \leftrightarrow \angle C$, which is a congruence by the SAS postulate. Another proof will be suggested as an exercise.

Corollary 4-5 Equilateral triangles are equiangular. (The proof is left to the reader.)

The converse of Theorem 4-5 is also true.

Theorem 4-6 If two angles of a triangle are congruent, the sides opposite these angles are congruent.

Given: $\triangle ABC$ with $\angle A \cong \angle B$
Prove: $\overline{AC} \cong \overline{BC}$

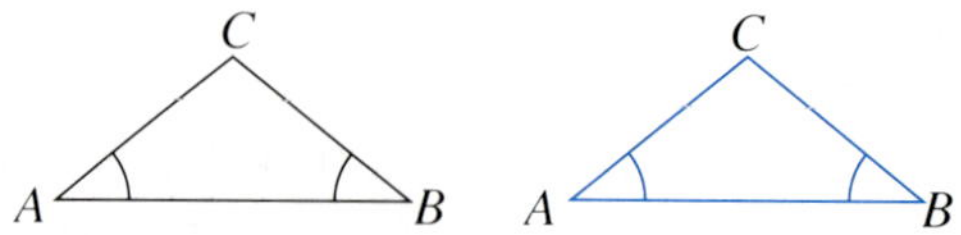

Proof As in the proof of Theorem 4-5, we see that $ABC \leftrightarrow BAC$ is the desired correspondence between the triangle and itself. Since $\angle A \cong \angle B$, $\overline{AB} \cong \overline{BA}$ and $\angle B \cong \angle A$, the correspondence is a congruence. Since $\triangle ABC \cong \triangle BAC$, it follows that $\overline{AC} \cong \overline{BC}$.

Corollary 4-6 An equiangular triangle is equilateral. (The proof is left to the reader.)

Example

Given: $\overleftrightarrow{AB}$ and $\overline{AC} \cong \overline{BC}$
Prove: $\angle 1 \cong \angle 2$

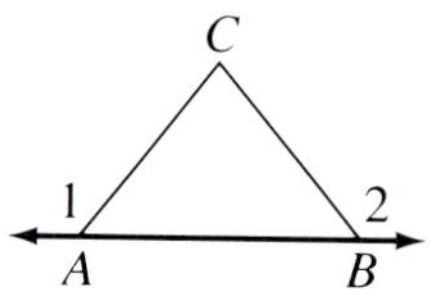

Proof

Statements	*Reasons*
1. $\overline{AC} \cong \overline{BC}$	1. Given
2. $\angle CAB \cong \angle CBA$	2. Isosceles Triangle Theorem (Theorem 4-5)
3. $\angle 1$ is the supplement of $\angle CAB$ $\angle 2$ is the supplement of $\angle CBA$	3. If the exterior sides of two adjacent angles form a straight angle, they are supplementary (Theorem 3-5)
4. $\angle 1 \cong \angle 2$	4. Supplements of congruent angles are congruent (Theorem 3-10)

Exercise Set 4.5

1. If $\overline{RT} \cong \overline{ST}$ and $\angle 1 \cong \angle 2$ in triangle RST, prove that $\triangle RQT \cong \triangle SQT$.

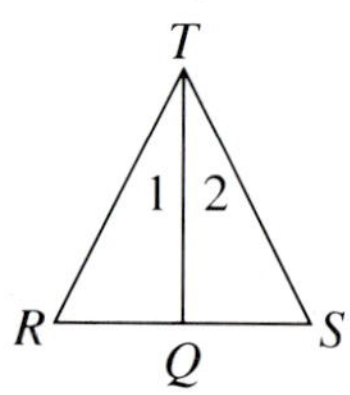

2. Using the figure in Exercise 1, if $\angle 1 \cong \angle 2$ and $\angle RQT \cong \angle SQT$, prove that $\triangle RQT \cong \triangle SQT$.

3. Using the figure in Exercise 1, if $\overline{RT} \cong \overline{ST}$ and Q is the midpoint of $\overline{RS}$, prove that $\triangle RQT \cong \triangle SQT$.
4. In triangle ABC, if $\angle A \cong \angle B$ and $\overline{AE} \cong \overline{BD}$, prove that
   (a) $\overline{CD} \cong \overline{CE}$
   (b) $\triangle ADC \cong \triangle BEC$

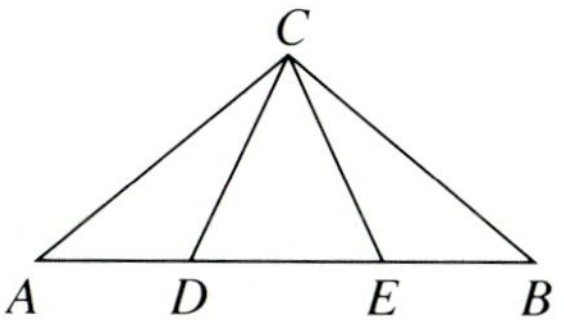

5. In triangle DEF, $\overline{DF} \cong \overline{EF}$ and $\angle 2 \cong \angle 3$. Prove that $\triangle DJG \cong \triangle EJH$.

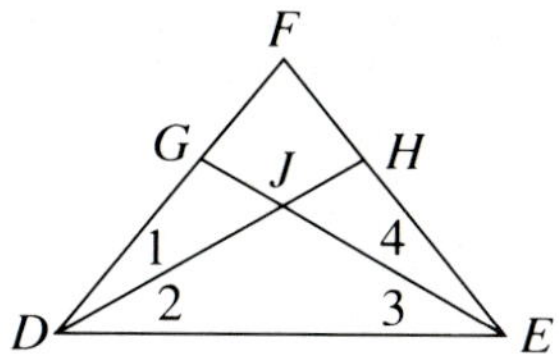

6. In triangle ACD, $\angle 1 \cong \angle 2$ and $\angle 3 \cong \angle 4$.
   Prove: **(a)** $\triangle BFC \cong \triangle EFD$
   (b) $\overline{AC} \cong \overline{AD}$

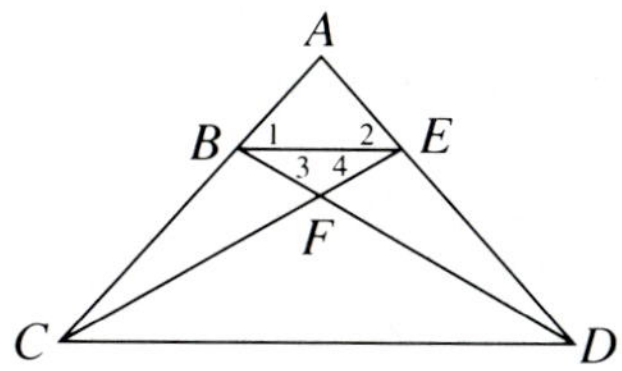

7. In the following figure, B, D, and E are collinear points, $\overline{AB} \cong \overline{CB}$, and $\overline{AE} \cong \overline{CE}$. Prove $\overline{AD} \cong \overline{CD}$.

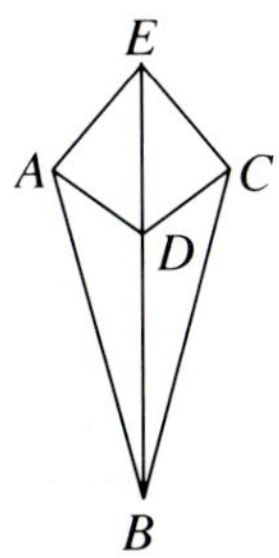

8. In the following figure, $\overrightarrow{YP}$ bisects $\angle XYZ$ and $\angle XWP \cong \angle ZWP$. Prove $\angle X \cong \angle Z$.

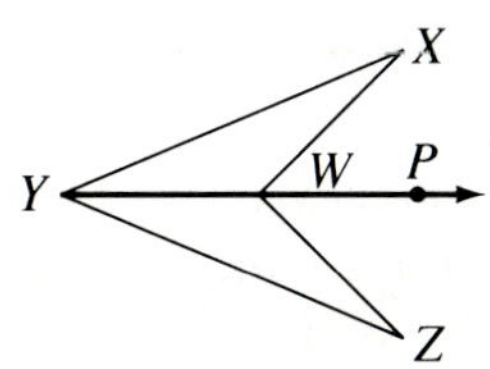

9. In the following figure, $\triangle ABC$ is isosceles with medians $\overline{AD}$ and $\overline{BE}$ to the congruent sides. Prove that $\overline{AD} \cong \overline{BE}$.

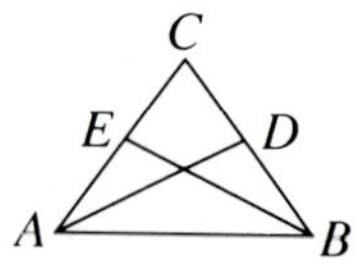

10. Prove that if two triangles are congruent, then the angle bisector of one triangle is congruent to the corresponding angle bisector of the other triangle.
11. Prove that the bisector of the vertex angle of an isosceles triangle is perpendicular to the base and bisects it.
12. Given the points $A(2, 0)$, $B(4, 0)$, $C(6, 0)$, $D(8, 0)$ and $E(5, 5)$, with $\angle A \cong \angle D$.
    Prove: **(a)** $\triangle BEC$ is isosceles
    (b) $\triangle ABE \cong \triangle DCE$

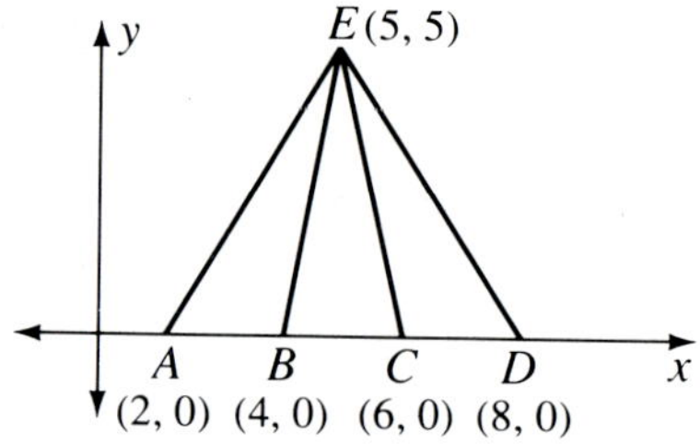

13. In a triangle RST with vertices $R(4, 0)$, $S(6, 6)$ and $T(0, 4)$, find the lengths of the medians.
14. **(a)** Prove Corollary 4-5.
    (b) Prove Corollary 4-6.

4-5 BASIC GEOMETRY INEQUALITIES

We have been chiefly concerned with the *interior* angles of a triangle; now we turn our attention to certain angles formed by extending a side (Fig. 4-14).

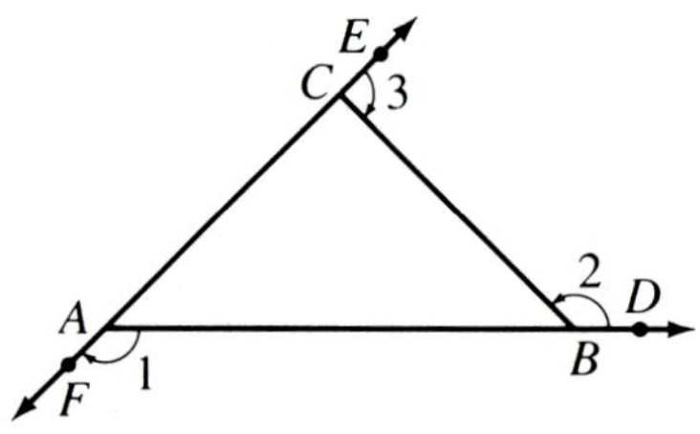

Figure 4-14

Definition 4-9 If A-B-D, then angle CBD is an **exterior angle** of triangle ABC.

Note that angles ABC and CBD are supplementary (Theorem 3-5). Since angles ABC and CBD are also adjacent to each other (Def. 3-5), we call $\angle CAB$ and $\angle ACB$ **nonadjacent interior angles** of triangle ABC. Figure 4-15 shows the six possible exterior angles of a triangle, three of which are vertical angles.

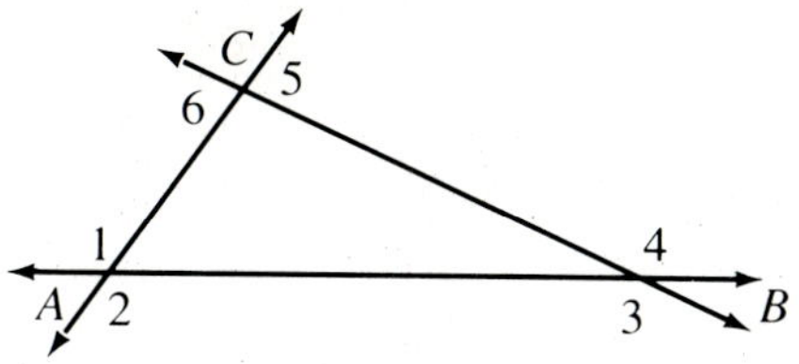

Figure 4-15

Since we have defined congruence of angles and segments in terms of their measures, it would be appropriate to define the inequality of segments and angles in terms of their measures also.

Definition 4-10 For segments AB and CD, $\overline{AB} < \overline{CD}$ if and only if $m\overline{AB} < m\overline{CD}$.

Definition 4-11 For angles ABC and DEF, $\angle ABC < \angle DEF$ if and only if $m\angle ABC < m\angle DEF$.

Theorem 4-7 (*Exterior Angle Theorem*) An exterior angle of a triangle is greater than either nonadjacent interior angle.

Given: $\triangle ABC$ with $\overrightarrow{AB}$ extending through D

Prove: $\angle C < \angle DBC$

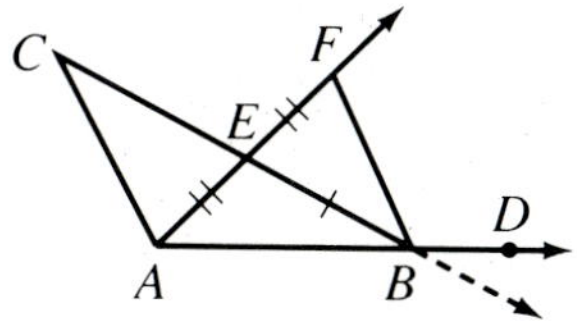

Proof

Statements	*Reasons*
1. Let E be the midpoint of $\overline{CB}$	1. Every segment has exactly one midpoint (Theorem 2-5)
2. On ray AE, let F be a point such that $\overline{AE} \cong \overline{FE}$	2. Segment Construction (Theorem 2-4)
3. Draw $\overline{FB}$	3. Two points determine a segment (Def. 2-2)
4. $\overline{EC} \cong \overline{EB}$	4. If P is the midpoint of $\overline{AB}$, $\overline{AP} \cong \overline{PB}$ (Theorem 2-6)
5. $\angle CEA \cong \angle BEF$	5. Vertical angles are congruent (Theorem 3-4)
6. $\triangle AEC \cong \triangle FEB$	6. SAS (Post. 4-1)
7. $\angle C \cong \angle FBE$	7. CPCTC
8. $m\angle C = m\angle FBE$	8. Definition of congruent angles (Def. 3-8)

Statements	Reasons
9. $m\angle DBC = m\angle DBF + m\angle FBE$	9. If P is in the interior of an $\angle ABC$, then $m\angle ABC = m\angle ABP + m\angle PBC$ (Post. 3-3)
10. $m\angle DBC = m\angle DBF + m\angle C$	10. Substitution Axiom for real numbers
11. $m\angle C < m\angle DBC$	11. Definition of less than for real numbers
12. $\therefore \angle C < \angle DBC$	12. Definition of less than for angles (Def. 4-11)

In similar fashion (extending $\overrightarrow{CB}$ through H, drawing $\overrightarrow{CG}$ through G, the midpoint of $\overline{AB}$, etc.) we can show that $\angle A < \angle DBC$.

Corollary 4-7 If a triangle has one right angle, the other two angles are acute.

In any scalene triangle RST, if we measure each angle with a protractor and each side with a ruler, the longest side will be opposite the largest angle. Thus in Fig. 4-16 we would find $\angle T$ is the largest angle and $\overline{RS}$ is the longest side.

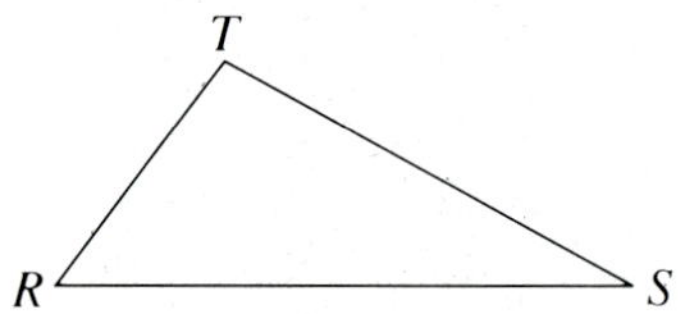

Figure 4-16

Theorem 4-8 If two sides of a triangle are not congruent, the angle opposite the longer side is the larger angle.

Given: In $\triangle ABC$, $\overline{AC} < \overline{BC}$

Prove: $\angle CBA < \angle CAB$

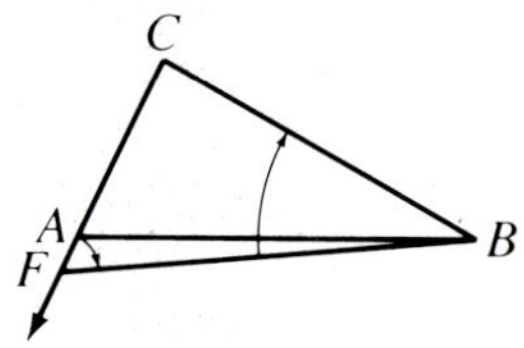

Proof

Statements	Reasons
1. Draw $\overrightarrow{CA}$	1. $\{\overrightarrow{CA}\} = \{\overline{CA}\} \cup \{X \mid C - A - X\}$ (Def. 2-3)
2. Locate F on $\overrightarrow{CA}$ so that $\overline{CF} \cong \overline{CB}$	2. Segment Construction (Theorem 2-4)
3. Draw $\overline{BF}$	3. Two points determine a segment (Def. 2-2)
4. BFC and BAF are triangles	4. Definition of triangle (Def. 4-1)

Statements	*Reasons*
5. $\angle CFB \cong \angle CBF$	5. Angles opposite congruent sides are congruent (Theorem 4-5)
6. $m\angle CBF = m\angle CBA + m\angle ABF$	6. Angle Addition Postulate (Post. 3-3)
7. $m\angle CBA < \angle CBF$	7. Definition of less than for real numbers
8. $m\angle CBA < m\angle CFB$	8. Step 5, substitution
9. $m\angle CFB < m\angle CAB$	9. An exterior angle of a triangle is greater than either nonadjacent angle (Theorem 4-7)
10. $m\angle CBA < m\angle CAB$	10. Transitivite Axiom of Order (Axiom 0-2)
11. $\angle CBA < \angle CAB$	11. Definition of less than for angles (Def. 4-11)

The converse of this theorem is also true:

Theorem 4-9 If two angles of a triangle are not congruent, the side opposite the larger angle is the longer side.

Exercise Set 4.6

1. In the following figure, name

(a) the exterior angles

(b) the nonadjacent interior angles of $\angle 1$, $\angle 4$, $\angle 9$

(c) the interior angle that is supplementary to $\angle 7$

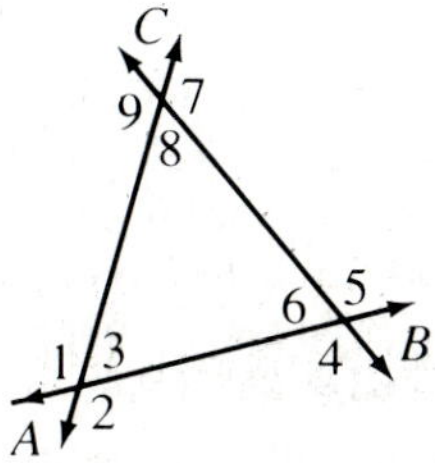

2. Referring to the following figure, replace the question mark with $<$ or $>$ in order to make each statement true.

(a) $\angle a$? $\angle c$ **(b)** $\angle b$? $\angle d$

(c) $\angle c$? $\angle b$ **(d)** $\angle a$? $\angle d$

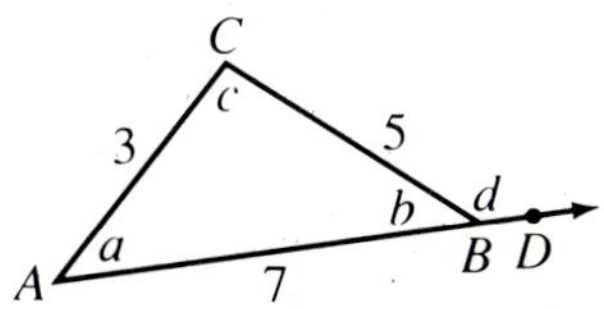

3. Referring to the following figure, complete each sentence:
   (a) If m$\angle A$ = 30, m$\angle B$ = 70, m$\angle 1$ > ______
   (b) If m$\angle A$ = 50, m$\angle B$ = 60, m$\angle 1$ > ______
   (c) If m$\angle A$ = 110, m$\angle B$ = 65, m$\angle 1$ > ______

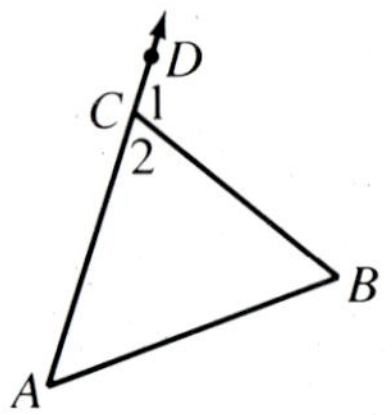

4. If, in $\triangle ABC$, m$\angle A$ = 42, m$\angle B$ = 63, and m$\angle C$ = 75, which is the longest side? The shortest side?
5. In $\triangle RST$, m$\angle R$ = 40, m$\angle S$ = 25, and m$\angle T$ = 115. List the sides of the triangle in order from shortest to longest.
6. In $\triangle ABC$ with m$\overline{AC}$ = 5, m$\overline{BC}$ = 7, and m$\overline{AB}$ = 10, the largest angle is angle ______ and the smallest angle is angle ______.
7. Prove Corollary 4-7.
8. Prove Theorem 4-9.
9. If ℓ_2 and ℓ_3 intersect ℓ_1 at points P and Q so that $\overline{RP} \perp \ell_1$, why can't $\overline{RQ}$ be perpendicular to ℓ_1?

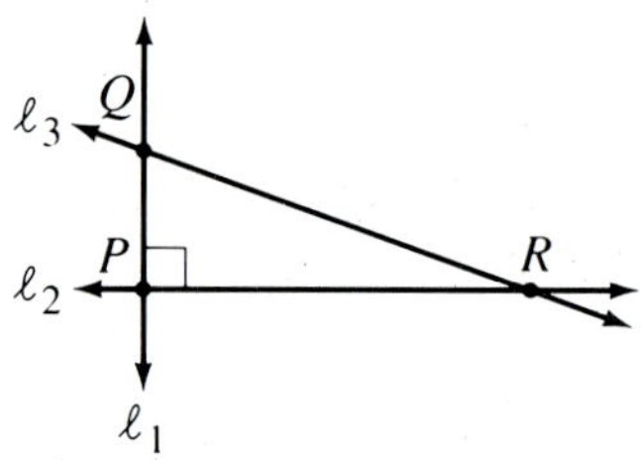

10. Prove that the hypotenuse of a right triangle is longer than either of its legs.
11. Write the converse of each of the following statements; then state whether or not the original statement is true and if the converse is true.
    (a) If a man lives in Reno, Nevada, he lives in Reno.
    (b) If it is snowing, it is cloudy outside.
    (c) If Mary has typhoid, she is ill.
    (d) If two angles are right angles, they are congruent.
    (e) If two triangles are congruent, their corresponding parts are congruent.
12. Complete the proof of Theorem 4-7.

REVIEW EXERCISES

1. In a right triangle, the median to the hypotenuse is $5\frac{1}{2}$ in. long. What is the length of the hypotenuse?

2. In the following figure, $\overline{RS} \cong \overline{RU}$ and $\angle S \cong \angle U$. Prove that $\overline{ST} \cong \overline{UT}$.

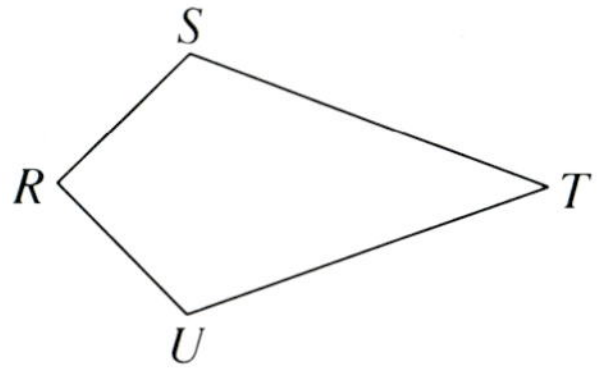

3. Prove Theorem 4-5 by drawing a median to the base opposite the vertex angle.

4. Construct a triangle RST with $m\overline{RS} = 3$ in., $m\overline{RT} = 4$ in., and $m\overline{ST} = 4$ in.
   (a) What kind of triangle is this?
   (b) Measure angles R and S with a protractor. How do they compare?

5. Construct an equilateral triangle with sides $3\frac{1}{2}$ in. in length.
   (a) Measure the three angles with a protractor. How do they compare?
   (b) Measure an exterior angle of the triangle.

6. Construct a triangle ABC with $m\overline{AC} = 2$ in., $m\overline{BC} = 3\frac{1}{2}$ in., and $m\overline{AB} = 4\frac{1}{2}$ in.
   (a) Classify this triangle with respect to its sides and then with respect to its angles.
   (b) Measure the angles with a protractor.
   (c) Measure $\angle CBD$ if A-B-D.
   (d) Discuss the results of your measurements as they relate to Theorems 4-7 and 4-8.

7. In the following figure, ABC is a triangle and A-D-B. Considering each set of additional conditions given below separately, indicate by SAS, SSS, or ASA if there is sufficient information to show the triangles ADC and BDC congruent. If not, write NO.

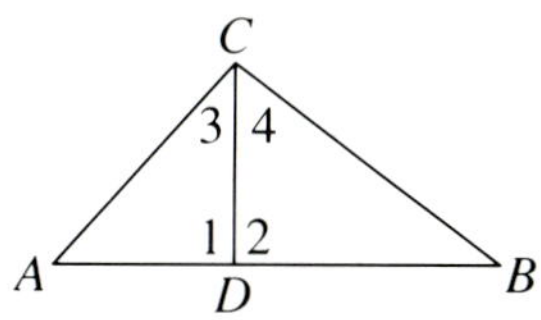

(a) $\angle A \cong \angle B$, $\overline{AD} \cong \overline{BD}$
(b) $\overline{AC} \cong \overline{BC}$, $\overline{AD} \cong \overline{BD}$
(c) $\angle 1 \cong \angle 2$, $\angle 3 \cong \angle 4$
(d) $\overline{CD}$ bisects $\angle ACB$
(e) $\overline{CD}$ is the median for $\overline{AB}$ and $\angle A \cong \angle B$
(f) $\angle B \cong \angle BCD$, $\angle A \cong \angle B$
(g) $\overline{CD} \perp \overline{AB}$
(h) $\overline{AD} \cong \overline{BC}$, $\angle 1 \cong \angle 2$

8. *Given:* $\overrightarrow{AB}$ bisects $\angle CAD$, $\overline{AC} \cong \overline{AD}$
   Prove: $\overline{BC} \cong \overline{BD}$

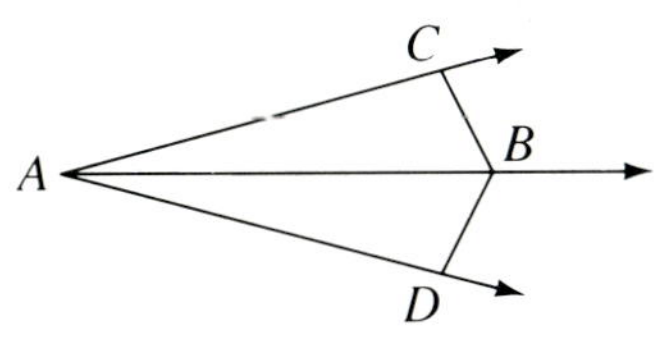

9. *Given:* $\angle RST \cong \angle UTS$, $\angle RSU \cong \angle UTR$
Prove: $\overline{RT} \cong \overline{US}$

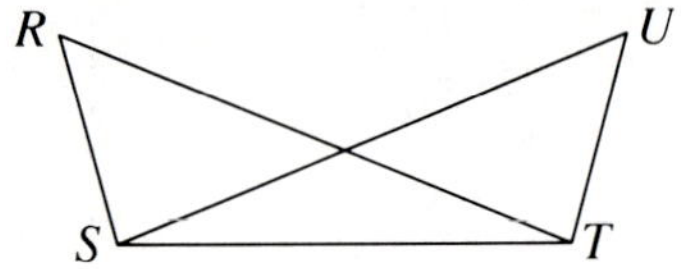

10. *Given:* $\triangle ABC$ with $\overline{AC} \cong \overline{BC}$, $\overline{BD} \perp \overline{AC}$, $\overline{AE} \perp \overline{BC}$
Prove: $\overline{AE} \cong \overline{BD}$, $\overline{AD} \cong \overline{BE}$

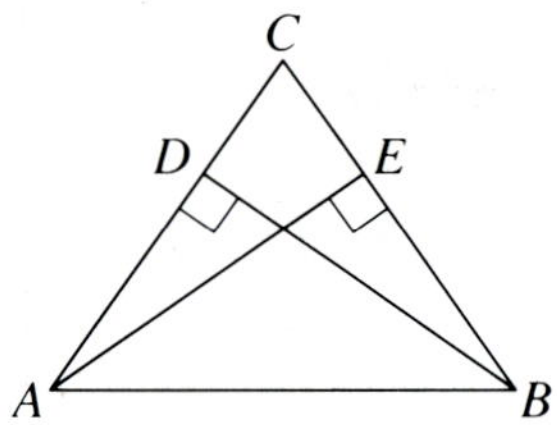

5

Perpendicular and Parallel Lines

Do not try to know everything if you wish to know anything.

DEMOCRITUS OF ABDERA (470 B.C.) Greek philosopher.

5-1 PERPENDICULARS

How many perpendiculars to a given line in a given plane can be drawn at a given point on the line? Using a protractor to do this it would seem that there is only one, but perhaps this is due to the inaccuracy of the drawing process. The following theorem settles the question.

Theorem 5-1 In a plane through a given point on a given line, there is one and only one perpendicular to the given line.

Given: Point R on line AB

Prove: There is one and only one line RS passing through R perpendicular to $\overleftrightarrow{AB}$

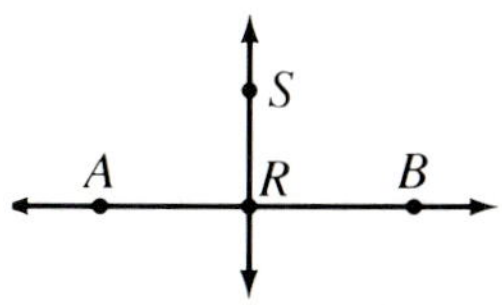

Existence Two lines are perpendicular if and only if their intersection determines a right angle. The angle construction postulate tells us that there is a ray RS such that $m\angle ARS = 90$. Hence the line RS determined by $\overrightarrow{RS}$ is perpendicular to $\overleftrightarrow{AB}$.

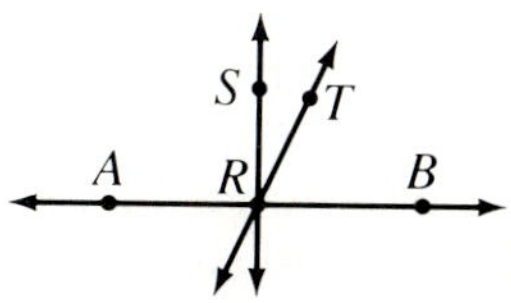

Uniqueness We can assume that there is only one perpendicular to $\overleftrightarrow{AB}$ at R, or that there are more than one. If we assume the latter, i.e., that $\overrightarrow{RS} \perp \overleftrightarrow{AB}$ and $\overrightarrow{RT} \perp \overleftrightarrow{AB}$, then $\angle SRA$ and $\angle TRA$ are right angles, hence congruent, and $m\angle SRA = m\angle TRA = 90$. The Angle Construction Postulate (Post. 3-2) states that there is only *one* ray on one side of a line that corresponds to a given measure. Therefore, the assumption that more than one line exists perpendicular to the given line in P is false, and there is only one.

Definition 5-1 In a plane, the **perpendicular bisector** of a segment is a line that passes through the midpoint of the segment and is perpendicular to the segment (Fig. 5-1).

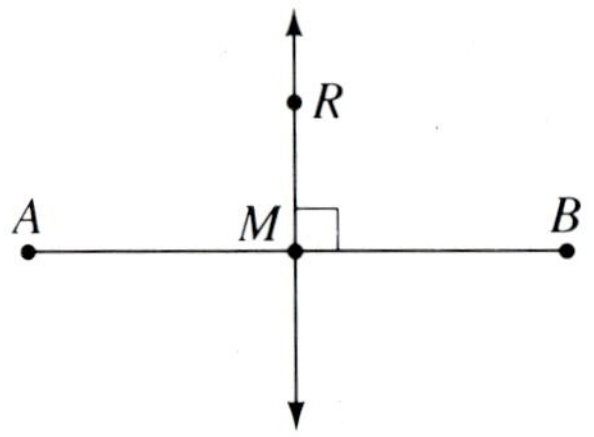

$\overleftrightarrow{MR} \perp \overline{AB}$ and
$\overline{AM} \cong \overline{MB}$

Figure 5-1

If a coordinate system is placed on $\overline{AB}$ so that M, its midpoint, has coordinates (0, 0) and $A(-a, 0)$, then B would have coordinates $(a, 0)$. Since $\overleftrightarrow{OY} \perp \overleftrightarrow{OX}$ in our coordinate system, $\overleftrightarrow{MY}$ is the perpendicular bisector of $\overline{AB}$ (Fig. 5-2).

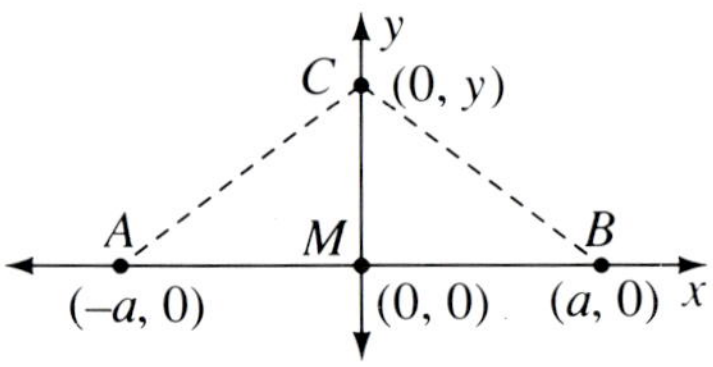

Figure 5-2

If we select any point $C(0, y)$ on line OY, we find $m\overline{AC} = \sqrt{(-a)^2 + y^2} = \sqrt{a^2 + y^2}$ and $m\overline{BC} = \sqrt{a^2 + y^2}$, or $m\overline{AC} = m\overline{BC}$. Thus:

Theorem 5-2 Any point on the perpendicular bisector of a segment is equidistant from the endpoints of the segment.

The converse of this theorem is as follows.

Theorem 5-3 If a point is equidistant from the endpoints of a segment, it lies on the perpendicular bisector of the segment.

We can prove this using coordinates also.

Given: $\overline{AB}$, a segment on $\overleftrightarrow{OX}$, with $A(a, 0)$, $B(-a, 0)$, $C(0, 0)$. Assume that $D(x, y)$ is equidistant from A and B, or $m\overline{AD} = m\overline{BD}$

Prove: $\overleftrightarrow{DC}$ is the perpendicular bisector of $\overline{AB}$

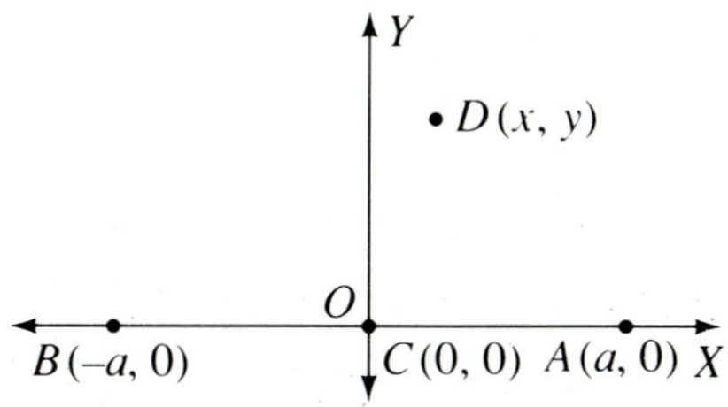

Proof (Since we know that $\overleftrightarrow{OY} \perp \overleftrightarrow{OX}$, the problem is to show that any point $D(x, y)$ that is equidistant from A and B is a point on $\overleftrightarrow{OY}$.) There are two possibilities:

1. D is on $\overleftrightarrow{OY}$
2. D is not on $\overleftrightarrow{OY}$.

Assume that the latter is true, so that $x \neq 0$. Then, since $m\overline{BD} = m\overline{AD}$, we obtain, upon substituting into the Distance Formula:

$$\sqrt{[x - (-a)]^2 + (y - 0)^2} = \sqrt{(x - a)^2 + (y - 0)^2}$$

Squaring both sides and simplifying,

$$(x + a)^2 + y^2 = (x - a)^2 + y^2$$

or

$$x^2 + 2ax + a^2 + y^2 = x^2 - 2ax + a^2 + y^2$$

Adding $(-x^2 - a^2 - y^2)$ to both members we obtain

$$2ax = -2ax, \text{ or}$$

$$4ax = 0$$

Since $a \neq 0$, x must be 0. But we assumed $x \neq 0$, which must therefore be false. If D has coordinates $(0, y)$ it must lie on the perpendicular to the segment and since C is the midpoint of $\overline{AB}$, $\overleftrightarrow{DC}$ is the perpendicular bisector of $\overline{AB}$.

Corollary 5-3 If two points are each equidistant from the endpoints of a line segment, they determine the perpendicular bisector of the segment.

The proofs of Theorems 5-1 and 5-3 relied on a special case of the Law of Disjunctive Syllogism (page 10) known as the *Law of the Excluded Middle*. The latter states, "If [P or not (P)] is true and not (P) is true, then P is true." This certainly seems obvious, but in the exercises that follow, you will be asked to verify this law by means of a truth table.

Example

Given the points $A(4, 2)$, $B(6, 6)$, and $C(2, \frac{11}{2})$.

(a) Does C lie on the perpendicular bisector of $\overline{AB}$?

(b) At what point M does the perpendicular bisector of $\overline{AB}$ intersect the segment?

(c) Determine the slopes of the segments $\overline{AB}$ and $\overline{CM}$.

(d) What is the product of the slopes found in part (c)?

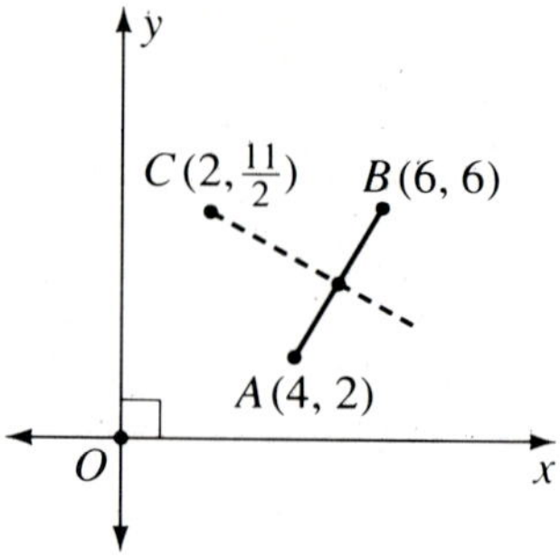

Solution

(a) $$m\overline{AC} = \sqrt{(4-2)^2 + \left(2 - \frac{11}{2}\right)^2} = \sqrt{2^2 + \left(-\frac{7}{2}\right)^2}$$

$$= \sqrt{4 + \frac{49}{4}} = \sqrt{\frac{65}{4}} = \frac{\sqrt{65}}{2}$$

$$m\overline{BC} = \sqrt{(6-2)^2 + \left(6 - \frac{11}{2}\right)^2} = \sqrt{4^2 + \left(\frac{1}{2}\right)^2}$$

$$= \sqrt{16 + \frac{1}{4}} = \sqrt{\frac{65}{4}} = \frac{\sqrt{65}}{2}$$

Since $m\overline{AC} = m\overline{BC}$, C lies on the perpendicular bisector of $\overline{AB}$ according to Theorem 5-2.

(b) The Midpoint Formula (Def. 3-14) gives

$$M = \left(\frac{4+6}{2}, \frac{2+6}{2}\right) = (5, 4)$$

(c) $m_{\overline{AB}} = \dfrac{6-2}{6-4} = \dfrac{4}{2} = 2$; $m_{\overline{CM}} = \dfrac{\frac{11}{2} - 4}{2 - 5} = \dfrac{1\frac{1}{2}}{-3} = -\dfrac{1}{2}$ from Def. 3-13.

(d) $m_{\overline{AB}} \cdot m_{\overline{CM}} = (2)(-\frac{1}{2}) = -1$ ◁

Example

It is given that $A(2, 1)$, $B(4, 3)$, and $C(2, y)$ are points in a plane with C equidistant from A and B. Find:

(a) the ordinate (y-coordinate) of C
(b) the lengths of the sides of $\triangle ABC$
(c) the slopes of the sides of $\triangle ABC$
(d) the midpoint M of $\overline{AB}$
(e) the slope of $\overline{CM}$
(f) the product of the slope of $\overline{AB}$ and the slope of $\overline{CM}$

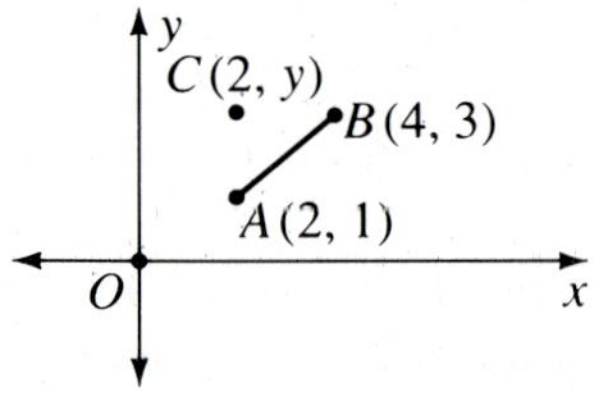

Solution

(a) Since C is equidistant from A and B, m$\overline{AC}$ = m$\overline{BC}$ and

$$\sqrt{(2-2)^2+(y-1)^2}=\sqrt{(4-2)^2+(3-y)^2}$$

or

$$(y-1)^2 = 2^2+(3-y)^2$$
$$y^2-2y+1 = 4+9-6y+y^2$$
$$4y = 12$$
$$y = 3$$

(b) m$\overline{AC} = \sqrt{(2-2)^2+(3-1)^2} = \sqrt{4} = 2$
m$\overline{BC} = \sqrt{(4-2)^2+(3-3)^2} = \sqrt{4} = 2$
m$\overline{AB} = \sqrt{(2-4)^2+(1-3)^2} = \sqrt{4+4} = \sqrt{8} = 2\sqrt{2}$

(c) $m_{\overline{AB}} = \dfrac{3-1}{4-2} = \dfrac{2}{2} = 1$

$m_{\overline{AC}} = \dfrac{3-1}{2-2} = \dfrac{2}{0}$, which is not a real number.

$m_{\overline{BC}} = \dfrac{3-3}{2-4} = \dfrac{0}{-2} = 0$

(Since the coordinates of B and C have the same ordinate value, how is this line related to the x-axis? Also, consider the segment AC for which the abscissas are equal. What relation does $\overline{AC}$ have to the y-axis? How are the x and y-axes related?)

(d) $M(\overline{AB}) = \left(\dfrac{2+4}{2}, \dfrac{1+3}{2}\right) = (3, 2)$

(e) $m_{\overline{CM}} = \dfrac{2-3}{3-2} = -1$

(f) $m_{\overline{AB}} \cdot m_{\overline{CM}} = (1)(-1) = -1$ ◁

Exercise Set 5.1

1. If $\overline{AD} \cong \overline{BD}$ and $\overline{CD} \perp \overline{AB}$, prove $\overline{AC} \cong \overline{BC}$ without the use of congruent triangles.

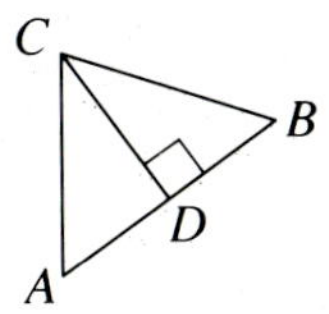

2. Draw the triangle determined by $A(3, 6)$, $B(9, 10)$ and $C(4, 11)$, and
   (a) Find the coordinates of the midpoint M of $\overline{AB}$.
   (b) Is $\overline{AC} \cong \overline{BC}$? Why?
   (c) Is $\angle CAB \cong \angle CBA$? Why?
   (d) Is $\angle CMB \cong \angle CMA$? Why?

3. If $\overline{AF} \cong \overline{FB}$, $\angle CAB \cong \angle CBA$, m$\overline{BC} = 6$, m$\overline{AF} = 4$ and m$\overline{AD} = 5$, find:
   (a) m$\overline{AC}$
   (b) m$\overline{BF}$
   (c) m$\overline{BD}$

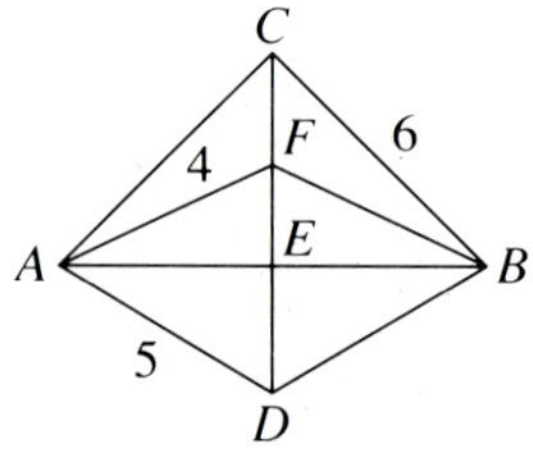

*4. Given the equilateral $\triangle A(0, 0)$, $B(4a, 0)$ and $C(2a, 2b)$. Prove analytically that the triangle joining the midpoints of the sides is also an equilateral triangle.

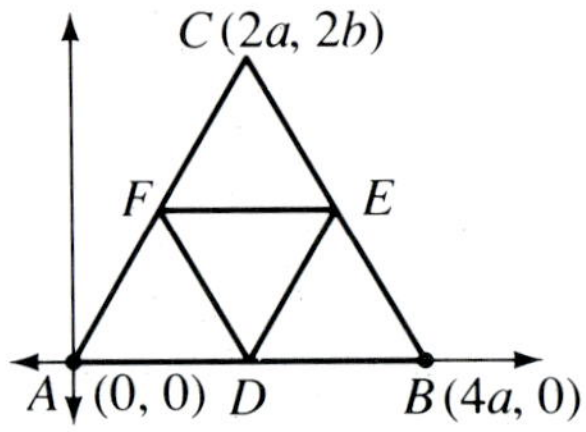

5. If $A(x, -3)$ is on the perpendicular bisector of $\overline{BC}$, $B(-2, 2)$ and $C(4, 4)$, determine the following:
   (a) abscissa of A
   (b) $d(\overline{AB})$ and $d(\overline{AC})$
   (c) coordinates of M, the midpoint of $\overline{BC}$
   (d) $m_{\overline{BC}}$
   (e) $m_{\overline{AM}}$
   (f) $(m_{\overline{BC}}) \cdot (m_{\overline{AM}})$
6. Use the distance formula to determine which of the following points lie on the perpendicular bisector of $\overline{PQ}$, given $P(-5, -1)$ and $Q(1, 1)$.
   (a) $A(-3, 3)$
   (b) $B(-4, 5)$
   (c) $C(-2, 0)$
   (d) $D(0, -3)$
7. If $T(6, y)$ and $U(x, 4)$ are on the perpendicular bisector of $\overline{GH}$, and given points $G(-1, -3)$, and $H(3, 7)$, determine the following:
   (a) ordinate of T
   (b) abscissa of U
   (c) $m_{\overline{GH}}$
   (d) $m_{\overline{TU}}$
   (e) $m_{\overline{GH}} \cdot m_{\overline{TU}}$
8. Given $K(-3, 3)$, $P(5, -1)$, $R(3, 5)$, $S(-1, -2)$ and $T(-1, -3)$, use the distance formula to determine whether $\overleftrightarrow{RS}$ or $\overleftrightarrow{RT}$ is the perpendicular bisector of $\overline{KP}$.
9. Prove Corollary 5-3.
10. Prove {[P or not(P)] and not(P)} $\rightarrow P$ is a law of logic (Law of the Excluded Middle) by means of a truth table.

Perpendiculars to a Line from a Point Not on the Line

In the previous discussion we were able to show that through a given point on a line there is exactly one perpendicular to a given line. We can also prove that in a

* The more difficult problems we indicated by an asterisk.

plane there is exactly one line perpendicular to a given line from a point *not* on the line.

Theorem 5-4 Through a point outside a line there is one and only one perpendicular to that line.

Given: Point P, not contained in line ℓ_1

Prove: There is one and only one line $\ell_2 \perp \ell_1$ such that $P \in \ell_2$

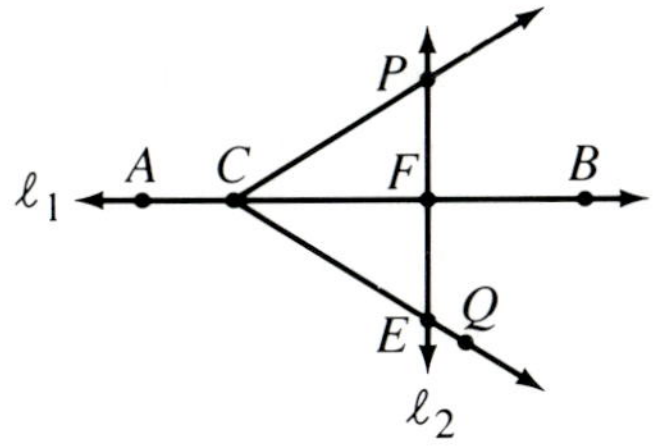

Existence Consider points A, B, and C on ℓ_1 such that A-C-B. By the definition of a ray (Def. 2-3), $\overrightarrow{CP}$ exists. Rays CP and CB form the angle BCP, hence there exists a point Q on the opposite side of ℓ_1 from P such that $m\angle BCP = m\angle BCQ$, by the Angle Construction Postulate (Post. 3-2). There exists on $\overrightarrow{CQ}$ a point E such that $m\overline{CP} = m\overline{CE}$, according to the Ruler Placement Postulate (Post. 2-5). Since P and E are on opposite sides of ℓ_1, $\overleftrightarrow{PE}$ intersects ℓ_1 at some point F. (F is the point of intersection; otherwise the points P and E would have to lie on the same side of $\overleftrightarrow{CB}$, by Def. 3-2(b).) By the SAS Postulate $\triangle CPF \cong \triangle CEF$, so $\angle PFC \cong \angle EFC$, which means that they are right angles, by the Right Angle Theorems (Cor. 3-6(b)). Thus $\overleftrightarrow{PE}$ is perpendicular to ℓ_1 at F.

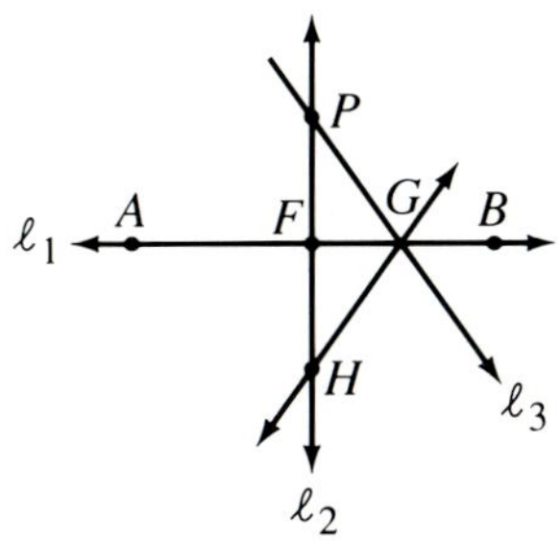

Uniqueness (We must show there is only one perpendicular from P to ℓ_1.) Suppose there is a second line ℓ_3 containing P and perpendicular to ℓ_1, which intersects ℓ_1 at G. Then there is a point H on ℓ_2 such that $\overline{PF} \cong \overline{HF}$, and $\triangle PFG \cong \triangle HFG$ by the SAS Postulate, since $\angle PFG \cong \angle HFG$. Now we have $\angle FGH \cong \angle FGP$, since they are corresponding parts of congruent triangles, but this cannot be true, since assuming $\ell_3 \perp \ell_1$ means $\angle PGF$ is a right angle. If $\angle HFG$ is also a right angle, $\overleftrightarrow{GH} \perp \ell_1$, and in Theorem 5-1 we proved that there is only one perpendicular to a line through a point on the line. Thus there is no other perpendicular from P to ℓ_1.

Auxiliary Lines

In proving the last few theorems we have used postulates and definitions to introduce lines, rays, and segments where needed. Such lines are called *auxiliary* lines, and if we are careful in using them they can make an otherwise impossible proof possible, and a difficult proof easier. What we must not do is place conditions on the auxiliary line which are not explicit in the postulate, definition or theorem that allow its use. For example, if we are given a scalene triangle ABC, as shown in Fig. 5-3(a), we cannot assume that the perpendicular bisector of $\overline{BC}$ will pass through the opposite vertex. In Fig. 5-3(b) we see that median $\overline{AD}$ is *not* perpendicular to $\overline{BC}$, and that segment $\overline{AE} \perp \overline{BC}$ does *not* bisect $\overline{BC}$. If the perpendicular bisector of the base *does* pass through the opposite vertex, the triangle is isosceles as a consequence of Theorem 5-2.

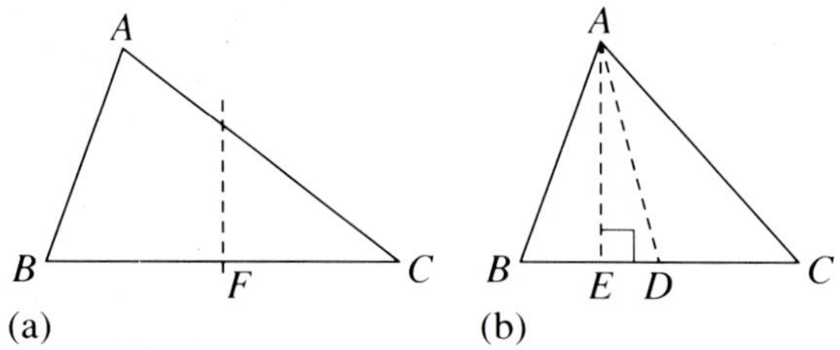

Figure 5-3

Examples

1. *Given:* The figure $ABCD$ with $\overline{AB} \cong \overline{CD}$ and $\overline{BC} \cong \overline{DA}$
Prove: $\angle A \cong \angle C$ and $\angle B \cong \angle D$

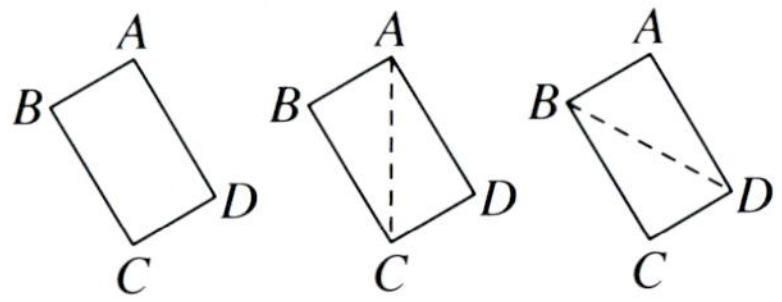

Solution If we first draw $\overline{AC}$, we have $\overline{AC} \cong \overline{AC}$, so $\triangle ABC \cong \triangle CDA$ by SSS, and $\angle B \cong \angle D$. If next we draw $\overline{BD}$, then $\overline{BD} \cong \overline{BD}$, $\triangle ABD \cong \triangle CDB$, and thus $\angle A \cong \angle C$.

2. If, in the following figure, $\overline{AD} \cong \overline{CB}$ and $\overline{AB} \cong \overline{CD}$, prove $\triangle AED \cong \triangle CEB$.

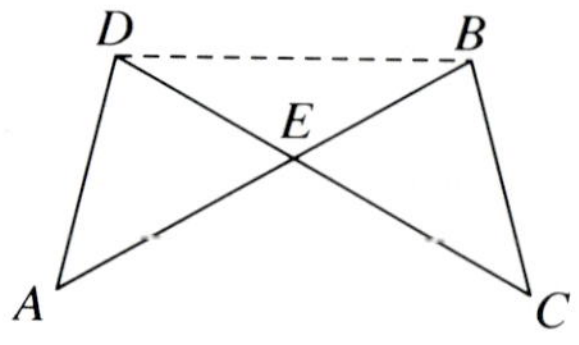

Proof

Statements	*Reasons*
1. Draw $\overline{DB}$	1. For any two distinct points there is exactly one line containing them (Post. 2-1(a))
2. $\overline{DB} \cong \overline{DB}$	2. Congruence of segments is reflexive. (Theorem 2-1(a))
3. $\overline{AB} \cong \overline{CD}$ and $\overline{AD} \cong \overline{CB}$	3. Given
4. $\triangle ADB \cong \triangle CBD$	4. SSS (Post. 4-3)
5. $\angle C \cong \angle A$, $\angle DBA \cong \angle BDC$ and $\angle ADB \cong \angle CBD$	5. CPCTC
6. $\angle ADE \cong \angle CBE$	6. If P is in the interior of $\angle ABC$ and Q is in the interior of $\angle RST$, so that $\angle ABC \cong \angle RST$ and $\angle ABP \cong \angle RSQ$, then $\angle PBC \cong \angle QST$ (Theorem 3-3(a))
7. $\triangle AED \cong \triangle CEB$	7. ASA (Post. 4-2) ◁

Distance from a Point to a Line

Sometimes when it is necessary to establish vertical lines through certain points, as in carpentry or changing wallpaper, a weight is tied to a string and dropped from the point. The line ℓ determined by the string will appear to be perpendicular to any line through Q, the intersection of ℓ with the horizontal plane of the floor or foundation. The segment from the given point P to Q will also seem to be the shortest segment which can be drawn from P to Q.

Theorem 5-5 The shortest segment joining a point to a line is the perpendicular segment.

Given: Point D not on $\overleftrightarrow{AC}$, $\overline{DB} \perp \overleftrightarrow{AC}$ with $B \in \overleftrightarrow{AC}$, and $E \in \overleftrightarrow{AC}$ with E different from B

Prove: $\overline{DB} < \overline{DE}$

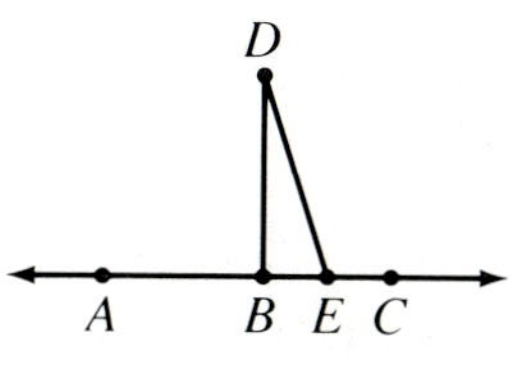

Proof

Statements	*Reasons*
1. For any two segments, $m\overline{DB} > m\overline{DE}$, $m\overline{DB} = m\overline{DE}$ or $m\overline{DB} < m\overline{DE}$	1. Trichotomy axiom (0-1)

Statements	*Reasons*
2. Consider $m\overline{DB} > m\overline{DE}$; then $\angle DEB > \angle DBC$	2. The angle opposite the longer side of a triangle is the larger angle (Theorem 4-8)
3. But $\angle DEB < \angle DBC$; hence, $m\overline{DB} \not> m\overline{DE}$	3. If a triangle has one right angle, the other two angles are acute (Corollary 4-7)
4. Consider $m\overline{DB} = m\overline{DE}$; then $\angle DBC \cong \angle DEB$	4. If two sides of a triangle are congruent, the angles opposite these sides are congruent (Theorem 4-5)
5. Again, since $\angle DEB < \angle DBC$, $m\overline{DB} \neq m\overline{DE}$	5. Same as 3

Therefore the last possibility must be correct, or $\overline{DB} < \overline{DE}$.

Definition 5-2 The **distance from a line to a point not on the line** is the length of the perpendicular segment from the point to the line.

Theorem 5-6 A point on the bisector of an angle is equidistant from the sides of the angle.

Definition 5-3 An **altitude** of a triangle is the perpendicular segment from a vertex to the line containing the opposite side.

Theorem 5-7 (*The Triangle Inequality Theorem*) The sum of the lengths of any two sides of a triangle is greater than the length of the third side.

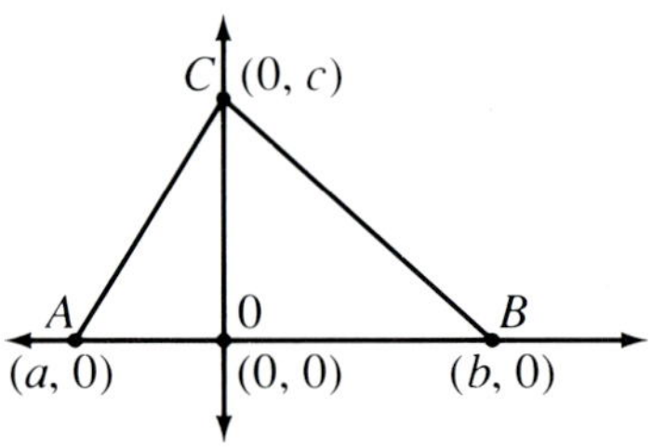

Proof A coordinate system can be placed on the triangle so that the coordinates of A are $(a, 0)$, of B are $(b, 0)$, and of C are $(0, c)$. (We have proved that there is a perpendicular from a point to a line, so let $\overleftrightarrow{CO}$ be perpendicular to $\overleftrightarrow{AB}$. There is no assumption here about *where* point O is other than that it is on line AB.) We want to show that $m\overline{AC} + m\overline{CB} > m\overline{BA}$, or since $m\overline{AC} = \sqrt{a^2 + c^2}$, $m\overline{CB} = \sqrt{c^2 + b^2}$ and $m\overline{BA} = \sqrt{(b - a)^2}$, that

$$\sqrt{a^2 + c^2} + \sqrt{c^2 + b^2} > \sqrt{(b - a)^2}$$

Since both sides of the inequality are positive, the order of the inequality will not change upon squaring them, and if we also apply Theorem 1-6 from Appendix I, we obtain the following:

$$(a^2 + c^2) + (c^2 + b^2) + 2\sqrt{(a^2 + c^2)(c^2 + b^2)} > (b - a)^2,$$
$$a^2 + 2c^2 + b^2 + 2\sqrt{a^2c^2 + a^2b^2 + b^2c^2 + c^4} > b^2 - 2ba + a^2,$$
$$2c^2 + 2\sqrt{a^2c^2 + a^2c^2 + b^2c^2 + c^4} > -2ba$$

Now if a and b are both positive or both negative, the left member is positive and the right member is negative, so the inequality holds. If a and b have opposite signs, then $-2ab$ is a positive number; but $2\sqrt{a^2b^2 + \cdots}$ is only *part* of the sum in the left member, so the left member must be greater than the right member and the inequality holds. Finally, if either a or b is zero (when would this occur?), the left member is still positive and the right member is 0, so the inequality is still satisfied. (Could a and b both be zero? Why?)

Theorem 5-8 If a and b are lengths of two sides of a triangle and $a > b$ then the length of the third side, x, is such that $a - b < x < a + b$.

Example

If $a = 17$ and $b = 12$, then $17 - 12 < x < 17 + 12$, or $5 < x < 29$.

Exercise Set 5.2

1. In each of the following, we would like to draw an auxiliary line with the given conditions. Decide which statement is true about each one:

   The line can be drawn as stated.
   There are not enough conditions to determine the line.
   There are too many conditions imposed on the line.

   (a) A line bisecting a given angle P
   (b) A line through a point R
   (c) In a triangle ABC, a line from B bisecting $\overline{AC}$
   (d) In a triangle ABC, a median from B perpendicular to $\overline{AC}$
   (e) In a triangle ABC, a line perpendicular to $\overline{AC}$ which bisects $\overline{BC}$

2. Prove that the base angles of an isosceles triangle are congruent. (Do not use Theorem 4-5. *Hint:* Draw the bisector of the angle that contains the two congruent sides.)

3. Referring to the following figure, put $\overline{TR}$, $\overline{TS}$, and $\overline{TQ}$ in correct order and state theorems for your conclusions.

   ________ < ________ < ________

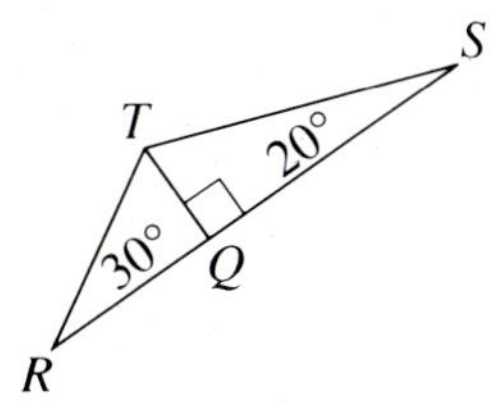

4. If the length of one side of a triangle is 7 and the second side is 4, what are the conditions that must be placed on the length of the third side? (i.e., ________ < 3rd side < ________)

5. Of the following sets of three numbers, which could be used for lengths of sides of a triangle? (All letters used represent positive numbers.)

   (a) 7, 11, 14 **(b)** 3, 5, 8 **(c)** 107, 165, 284 **(d)** $x, y, x + y$ **(e)** $e, 4, 5$

6. Prove that $m\overline{DB} < \frac{1}{2}[m\overline{AD} + m\overline{AB} + m\overline{DC} + m\overline{BC}]$ in the figure below.

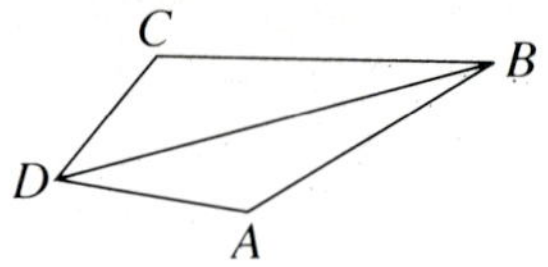

7. Draw triangles like those in the following figures; then draw the three altitudes of each.

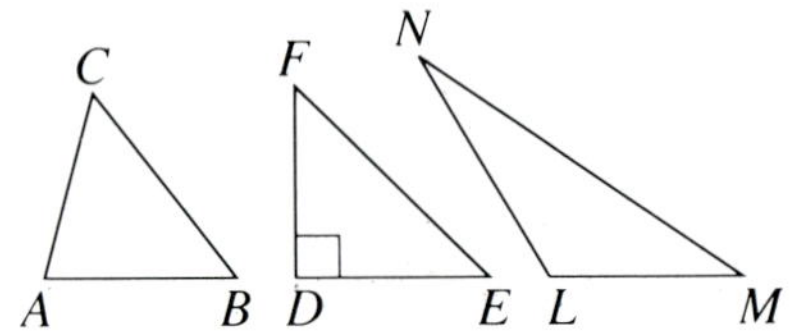

8. In the following figure, what is the measure of the altitude of
   (a) $\triangle ABP$ **(b)** $\triangle DCP$ **(c)** $\triangle ADP$ **(d)** $\triangle BCP$

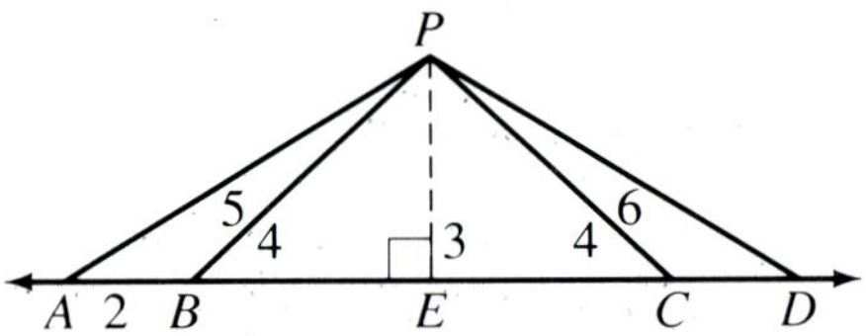

9. Prove that altitudes to the sides of an equilateral triangle are congruent.
10. Prove Theorem 5-8.
11. If two sides of a triangle measure 9 and 11, respectively, then what are the smallest and the largest lengths that the third side could have?

Construction of Perpendiculars to a Line

Construction 7: Construct the perpendicular to a line from a point not on the line.

If P is the given point, set a distance on your compass so that with P as center and the distance as a radius the arc will intersect ℓ in two points, say H and G. Using H and G as centers select a radius so that Q, the intersection of the arcs, will be on the opposite side of ℓ from P. Draw $\overline{PQ}$. $\overline{PQ}$ is perpendicular to ℓ.

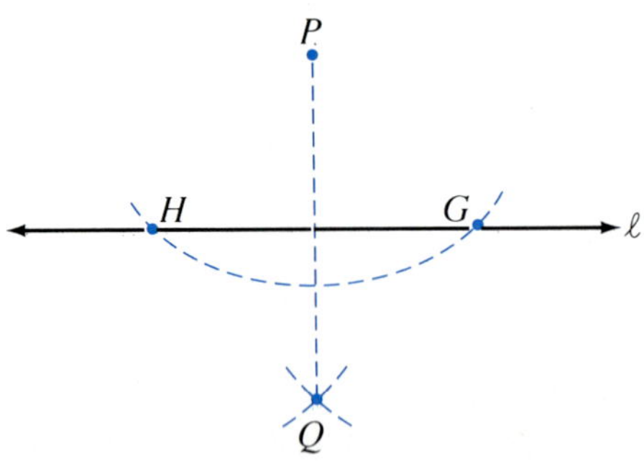

Proof $\overline{PH} \cong \overline{PG}$ and $\overline{HQ} \cong \overline{GQ}$. Thus P and Q are each equidistant from the end points of $\overline{HG}$ and determine the perpendicular bisector of $\overline{HG}$ by Corollary 4-12.

Construction 8: Construct the perpendicular to a line from a point on the line. See the following figure. A description and proof of the construction is left as an exercise.

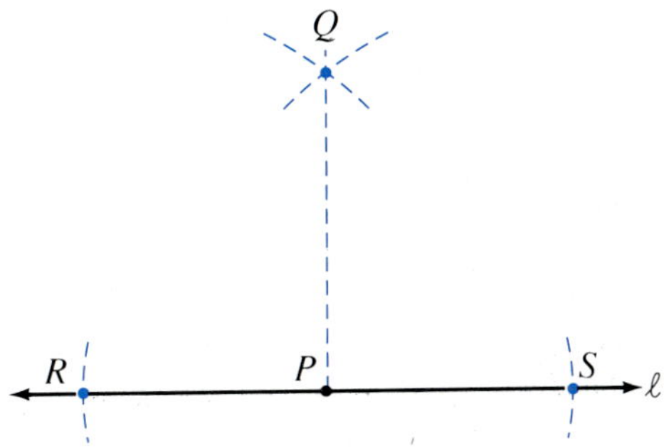

Construction 9: Construct the perpendicular bisector of a line segment.

If the segment is $\overline{MN}$, draw circles with centers M and N with radius greater than $\frac{1}{2}m\overline{MN}$. The arcs will intersect in two points P and Q on opposite sides of $\overline{MN}$. Draw a line through P and Q. $\overleftrightarrow{PQ} \perp \overline{MN}$, and if T is the point of intersection of $\overleftrightarrow{PQ}$ and $\overline{MN}$, then $\overline{MT} \cong \overline{NT}$. (Proof is left to the student as an exercise.)

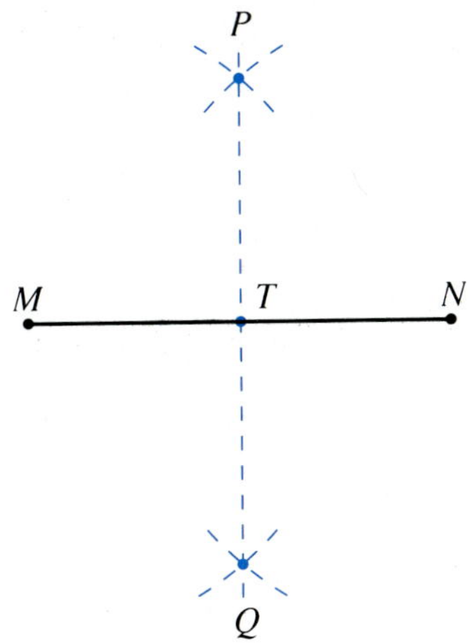

Exercise Set 5.3

1. Draw a triangle with sides 4 in., 5 in., and 6 in. Construct the perpendicular bisectors of the sides.
2. Draw a triangle with sides 4 in., 3 in., and 2 in. Construct the altitude to each side.
3. Draw a triangle with sides $5\frac{1}{4}$ in., $3\frac{1}{2}$ in., and $2\frac{1}{4}$ in. Construct the altitude to each side.
4. Describe and prove Construction 8. (Note that we are bisecting a straight angle.)
5. Construct a right triangle ABC given $m\angle A = 20$ and $c = 4$ in.

6. Suppose that a line is near the bottom edge of a sheet of paper and you would like to construct a perpendicular to it from a point above the line. Show a possible construction and prove that it is valid.

5-2 Parallel Lines

The intersection of sets of points in a plane may be a nonempty set of points or the empty set. Theorem 2-8 stated that two different lines intersect in at *most* one point. The only other possibility would be that two different lines in a plane do not intersect at all and then the lines are said to be *parallel*.

Definition 5-4 **Parallel lines** are two lines in the same plane that do not intersect. Any two segments, rays or half-lines are said to be **parallel** if the lines containing them are parallel. (The symbol $\|$ means "is parallel to.")

Thus we see that if $A, B \in \ell_1$ and $C, D \in \ell_2$, then $\ell_1 \| \ell_2$, then $\overline{AB} \| \overline{CD}$, $\ell_1 \| \overline{CD}$, $\ell_2 \| \overline{AB}$, $\overline{CD} \| \overleftrightarrow{AB}$, etc., all as a consequence of Def. 5-4 (see Fig. 5-4).

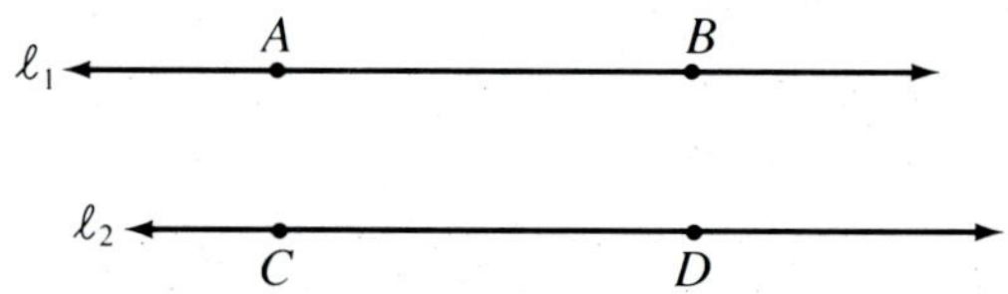

Figure 5-4

Although we have defined parallel lines, our definition does not tell us how to determine whether two given lines are parallel or not parallel. It is impossible for us to examine the infinite extent of two lines to decide whether or not their intersection is the empty set. We need, therefore, to consider other conditions which guarantee parallelism.

Theorem 5-9 Two coplanar lines that are perpendicular to a given line in the same plane are parallel.

Given: $\ell_1 \perp \ell_3$ at B, $\ell_2 \perp \ell_3$ at A

Prove: $\ell_1 \| \ell_2$

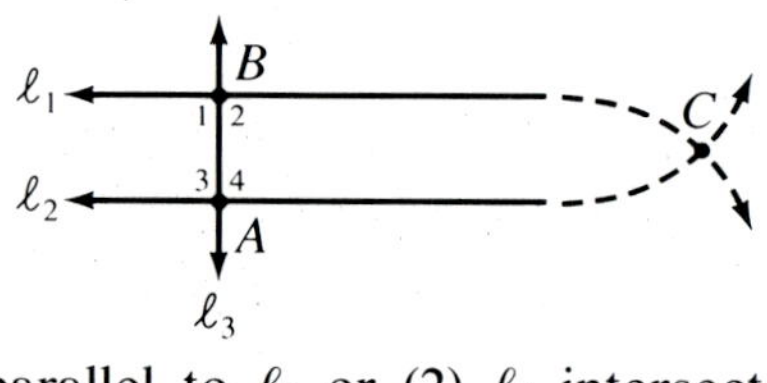

Proof Either (1) ℓ_1 is parallel to ℓ_2 or (2) ℓ_1 intersects ℓ_2 at some point C. Assuming the latter to be true, the figure ABC is a triangle and $\angle 1 > \angle 4$, since the exterior angle of a triangle is greater than either nonadjacent interior angle (Theorem 4-7). But $\angle 1$ and $\angle 4$ are right angles and all right angles are congruent (Theorem 3-6). Thus assumption (2) is false and the lines must be parallel.

This theorem allows us to prove an important theorem concerning the existence of parallel lines.

Theorem 5-10 Through a point P not on a line ℓ_1, there is a line ℓ_2 parallel to ℓ_1.

Given: Line ℓ_1 and P not on ℓ_1

Prove: There is a line ℓ_2 through P such that $\ell_2 \parallel \ell_1$

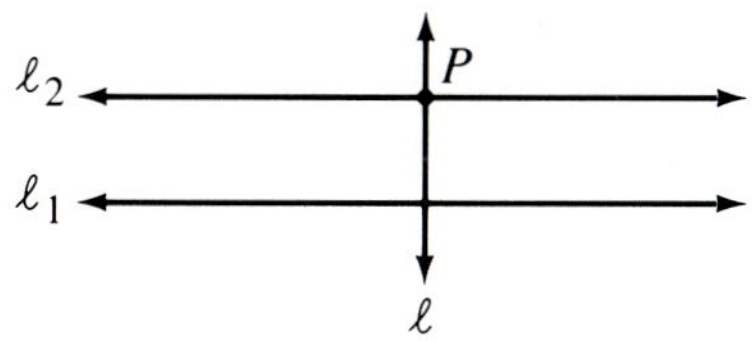

Proof

Statements	*Reasons*
1. Let ℓ be a line through P and perpendicular to ℓ_1	1. Through a point outside a line there is one and only one perpendicular to the line (Theorem 5-4)
2. There is a line ℓ_2 perpendicular to ℓ through P	2. Through a given point on a line there is one and only one perpendicular to a given line (Theorem 5-1)
3. Hence $\ell_1 \parallel \ell_2$	3. Two coplanar lines that are perpendicular to the same line are parallel (Theorem 5-9)

It would seem appropriate at this point to try to prove that there is only one line parallel to a given line at a given point. It *cannot* be proved, based on the postulates we have stated thus far, so we must postulate the following statement.

NIKOLAI IVANOVICH LOBACHEVSKI (1793–1856) This Russian mathematician was one of the first to recognize the merits of a geometry that denied Euclid's parallel postulate. In his geometry, more than one parallel can be drawn to a given line through a point not on that line. Unfortunately, his work in plane hyperbolic geometry was largely ignored by others until after his death. Non-Euclidean geometries such as Lobachevski's provided Albert Einstein with a basis for his theory of relativity. The work of Lobachevski and others made it apparent that Euclidean geometry is only one of many possible geometries. (*New York Public Library Picture Collection*)

Postulate 5-1 (*The Parallel Postulate*) Through a point not on a given line there is at most one line parallel to the given line.

In the diagrams that we used in the last two theorems we have seen that the parallel lines were intersected by a third line in each case.

Definition 5-5 A **transversal** is a line that intersects two coplanar lines in two distinct points. (We often say that a transversal "cuts" the two lines.)

Referring to Figure 5-5, we will name the angles formed by two lines and a transversal as follows:

(a) Angles a, b, g, and h are called *exterior* angles.
(b) Angles c, d, e, and f are called *interior* angles.
(c) Pairs of angles with interiors that lie on the same side of the transversal and on corresponding sides of the lines are called *corresponding* angles ($\angle a$ and $\angle e$, $\angle b$ and $\angle f$, $\angle d$ and $\angle h$, $\angle g$ and $\angle c$).

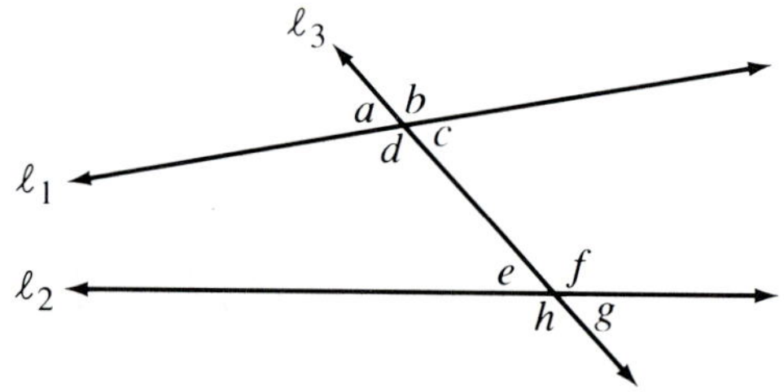

Figure 5-5

(d) Pairs of interior angles that are nonadjacent and lie on opposite sides of the transversal are called *alternate-interior* angles ($\angle d$ and $\angle f$, $\angle c$ and $\angle e$).
(e) Pairs of exterior angles that are nonadjacent and lie on opposite sides of the transversal are called *alternate-exterior* angles ($\angle a$ and $\angle g$, $\angle b$ and $\angle h$).

Example

Given $\overleftrightarrow{AB}$ and $\overleftrightarrow{CD}$ with transversal $\overleftrightarrow{EF}$, name
(a) four pairs of corresponding angles;
(b) two pairs of interior angles on the same side of the transversal;
(c) two pairs of alternate-interior angles;
(d) two pairs of alternate-exterior angles.

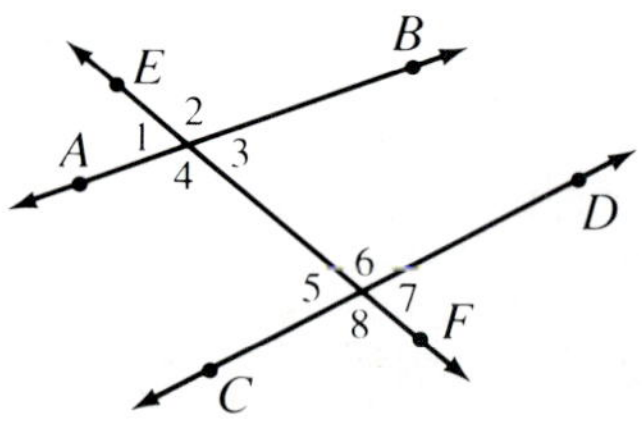

Solution

(a) $\angle 1$ and $\angle 5$, $\angle 2$ and $\angle 6$, $\angle 3$ and $\angle 7$, $\angle 4$ and $\angle 8$;
(b) $\angle 3$ and $\angle 6$, $\angle 4$ and $\angle 5$;
(c) $\angle 3$ and $\angle 5$, $\angle 4$ and $\angle 6$;
(d) $\angle 1$ and $\angle 7$, $\angle 2$ and $\angle 8$. ◁

Exercise Set 5.4

1. In the following figure, $\overleftrightarrow{AC}$ and $\overleftrightarrow{BC}$ are transversals of $\overleftrightarrow{AG}$ and $\overleftrightarrow{DF}$.
   (a) Name two pairs of alternate interior angles.
   (b) Name two pairs of interior angles on the same side of the transversals.
   (c) Name two pairs of corresponding angles.

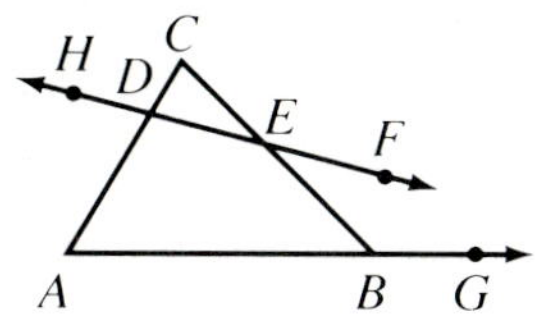

2. In the following figure,
   (a) What are the transversals of $\overline{AD}$ and $\overleftrightarrow{EF}$?
   (b) What are the transversals of $\overrightarrow{AC}$ and $\overrightarrow{DB}$?
   (c) What are the transversals of $\overrightarrow{DB}$ and $\overleftrightarrow{EF}$?
   (d) What are the transversals of $\overleftrightarrow{EF}$ and $\overrightarrow{AC}$?

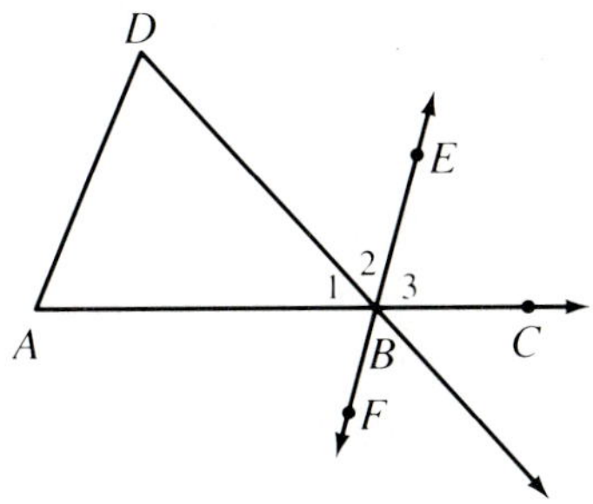

3. In the following figure, $\overline{AE} \perp \overline{AB}$, $\overline{DB} \cong \overline{DC} \cong \overline{AD}$, $\angle 2 \cong \angle 5$. Prove $\overline{AE} \parallel \overline{BC}$. (Give reasons for each statement.)

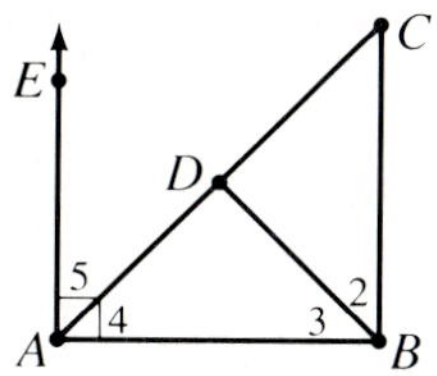

Proof

Statements	*Reasons*
1. $\overline{AD} \cong \overline{BD}$ and $\overline{DB} \cong \overline{DC}$ $\overline{AE} \perp \overline{AB}$, $\angle 2 \cong \angle 5$	1.

Statements	*Reasons*
2. $\angle 4 \cong \angle 3$	2.
3. $\angle EAB$ is a right angle and its measure is 90.	3.
4. $m\angle 4 + m\angle 5 = m\angle EAB = 90$	4.
5. $m\angle 3 = m\angle 4$ and $m\angle 2 = m\angle 5$	5.
6. $m\angle 3 + m\angle 2 = 90$	6.
7. $m\angle 3 + m\angle 2 = m\angle CBA$	7.
8. $m\angle CBA = 90$, so $\angle CBA$ is a right angle and $\overline{AB} \perp \overline{BC}$	8.
9. $\overline{AE} \parallel \overline{BC}$	9.

Now we are able to prove some very useful theorems containing parallel lines and the angles determined by their transversals.

Theorem 5-11 If two lines are cut by a transversal so that the angles in one pair of alternate-interior angles are congruent, then the angles in the other pair of alternate-interior angles are congruent.

Given: ℓ_1 and ℓ_2 cut by ℓ_3 such that $\angle a \cong \angle d$
Prove: $\angle b \cong \angle c$

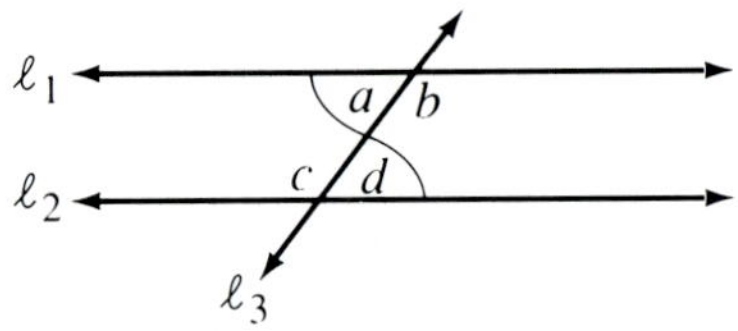

(The proof of this theorem is left to the student.)

Theorem 5-12 If two lines are cut by a transversal so that the alternate-interior angles are congruent, the lines are parallel.

Given: ℓ_1 and ℓ_2 cut by ℓ_3 such that $\angle a \cong \angle b$
Prove: $\ell_1 \parallel \ell_2$

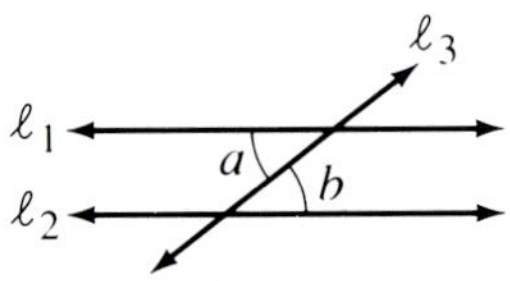

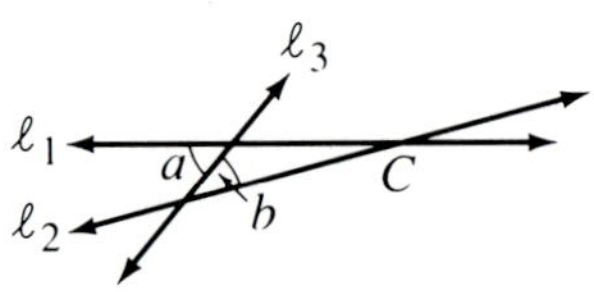

Proof Either (1) $\ell_1 \parallel \ell_2$, or (2) ℓ_1 intersects ℓ_2 at some point C. If ℓ_1 intersects ℓ_2 at C, then a triangle is formed (Def. 4-1); hence $\angle a$ is an exterior angle and would be greater than $\angle b$ (Theorem 4-7), contradicting the hypothesis. Therefore, assumption (2) is false and $\ell_1 \parallel \ell_2$.

The converse of Theorem 5-12 can be stated as follows.

Theorem 5-13 If two parallel lines are cut by a transversal, then the angles in each pair of alternate-interior angles are congruent.

This theorem and the set to follow will be left to the student to prove as exercises.

Theorem 5-14 If two lines are cut by a transversal so that the corresponding angles are congruent, the lines are parallel.

Theorem 5-15 (*Converse of Theorem 5-14*) If two parallel lines are cut by a transversal, then the angles of each pair of corresponding angles are congruent.

Theorem 5-16 If parallel lines are cut by a transversal, the interior angles on the same side of the transversal are supplementary.

Theorem 5-17 In a plane, two lines parallel to a third line are parallel to each other.

Theorem 5-18 In a plane, if a line is perpendicular to one of two parallel lines, it is perpendicular to the other.

Theorem 5-19 If two parallel lines are cut by two parallel lines, the opposite pairs of segments determined are congruent.

Given: $\ell_1 \parallel \ell_2$, $\ell_3 \parallel \ell_4$, $\ell_1 \neq \ell_3$
Prove: $\overline{AB} \cong \overline{CD}$ and $\overline{AC} \cong \overline{BD}$

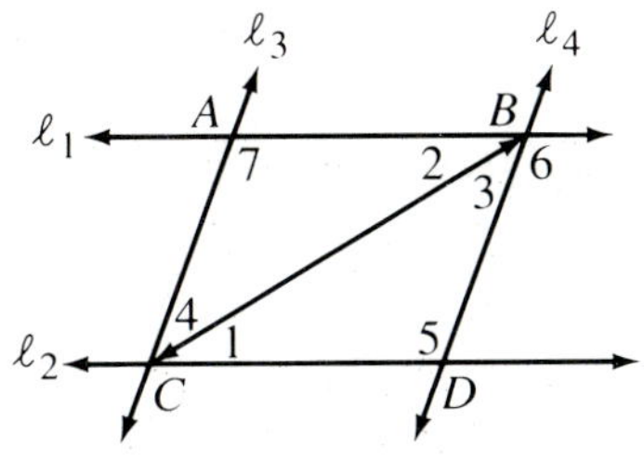

Proof

Statements	*Reasons* (Left as an exercise)
1. Draw $\overline{CB}$	1. Why possible?
2. $\angle 4 \cong \angle 3$, $\angle 1 \cong \angle 2$	2. Why?
3. $\overline{CB} \cong \overline{CB}$	3. Why?
4. $\triangle ABC \cong \triangle DCB$	4. Why?
5. $\overline{AB} \cong \overline{CD}$ and $\overline{AC} \cong \overline{BD}$	5. Why?

Definition 5-6 The **distance between two parallel lines** is the perpendicular distance from a point on one line to the other line.

Corollary 5-19 Two parallel lines are everywhere equidistant.

Theorem 5-20 If parallel lines cut congruent segments on one transversal, they cut congruent segments on any transversal.

Given: $\ell_1 \parallel \ell_2 \parallel \ell_3$, transversals t_1 and t_2; $\overline{AB} \cong \overline{BC}$
Prove: $\overline{DE} \cong \overline{EF}$

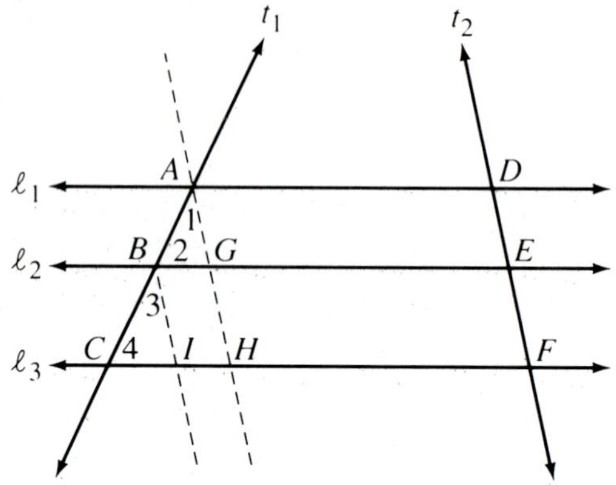

(The proof is for the particular case of three parallels but could apply as well to any number greater than three.)

Statements	*Reasons* (Left to student)
1. $\ell_1 \parallel \ell_2 \parallel \ell_3$ and $\overline{AB} \cong \overline{BC}$	1. Given
2. Through A draw the line parallel to t_2 and through B draw the line parallel to t_2 intersecting ℓ_2 and ℓ_3 as shown	2. Why possible?
3. $\overline{AG} \cong \overline{DE}$ and $\overline{GH} \cong \overline{EF}$; $\overline{BI} \cong \overline{GH} \cong \overline{EF}$	3. Theorem 5-19
4. $\angle 1 \cong \angle 3$ and $\angle 2 \cong \angle 4$	4. Why?
5. $\triangle ABG \cong \triangle BCI$	5. Why?
6. $\overline{AG} \cong \overline{BI}$	6. Why?
7. $\overline{DE} \cong \overline{EF}$	7. Why?

Exercise Set 5.5

1. In the figure, identify the following sets of angles:

(a) $\angle 1$ and $\angle 7$
(b) $\angle 4$ and $\angle 6$
(c) $\angle 8$ and $\angle 2$
(d) $\angle 5$ and $\angle 3$
(e) $\angle 3$ and $\angle 6$
(f) $\angle 7$ and $\angle 3$
(g) $\angle 8$ and $\angle 4$
(h) $\angle 1$ and $\angle 5$
(i) $\angle 4$ and $\angle 5$

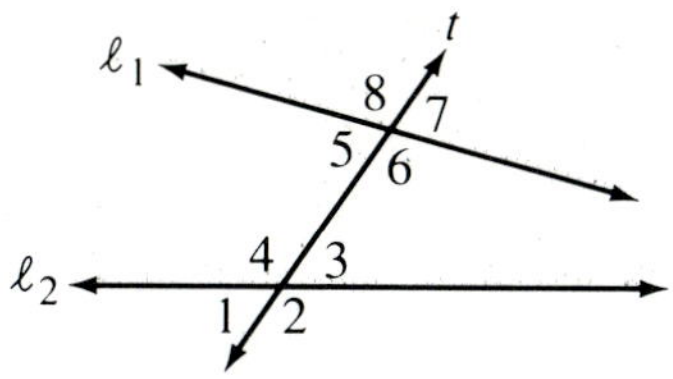

2. In the following figure, classify

(a) $\angle 1$ and $\angle 5$ **(b)** $\angle 2$ and $\angle 6$

(c) $\angle 3$ and $\angle 7$ **(d)** $\angle 4$ and $\angle 8$

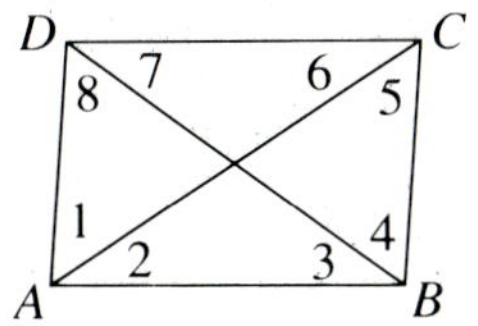

3. In the figure in Exercise 2,

(a) Is $\angle 1 \cong \angle 5$? Why?

(b) Is $\overline{AB} \parallel \overline{DC}$? Why?

(c) Are $\angle 7$ and $\angle 3$ alternate-interior angles?

4. **(a)** In the following figure $\overrightarrow{BA} \parallel \overrightarrow{DC}$ and $\overrightarrow{BC} \parallel \overrightarrow{DE}$. If $m\angle 3 = 50$, find $m\angle 1$, $m\angle 2$, $m\angle 4$ and $m\angle 5$.

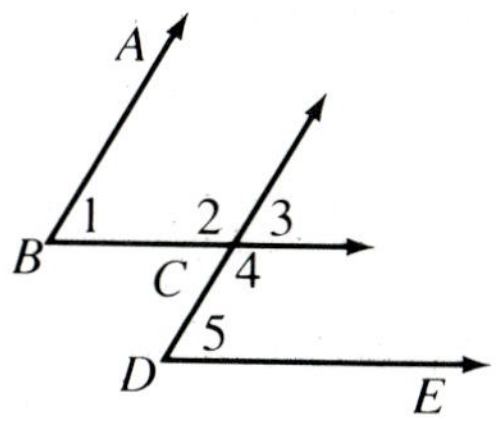

(b) Prove the following theorem:
If two angles have their sides parallel right side to right side and left side to left side, the angles are congruent. (The right side of an angle is the side to the right if looking into the interior of an angle from the vertex.)

5. Prove Theorem 5-13. (*Hint:* Use an indirect proof.)

6. Prove Theorem 5-14.

7. Prove Theorem 5-15.

8. Prove Theorem 5-16.

9. Prove Theorem 5-17.

10. Prove Theorem 5-18.

11. **(a)** Give reasons for the proof of Theorem 5-19.

(b) Prove Corollary 5-19.

12. Give reasons for the proof of Theorem 5-20.

13. What segments may be proved parallel if

(a) $\angle 1 \cong \angle 5$?

(b) $\angle 4 \cong \angle 8$?

(c) $\angle 7 \cong \angle 3$?

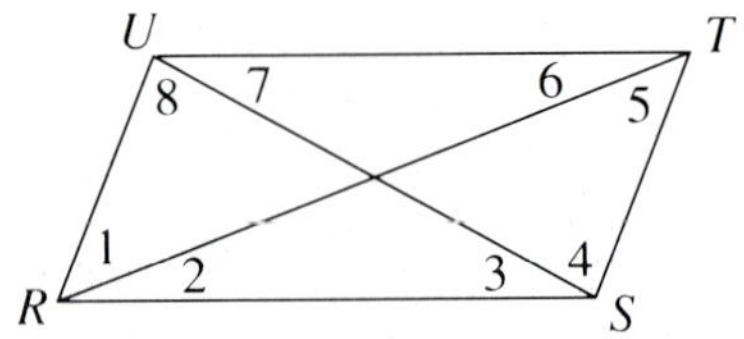

14. In the following figure, $\overline{AB}$, a side of $\triangle ABC$, is extended through D, and $\overrightarrow{BE} \parallel \overline{AC}$.

(a) If $m\angle 1 = 60$, then $m\angle A =$ ________

(b) If $m\angle 2 = 50$, then $m\angle C =$ ________

(c) If $m\angle 1 + m\angle 2 = 110$, then $m\angle 3 =$ ________

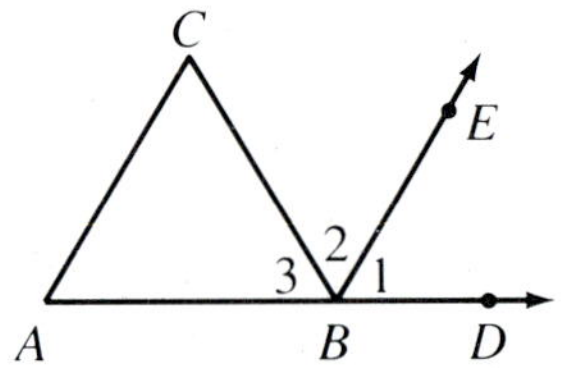

15. If $\triangle ABC$ is isosceles, with $\overline{AC} \cong \overline{BC}$ and $\overline{DE} \parallel \overline{AB}$, prove that $\triangle DCE$ is isosceles.

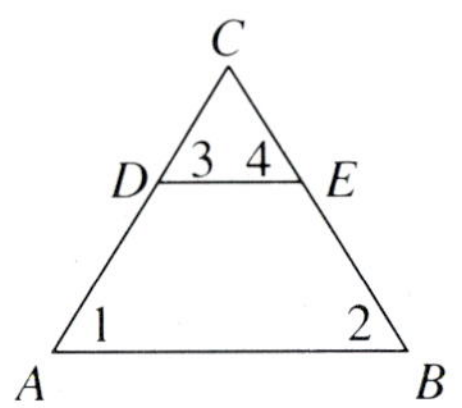

16. Given that $\overleftrightarrow{TU}$ is parallel to $\overline{RS}$, $m\angle 1 = 70$ and $m\angle 3 = 60$, find $m\angle R + m\angle 2 + m\angle S$.

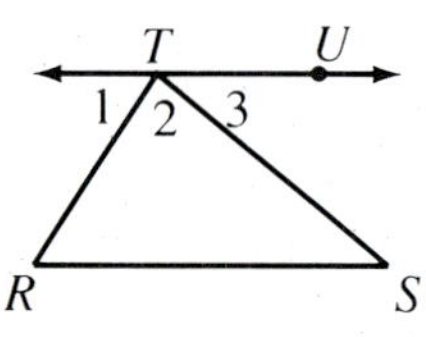

17. If $m\overline{AC} = 15$, $m\overline{BC} = 21$, and lines ℓ_1 through ℓ_5 are parallel to $\overline{AB}$ and cut congruent segments on $\overline{AC}$, what is the measure of each segment cut on $\overline{BC}$?

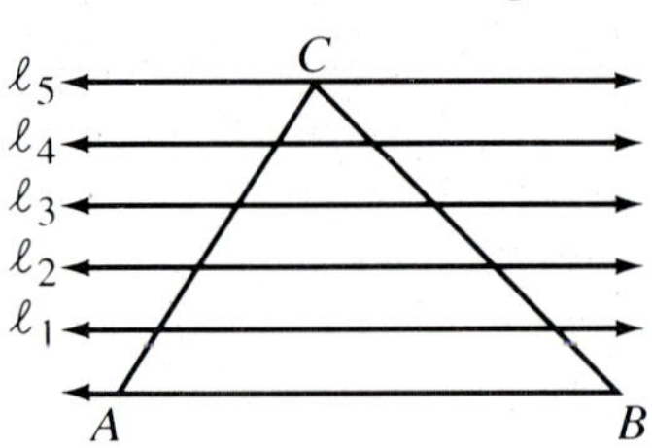

18. If $\overline{AB} \parallel \overline{FG} \parallel \overline{EH} \parallel \overline{DC}$, $m\overline{AF} = m\overline{FE} = m\overline{ED} = 2$, $m\overline{AC} = 12$ and $m\overline{BC} = 8$, find
(a) $m\overline{AK}$ **(b)** $m\overline{BG}$

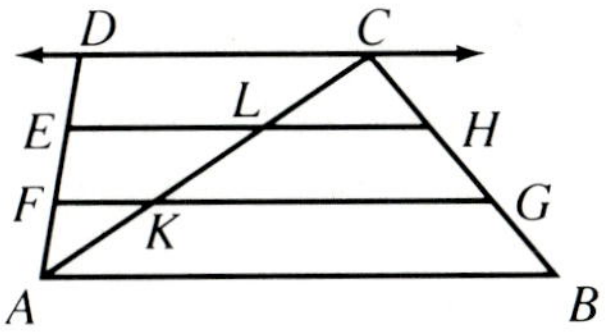

19. Lines ℓ_1 through ℓ_7 in the following figure are parallel, $\overline{AB}$, $\overline{BC}$, $\overline{CD}$, $\overline{DE}$, $\overline{EF}$, and $\overline{FG}$ are congruent, and $m\overline{AG} = m\overline{AH} = m\overline{AI} = m\overline{AJ} = m\overline{AP} = 12$. Determine each of the following measures:
(a) $m\overline{FG}$ **(b)** $m\overline{KH}$
(c) $m\overline{LI}$ **(d)** $m\overline{MJ}$
(e) $m\overline{EG}$ **(f)** $m\overline{NH}$

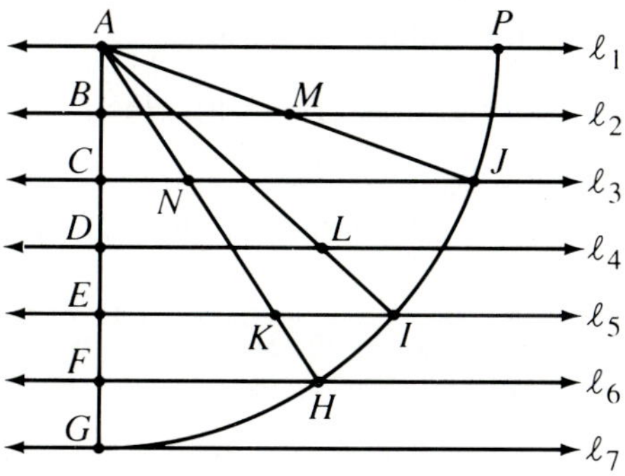

20. Prove that in a plane if a line intersects one of two parallel lines, then it intersects the other. (*Hint:* Use an indirect proof. Refer to Theorem 5-17.)

5-3 SUM OF MEASURES OF THE ANGLES OF A TRIANGLE

As a consequence of Theorems 5-10 and 5-13, we may prove another theorem of considerable practical value.

Theorem 5-21 The sum of the measures of the angles of a triangle is 180.
Given: $\triangle ABC$
Prove: $m\angle A + m\angle B + m\angle C = 180$

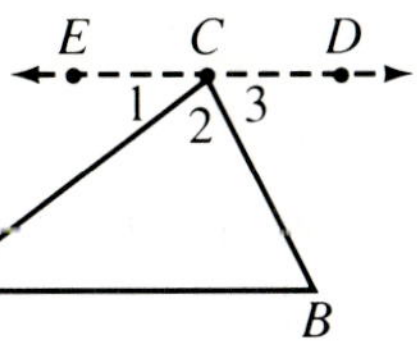

Proof

Statements	*Reasons*
1. Through C, draw $\overleftrightarrow{CD} \parallel \overleftrightarrow{AB}$	1. Why possible?
2. $\angle ECD$ is a straight angle and $m\angle ECD = 180$	2. Why?
3. $m\angle 1 + m\angle 2 + m\angle 3 = 180$	3. Why?
4. $\angle 1 \cong \angle A$ and $\angle 3 \cong \angle B$	4. Why?
5. $m\angle 1 = m\angle A$ and $m\angle 3 = m\angle B$	5. Why?
6. $m\angle A + m\angle B + m\angle C = 180$	6. Why?

The proofs of the following corollaries are left to the reader.

Corollary 5-21(a) If two angles of a triangle are congruent to two angles of a second triangle, then the third angles are congruent.

Corollary 5-21(b) The acute angles of a right triangle are complementary.

Corollary 5-21(c) If the hypotenuse and an acute angle of one right triangle are congruent to the hypotenuse and corresponding acute angle of another right triangle, the triangles are congruent.

Corollary 5-21(d) The measure of an exterior angle of a triangle is equal to the sum of the measures of the two nonadjacent interior angles of the triangle.

Corollary 5-21(e) The measure of each angle of an equilateral triangle is 60.

Theorem 5-22 If two nonvertical lines are parallel, their slopes are equal.

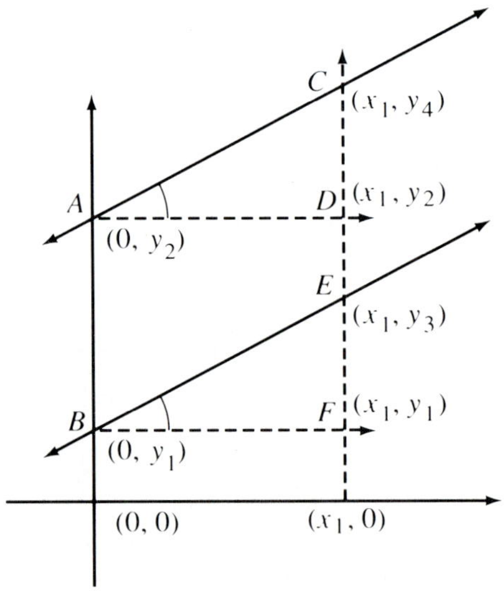

Proof If the lines are not vertical, they will intersect the y-axis at two distinct points, A and B. If the lines are horizontal, their slopes are both equal to zero and the theorem is proved. If they are not horizontal, we may draw lines through A and B parallel to the x-axis. These lines are parallel to each other, since two lines parallel to a third line are parallel to each other (Theorem 5-17). At any point $(x_1, 0)$ with $x_1 \neq 0$, draw a perpendicular to the x-axis intersecting the given parallels and the others as indicated. The slope of $\overleftrightarrow{AC}$ is $(y_4 - y_2)/x_1$ and the slope of $\overleftrightarrow{BE}$ is $(y_3 - y_1)/x_1$. Thus it is necessary to show that $y_4 - y_2 = y_3 - y_1$ in order to have their slopes equal. We see that $y_4 - y_2 = \text{m}\overline{CD}$ and $y_3 - y_1 = \text{m}\overline{EF}$, so it would suffice to show that $\overline{CD} \cong \overline{EF}$.

Line CF is perpendicular to $\overleftrightarrow{AD}$ and $\overleftrightarrow{BF}$; hence $\triangle ADC$ and $\triangle BFE$ are right triangles. $\angle ACD \cong \angle BEF$, since if two parallel lines are cut by a transversal, the corresponding angles formed are congruent (Theorem 5-15). $\overline{AC} \cong \overline{BE}$, since if two parallel lines are cut by two parallel lines, the segments formed are congruent (Theorem 5-19). $\overleftrightarrow{AB} \parallel \overleftrightarrow{CF}$; lines perpendicular to the same line are parallel (Theorem 5-9). Thus $\triangle ACD \cong \triangle BEF$, since we have shown that the hypotenuse and an acute angle of one triangle are congruent to the hypotenuse and acute angle of the other triangle (Corollary 5-21(c)). Thus $\overline{CD} \cong \overline{EF}$ (CPCTC).

Theorem 5-23 If the slopes of two nonvertical lines are equal, then they are parallel. (The proof is left to the reader.)

Exercise Set 5.6

1. If two angles of a triangle have the measures given, find the measure of the third angle.

(a) 30 and 70 **(b)** 45 and 85
(c) 59 and 115 **(d)** x and $2x$
(e) 90 and 30

2. In the following figure,

(a) if $\text{m}\angle 1 = 80$ and $\text{m}\angle 3 = 50$, $\text{m}\angle 4 =$ ________
(b) if $\text{m}\angle 1 = 20$ and $\text{m}\angle 2 = 40$, $\text{m}\angle 3 =$ ________
(c) if $\text{m}\angle 1 = \text{m}\angle 3$ and $\text{m}\angle 4 = 120$, then $\text{m}\angle 1 =$ ________ and $\text{m}\angle 2 =$ ________

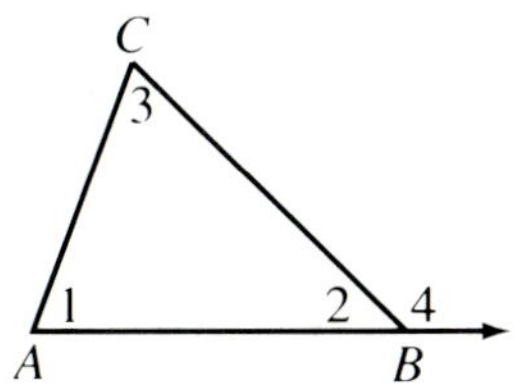

3. Prove Corollary 5-21(d).

4. In the following figure, $\overleftrightarrow{CD} \parallel \overleftrightarrow{AB}$. Find the measures of the remaining numbered angles if:

(a) $\text{m}\angle 2 = 40$ and $\text{m}\angle 6 = 60$
(b) $\text{m}\angle 5 = \text{m}\angle 7 = 60$
(c) $\text{m}\angle 2 = 130$ and $\text{m}\angle 3 = 40$

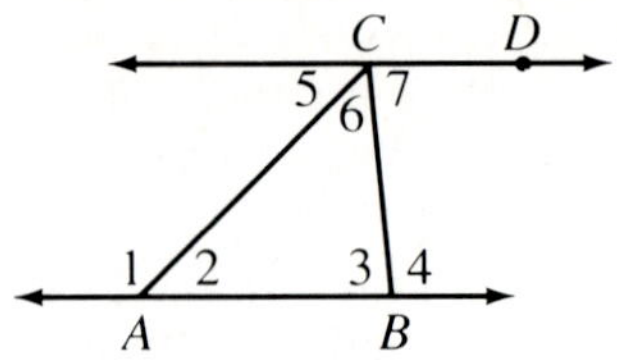

5. Given $\triangle ABC$ with $A(1, 10)$, $B(5, 5)$, slope of $\overline{AB} = -\frac{5}{4}$, slope of $\overline{BC} = 2$ and slope of $\overline{AC} = \frac{3}{8}$. $\triangle DEF$ with $D(11, 6)$, $E(15, 1)$, slope of $\overline{DE} = -\frac{5}{4}$, slope of $\overline{DF} = 2$ and slope of $\overline{EF} = \frac{3}{8}$. Prove that the triangles are congruent.

6. Prove that if two triangles have their bases parallel, then the altitudes to the bases are parallel or lie on the same line.

7. Prove Theorem 5-23.

8. Prove that the measure of the complement of a base angle of an isosceles triangle is half that of the vertex angle.

*9. If the measure of angle B is 10 more than the measure of angle A and the measure of angle BCD is 20 more than the sum of the measures of angles A and ACB, find the measures of angles A and ACB.

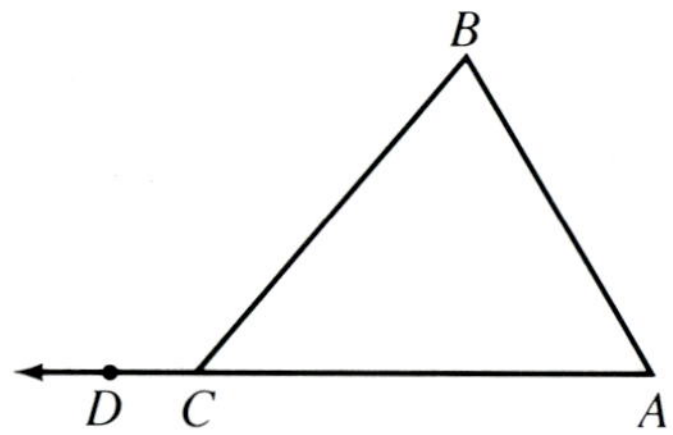

10. Given the points ($R(3, 5)$, $S(7, 2)$ and $T(6, 9)$, which are the vertices of $\triangle RST$, and that angle R has measure 50. Find the measure and/or the relative sizes of angle T and angle S.

11. If $\ell_1 \parallel \ell_2$ and $\overline{BC} \parallel \overline{AD}$, find
    (a) the slope of ℓ_1
    (b) the slope of $\overline{BC}$
    (c) the coordinates of point C

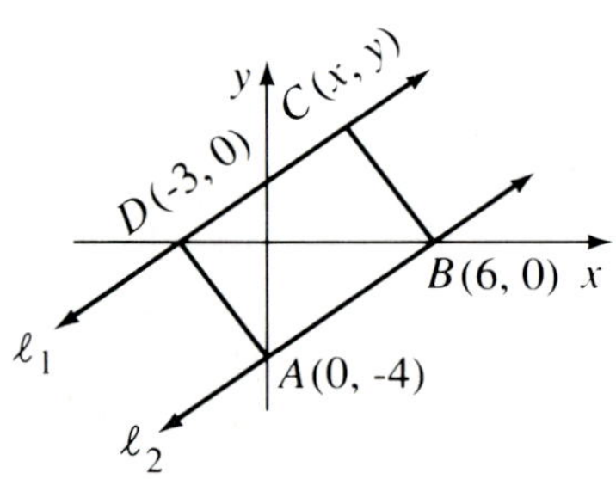

12. If, in the following figure, $\overline{AB} \cong \overline{AD}$ and $m\angle DBC = 125$, find the measure of
    (a) $\angle DBA$
    (b) $\angle A$
    (c) $\angle D$

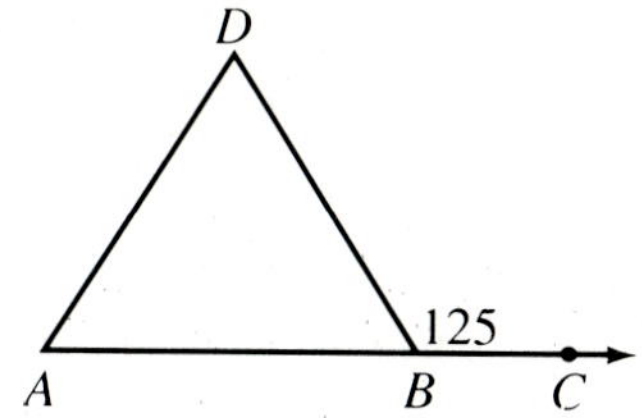

13. In the following figure $m\angle CBD = 130$. Find
   (a) $m\angle ABC$
   (b) $m\angle A + m\angle C + m\angle ABC$
   (c) $m\angle ABC + m\angle CBD$
   (d) $m\angle A + m\angle C$
   (e) How are the measures of $\angle A$ and $\angle C$ related to $m\angle CBD$?

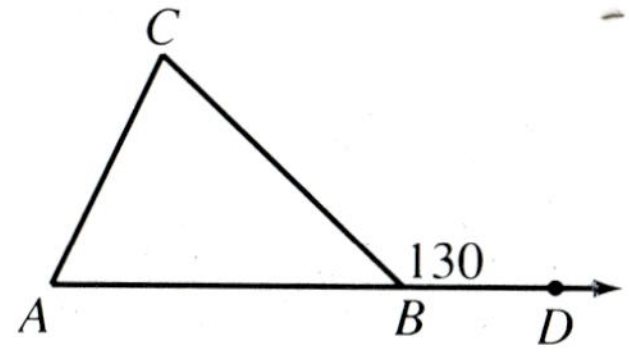

14. *Given:* $\overrightarrow{AB}$ bisects $\angle CAD$, $\overline{BC} \perp \overrightarrow{AC}$, $\overline{BD} \perp \overrightarrow{AD}$
   Prove: $\overline{BC} \cong \overline{BD}$

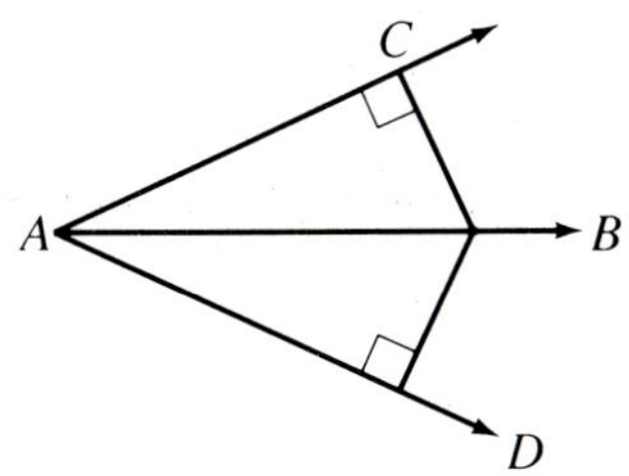

Construction of Parallels

Construction 10: Construct a line parallel to a given line ℓ through a given point P. Draw a line from P through any point Q of ℓ. Construct $\angle TPQ \cong \angle PQR$, with T and R on opposite sides of $\overleftrightarrow{PQ}$. Line TP is parallel to ℓ.

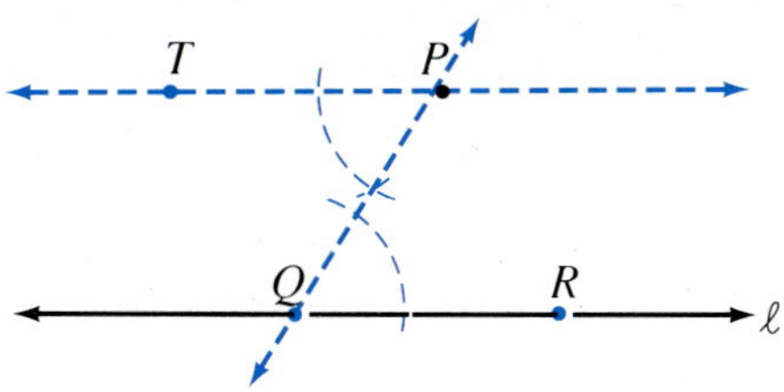

Construction 11: Divide a given segment AB into n congruent parts.

Draw a ray AC, where C is any point not on $\overleftrightarrow{AB}$. Choose an arbitrary point P_1 on $\overrightarrow{AC}$; then construct points $P_2, P_3, \ldots, P_n$ such that $\overline{AP_1} \cong \overline{P_1P_2} \cong \overline{P_2P_3} \cong \cdots \cong \overline{P_{n-1}P_n}$. Draw $\overline{P_nB}$. Then construct $\angle AP_1Q_x \cong \angle AP_2Q_2 \cong \angle AP_3Q_3 \cong \cdots \cong \angle AP_{n-1}Q_{n-1} \cong \angle AP_nB$ with $Q_1, Q_2, Q_3, \ldots, Q_{n-1}$ the intersections of the angles with $\overline{AB}$. Then $\overline{AQ_1} \cong \overline{Q_1Q_2} \cong \overline{Q_2Q_3} \cong \cdots \cong \overline{Q_{n-1}B}$, and $\overline{AB}$ is divided into n congruent parts.

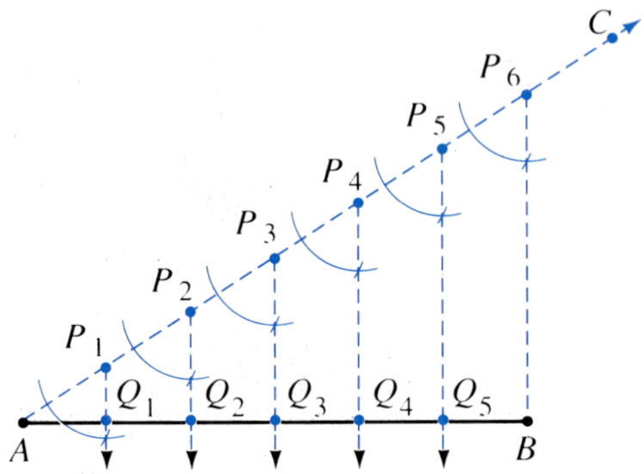

Construction of A Parabola

A *parabola* is a geometric figure consisting of a set of points equidistant from a fixed point, the *focus*, and a fixed line, the *directrix*. In Fig. 5-6, the fixed point is F, the fixed line is ℓ, and the lines through A, B, C, etc. are perpendicular to ℓ. You will be asked to construct a parabola in the exercise set that follows.

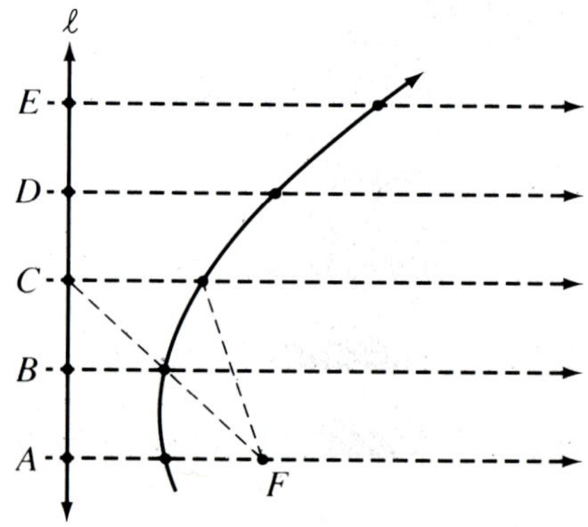

Figure 5-6

Exercise Set 5.7

1. Draw a line AB. Through a point P, approximately 2 in. above $\overleftrightarrow{AB}$, construct a line PC parallel to $\overleftrightarrow{AB}$.
2. Show a construction of a line parallel to a given line that makes use of corresponding angles instead of alternate-interior angles.

3. Draw a segment that is 72 mm (7.2 cm) long. Divide the segment into eight equal parts by means of Construction 11. Check your results by measuring the lengths of the parts.
4. Construct a parabola:
   (a) Construct a line ℓ perpendicular to the ruled lines on a piece of notebook paper.
   (b) Locate points A, B, C, . . . at the intersections of ℓ and the ruled lines.
   (c) Choose a point F about 1 in. to the right of A on line AF.
   (d) Construct the perpendicular bisectors of $\overline{AF}$, $\overline{BF}$, $\overline{CF}$, $\overline{DF}$, and $\overline{EF}$.
   (e) The points V, W, X, Y, and Z at which the perpendicular bisectors intersect the ruled lines through A, B, C, . . . are points on the parabola. Draw a smooth curve through these points. This will be only one *branch* of the parabola. You can construct the other branch in a similar fashion.

REVIEW EXERCISES

1. In the following figure t is a transversal for ℓ_1 and ℓ_2. Name
   (a) A pair of corresponding angles
   (b) A pair of alternate-interior angles.

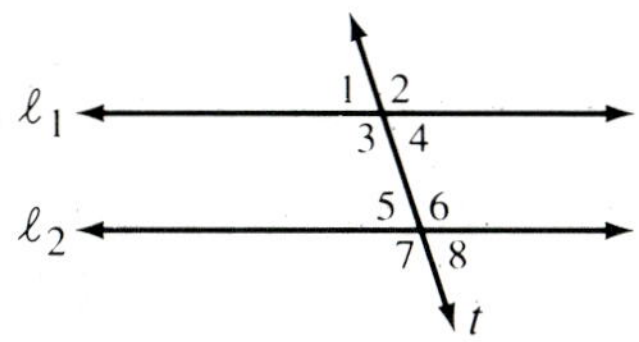

2. In triangle XYZ, $m\angle X = 50$, $m\angle Y = 60$ and $m\angle Z = 70$. The longest side is side ________ and the shortest side is side ________.
3. If $\ell_1 \parallel \ell_2$ and t is a transversal, and
   (a) if $m\angle 1 = 100$, then $m\angle 6 =$ ________
   (b) if $m\angle 3 = 70$, then $m\angle 5 =$ ________
   (c) if $m\angle 4 = 105$, then $m\angle 5 =$ ________

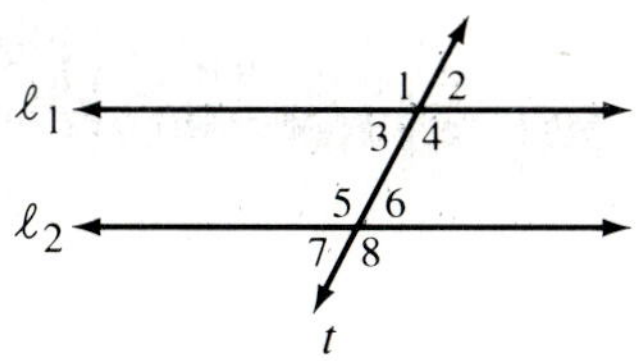

4. If, in triangle ABC, $m\overline{AB} = 12$, $m\overline{BC} = 12$ and $m\angle A = 70$, then $m\angle B =$ ________ and $m\angle C =$ ________.
5. In the given triangle, $m\angle GOP = 50$, $m\angle K = x$ and $m\angle G = x + 20$. What is the value of x?

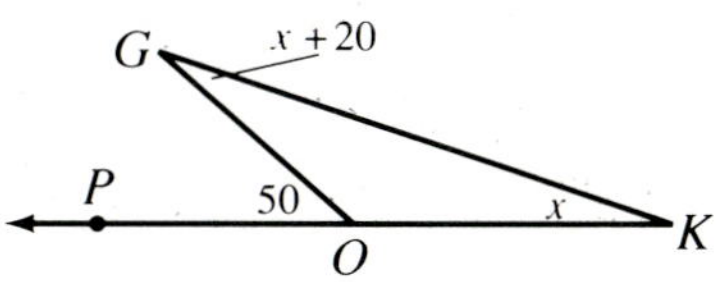

6. Lines ℓ_1 and ℓ_2 are parallel and t is a transversal. Find the value of y.

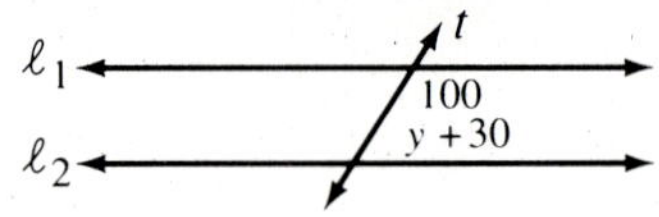

7. A, B, and C are on line ℓ and P is not on ℓ; $m\overline{PA} = 12$, $m\overline{PB} = 14$ and $m\overline{PC} = 15$. If one of these segments is perpendicular to ℓ, which one must it be? Why?

8. If $U(-4, 1)$ and $S(2, -2)$ are the endpoints of a segment US, does either $K(2, 5)$ or $L(1, 3\frac{1}{2})$ lie on the perpendicular-bisector of $\overline{US}$? Prove your answer.

9. Given $R(-2, -2)$, $K(1, 2)$, $A(-1, -1)$, $B(3, 4)$, $C(4, 6)$ and $D(5, 7)$, which of the segments AB, AC, or AD are parallel to $\overline{RT}$, if any? Prove your answer.

10. Prove Corollary 5-21(c).

11. *Given:* $\triangle UCB \cong \triangle UCS$ and $\overline{BS}$ intersects $\overline{UC}$ at T
Prove: **(a)** $\triangle BTU \cong \triangle STU$
(b) $\overline{BS} \perp \overline{UC}$

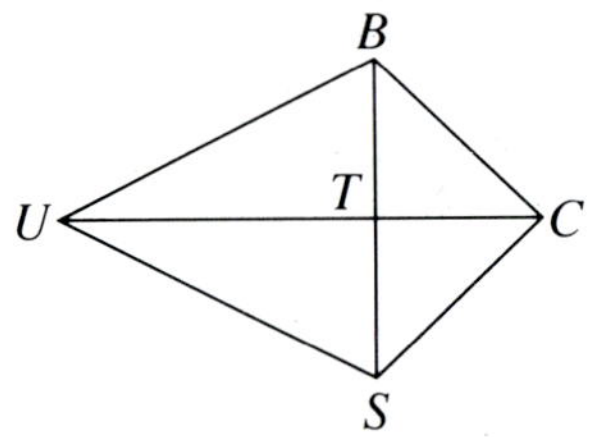

12. If A, B, D, and F are collinear, $\overline{AB} \cong \overline{DF}$, $\overline{AC} \parallel \overline{BE}$, and $\overline{AC} \cong \overline{BE}$, prove that $\overline{CD} \parallel \overline{EF}$.

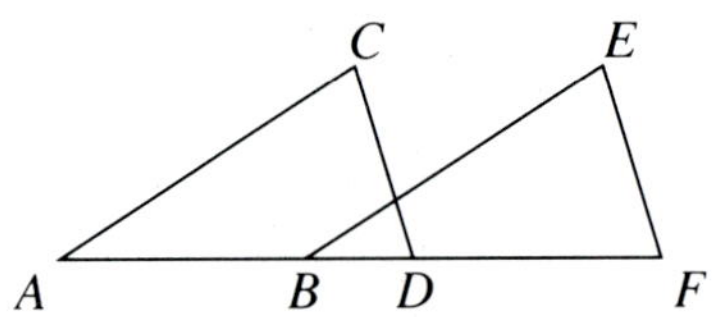

13. Using the information given for Exercise 12, that $m\angle A$ is 20 less than $m\angle C$, and also that $m\angle CDF = 110$, determine the measure of

(a) $\angle A$	**(b)** $\angle C$
(c) $\angle E$	**(d)** $\angle F$
(e) $\angle CDA$	**(f)** $\angle EBD$

14. Prove that the altitudes on the legs of an isosceles triangle are congruent.

6

Similar Triangles and Numerical Trigonometry

The chief forms of beauty are order and symmetry and precision, which the mathematical sciences demonstrate in a special degree.

ARISTOTLE (384–322 B.C.) Greek philosopher.

6-1 SIMILAR TRIANGLES

When we refer to two objects as being similar we mean that they have the same shape but are not necessarily the same size. Two congruent triangles have the same shape and the same size, but a set of equilateral triangles, for example, can all have the same shape and yet not be congruent (Fig. 6-1).

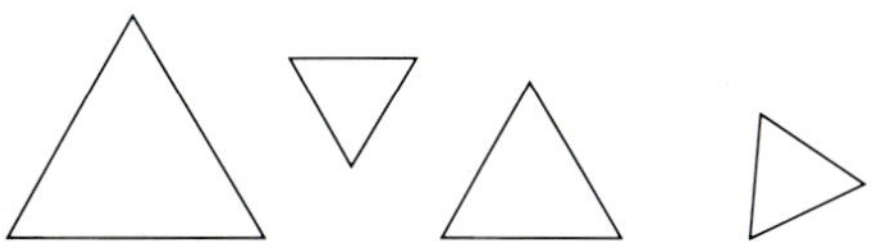

Figure 6-1

Before we define this special relation that we call similarity, we need to develop the idea of *proportionality*. The simplest of the proportions are those involving only four numbers.

Definition 6-1 The numbers a, b, c, and d (where none equal zero) are said to be **proportional** if $a/b = c/d$. The expressions a/b and c/d are called **ratios.**

The numbers a, b, c, and d are the *terms* of the proportion. Numbers in the positions of a and d are the *extremes* and those in the positions of b and c are the *means*. The definition can be extended to include sequences of numbers a, b, c, . . . and r, s, t, . . . , which are proportional if $a/r = b/s = c/t = \ldots = k$ (k is the *constant of proportionality*).

Since $a/b = c/d$ if and only if $ad = bc$, as can be proved with the axioms for real numbers, we can establish the following, which are all *equivalent* to the original proportion, although they are not the *same*.

1. $a/c = b/d$ or $d/b = c/a$ (equivalent by *alternation*)
2. $b/a = d/c$ (equivalent by *inversion*)

Note that in each case $ad = bc$. If the original proportion is $\frac{3}{4} = \frac{15}{20}$, for example, $\frac{3}{15} = \frac{4}{20}$ by alternation.

If $b = c$ in the equation $\frac{a}{b} = \frac{c}{d}$, we can substitute to obtain $\frac{a}{b} = \frac{b}{d}$; then $ad = b(b)$, or $b^2 = ad$; hence $b = \pm\sqrt{ad}$.

Definition 6-2 If a, b and d are positive real numbers and if $\frac{a}{b} = \frac{b}{d}$, then $b = \sqrt{ad}$ is the **geometric mean** between a and d.

Examples

1. For what value of c will $\frac{3}{5} = \frac{c}{20}$ be a proportion?

Solution $5c = 3(20) = 60$; therefore, $c = 12$.

2. If $\frac{2}{3} = \frac{6}{9}$, then $\frac{2}{?} = \frac{3}{?}$ is an equivalent proportion.

Solution $\frac{2}{6} = \frac{3}{9}$ by alternation.

3. If $\frac{c}{d} = \frac{4}{9}$, then $9c = ?$

Solution $9c = 4d$, since $c(9) = 9c$ and $d(4) = 4d$.

4. If $\frac{x}{y} = \frac{p}{q}$, does $\frac{(x + y)}{y} = \frac{(p + q)}{q}$?

Solution Yes, since $(x + y)q = y(p + q)$, $xq + yq = yp + yq$, and $xq = yp$ are all equivalent statements. (Here we say the proportions are equivalent by *addition*.)

5. Find the ratio of x to y if $5x = 6y$.

Solution $\frac{x}{y} = \frac{6}{5}$, since $5x = x(5)$ and $6y = y(6)$.

6. If 1 in. = 2.54 cm, how many inches are there in 40 cm?

Solution Let x be the number of inches required. Then $1/x = 2.54/40$, or $1/2.54 = x/40$, so that $2.54x = 1(40)$, or $x = 40/2.54 = 15.75$ in. (correct to the nearest hundredth inch).

Exercise Set 6.1

1. In each of the following proportions, find the value of a.

   (a) $\frac{a}{4} = \frac{6}{8}$ (b) $\frac{2}{5} = \frac{a}{15}$

   (c) $\frac{3}{a} = \frac{4}{7}$ (d) $\frac{5}{6} = \frac{4}{a}$

   (e) $\frac{a}{5 - a} = \frac{7}{8}$ (f) $\frac{2a - 1}{3} = \frac{3}{1}$

   (g) $\frac{a}{b} = \frac{c}{d}$ (h) $\frac{ab}{4} = \frac{b}{8}$

2. Prove that a mean of a proportion is equal to the product of the extremes divided by the other mean and that an extreme is equal to the product of the means divided by the other extreme.

3. Find the ratio of x to y if

   (a) $3x = 4y$ (b) $ax = 2y$

   (c) $x = 3y$ (d) $y = 5x$

   (e) $3y - x = 0$ (f) $\frac{x - y}{x + y} = \frac{4}{8}$

4. In each of the following, find x in terms of w, y, and z.

   (a) $\frac{5x}{6z} = \frac{2w}{y}$ (b) $\frac{7y}{12z} = \frac{3x}{5w}$

   (c) $\frac{y}{3w} = \frac{5z}{4x}$ (d) $\frac{3y}{2x} = \frac{6z}{4w}$

5. Which of the following are true?

   (a) $\frac{13}{14} = \frac{15}{16}$ (b) $\frac{a}{3a} = \frac{b + b}{6b}$

   (c) $\frac{a^2}{ab} = \frac{2a}{2b} = \frac{a^2b}{ab^2}$ (d) $\frac{x^2 + y^2}{x + y} = \frac{x + y}{1}$

   (e) $\frac{1}{r + s} = \frac{r - s}{r^2 - s^2}$

6. Find the value of n in each of the following:

   (a) $\frac{2n}{5} = \frac{3}{4}$ (b) $\frac{n + 3}{3} = \frac{n + 5}{5}$

   (c) $\frac{2}{n} = \frac{n}{8}$ (d) $\frac{9}{2n} = \frac{18}{n + 9}$

7. Since two sequences of nonzero numbers form a proportion if for a, b, c . . . and r, s, t . . . it is true that $a/r = b/s = c/t$ . . . , decide which pairs of the following sequences are proportional and list these pairs. (*Hint: a* and *d* are one such pair, such $\frac{3}{9} = \frac{5}{15} = \frac{7}{21}$)

   (a) 3, 5, 7 (b) 5, 4, 8

   (c) $\frac{1}{2}, \frac{2}{3}, \frac{3}{4}$ (d) 9, 15, 21

   (e) 6, 4, 9 (f) 20, 16, 32

   (g) $2, \frac{10}{3}, \frac{14}{3}$ (h) $\frac{1}{4}, \frac{3}{4}, 2$

   (i) 15, 12, 24 (j) 2, 6, 16

8. Find the geometric mean of
   (a) 6 and 8 (b) 4 and 9
   (c) 8 and 12 (d) 3 and 12
   (e) 3 and 5
9. If 1 oz is equivalent to 28.35 g, how many grams does 48 oz equal?
10. A model of a boat is to be constructed on a scale of 1 : 50. If the boat is 12.5 m long, how long will the model be?
11. If a picture that measures 9 in. × 15 in. is to be enlarged so that the larger dimension will be 38 in., what will be the size of the smaller dimension?
12. Use the laws and axioms of real numbers to prove that $a/b = c/d$ if and only if $ad = bc$.
13. (a) Show how to derive $a/c = b/d$ from $a/b = c/d$.
    (b) Show how to derive $b/a = d/c$ from $a/b = c/d$.
14. If a certain amount of gold is valued at \$468 and a similar but smaller amount that weighs 0.46 oz is valued at \$161, what is the weight of the larger piece?

Similarity and Correspondence

We are now ready to define similarity between two triangles. Given two triangles ABC and DEF (Fig. 6-2) for which there is a correspondence

$$\overline{AB} \leftrightarrow \overline{DE} \quad \text{and} \quad \angle A \leftrightarrow \angle D$$
$$\overline{BC} \leftrightarrow \overline{EF} \qquad \angle B \leftrightarrow \angle E$$
$$\overline{AC} \leftrightarrow \overline{DF} \qquad \angle C \leftrightarrow \angle F$$

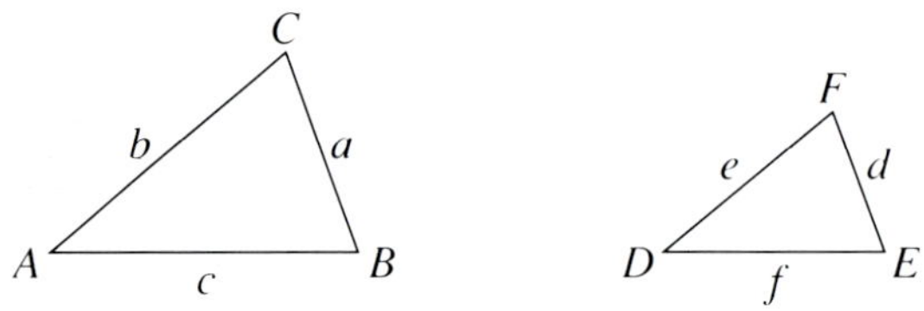

Figure 6-2

Let a be the measure of the side opposite $\angle A$, b the measure of the side opposite $\angle B$, etc. If this correspondence is such that the corresponding angles are congruent and $a/d = b/e = c/f$, then we can say $\triangle ABC$ is similar to $\triangle DEF$. We symbolize this relation by $\triangle ABC \sim \triangle DEF$.

Definition 6-3 **Similar triangles** are triangles that have corresponding angles congruent and corresponding sides proportional.

Actually it will be necessary to demonstrate only one or the other or certain combinations of the preceding conditions in order to prove similarity, as was the case in proving congruency. First let us consider some relations between triangles, however.

Our definition of similar triangles depends on the proportionality of corresponding sides. Since these segments have lengths, when we speak of the ratio of two segments we will mean the ratio of the numbers that are the measures or lengths of the segments.

Definition 6-4 If AB and CD are segments, then the ratio of $m\overline{AB}$ to $m\overline{CD}$ is a number $m\overline{AB}/m\overline{CD} = r$. The number r is called the **ratio of similitude** between the measures of the corresponding sides of similar triangles.

Example

If $\triangle ABC \sim \triangle DEF$ and the measures of their sides are as indicated in the figure,

(a) what is the ratio of similitude between the two triangles?

(b) what is the length of $\overline{DE}$? of $\overline{EF}$?

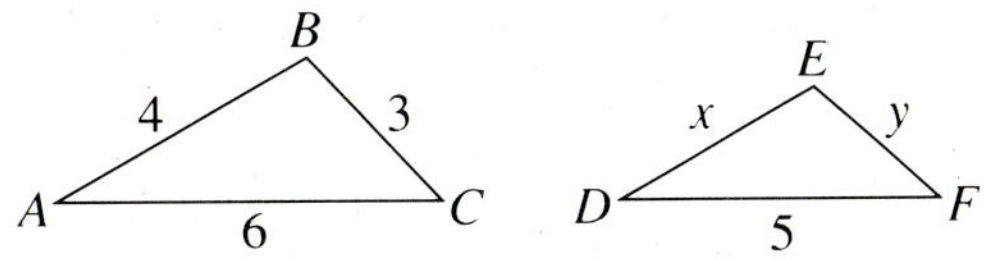

Solution

(a) By the definition of the ratio of similitude $m\overline{AC}/m\overline{DF} = \frac{6}{5} = r$.

(b) From the definition of similar triangles, we have $m\overline{AB}/m\overline{DE} = m\overline{BC}/m\overline{EF} = m\overline{CA}/m\overline{FD}$; therefore

$$\frac{6}{5} = \frac{4}{x} = \frac{3}{y}; \qquad x = \frac{20}{6} = 3\tfrac{1}{3}, \qquad y = \frac{15}{6} = 2\tfrac{1}{2}$$

Exercise Set 6.2

1. If two triangles are similar and the first triangle has sides of length 3, 7, and 9, and the second has sides of 6, 14, and x, respectively, find the length of side x of the second triangle.
2. Write the proportionality between the corresponding sides when
$$\triangle ABC \sim \triangle RST$$
3. If $\triangle ABC \sim \triangle DEF$ and $m\overline{AB} = 15$, $m\overline{BC} = 12$, $m\overline{AC} = 21$, and $m\overline{DE} = 12$, find the lengths of the other sides of $\triangle DEF$.
4. If $\triangle ABC \sim \triangle DEF$, $m\overline{AB} = 4$, and $m\overline{DE} = 7$, what is the ratio of similitude? What is the length of $\overline{AC}$ if $m\overline{DF} = 8$?
5. If the triangle on the left is similar to the triangle on the right, write the proportion for corresponding sides.

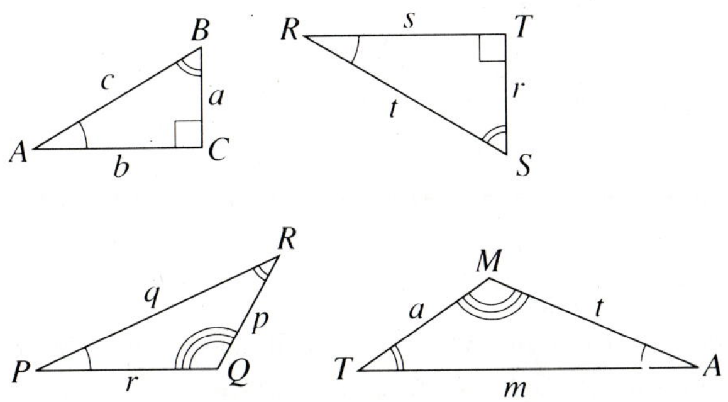

6. To find the distance across a lake, a developer measured the angles and distances as indicated in the figure so that the triangles shown are similar. What is the distance across the lake?

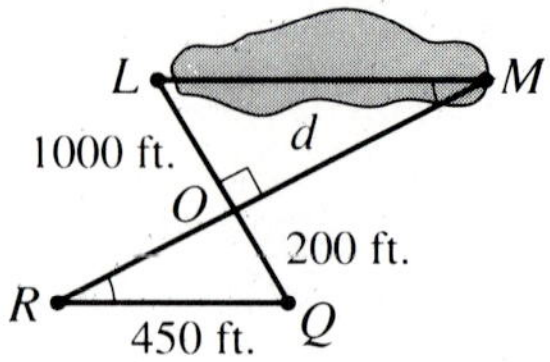

7. If the pairs of triangles shown are similar, find the measures x and y.

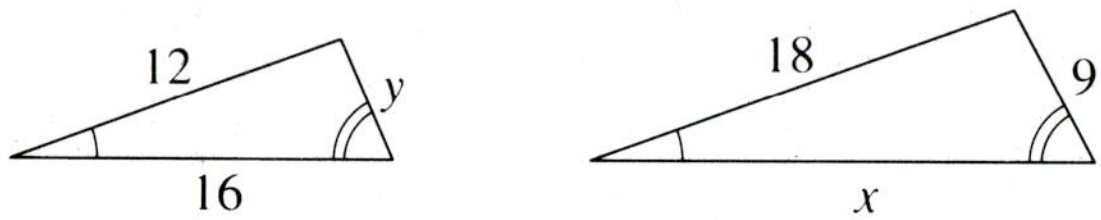

(a)

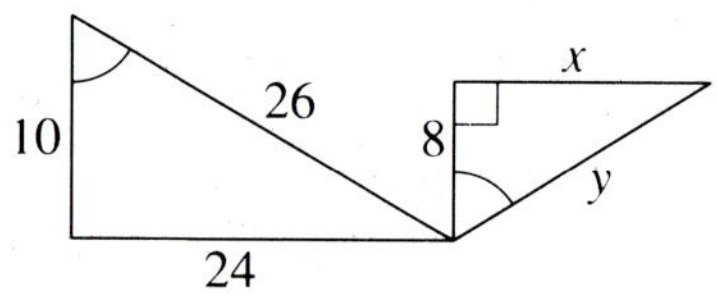

(b)

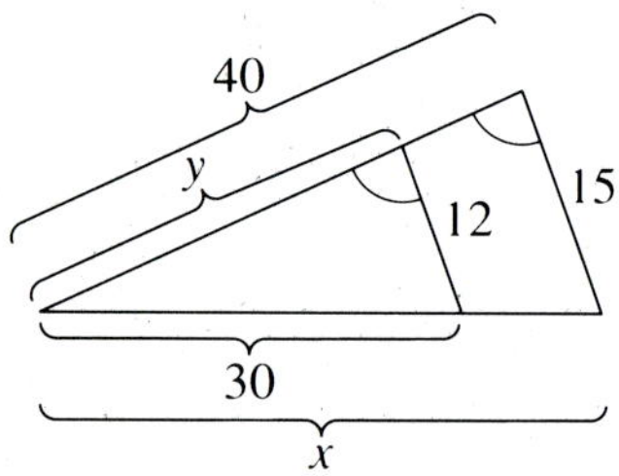

(c)

8. In the following problems, triangles ABC and XYZ have sides a, b, and c, and x, y, and z opposite the respective angles. Draw as indicated and state your conclusion.
   (a) $m\overline{AB} = 2$ in., $m\angle A = 45$, $m\angle B = 30$;
   $m\overline{XY} = 2\frac{1}{2}$ in., $m\angle X = 45$, $m\angle Y = 30$.
   Are the triangles similar?
   (b) $m\overline{AB} = 3$ in., $m\angle A = 60$, $m\overline{AC} = 1\frac{1}{2}$ in. Draw triangle XYZ with $X = A$, A-Y-B and A-Z-C so that $m\overline{XZ} = 1$ in. and $m\overline{XY} = 2$ in. Are the triangles similar? What can you say about $\overline{BC}$ and $\overline{YZ}$?
   (c) $m\overline{AB} = 1\frac{7}{8}$ in., $m\overline{AC} = 1\frac{1}{2}$ in., $m\overline{BC} = 1\frac{1}{8}$ in., $m\overline{XY} = 3\frac{3}{8}$ in., $m\overline{XZ} = 1\frac{7}{8}$ in., and $m\overline{YZ} = 2\frac{1}{2}$ in. (Use the method described in the first exercise set of Chapter 1.) Measure the angles. Are the triangles similar?

Theorems on Similar Triangles

Theorem 6-1 (*AAA Similarity Theorem*) If the corresponding angles of two triangles are congruent, the triangles are similar. (Proof of a special case of this theorem appears in Appendix III.)

Corollary 6-1(a) If two angles of a triangle are congruent to two corresponding angles of a second triangle, the triangles are similar.

Given: $\angle A \cong \angle X$, $\angle B \cong \angle Y$
Prove: $\triangle ABC \sim \triangle XYZ$

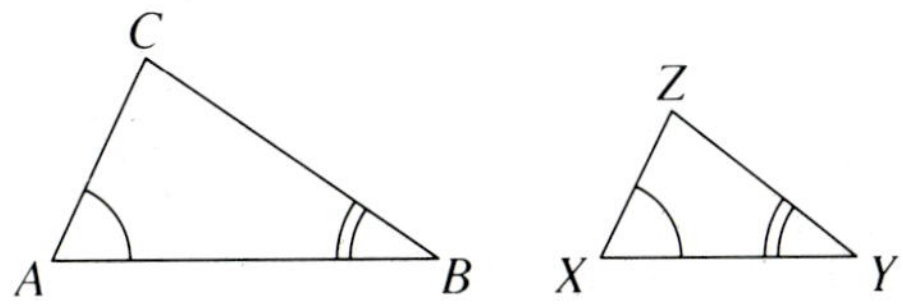

Here is a case in which a corollary is more useful than the original theorem, since we will usually be content to show only two pairs of angles congruent in a proof.

Corollary 6-1(b) If an acute angle of a right triangle is congruent to an acute angle of another right triangle, the triangles are similar.
Given: Right $\triangle$s ABC and RST, $\angle A \cong \angle R$
Prove: $\triangle ABC \sim \triangle RST$

(The proof is left to the reader.)

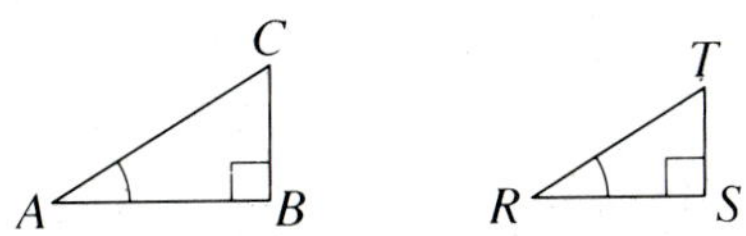

Next let us consider the basic proportionality theorem.

Theorem 6-2 If a line parallel to one side of a triangle intersects the other two sides in distinct points, it divides these sides proportionally.
Given: $\triangle ABC$ with $DE \parallel AB$

Prove: (a) $\dfrac{m\overline{AC}}{m\overline{CD}} = \dfrac{m\overline{BC}}{m\overline{CE}}$

(b) $\dfrac{m\overline{AD}}{m\overline{CD}} = \dfrac{m\overline{BE}}{m\overline{CE}}$

(c) $\dfrac{m\overline{AC}}{m\overline{AD}} = \dfrac{m\overline{BC}}{m\overline{BE}}$

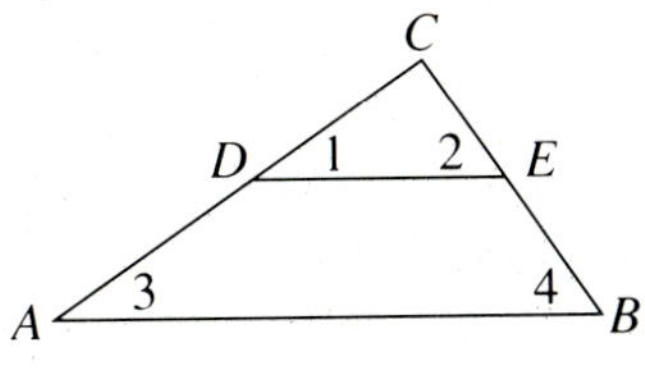

Proof

Statements	*Reasons*
1. $\overline{DE} \parallel \overline{AB}$	1. Given
2. $\angle 1 \cong \angle 3$ and $\angle 2 \cong \angle 4$	2. Corresponding angles are congruent (Theorem 5-15)

Statements	*Reasons*
3. $\triangle ABC \sim \triangle DEC$	3. Two angles of one triangle are congruent to two angles of a second triangle (Corollary 6-1(a))
4. $\dfrac{m\overline{AC}}{m\overline{CD}} = \dfrac{m\overline{BC}}{m\overline{CE}}$	4. Similar triangles have corresponding sides proportional (Def. 6-3)
5. $m\overline{CD} + m\overline{AD} = m\overline{AC}$ and $m\overline{CE} + m\overline{BE} = m\overline{BC}$	5. If C is a point on $\overline{AB}$ such that A-C-B, then $m\overline{AC} + m\overline{CB} = m\overline{AB}$. (Post 2-7)
6. $\dfrac{m\overline{CD} + m\overline{AD}}{m\overline{CD}} = \dfrac{m\overline{CE} + m\overline{BE}}{m\overline{CE}}$	6. Substitution axiom (E-4)
7. $\dfrac{m\overline{AD}}{m\overline{CD}} = \dfrac{m\overline{BE}}{m\overline{CE}}$	7. Equivalent proportions (By addition)

The proof of part (c) is left to the reader.

Theorem 6-3 (Converse of Theorem 6-2) If a line intersects two sides of a triangle and divides them proportionally, the line is parallel to the third side.

Given: $\triangle ABC$ with $\overline{DE}$ such that $m\overline{AC}/m\overline{DC} = m\overline{BC}/m\overline{EC}$
Prove: $\overline{DE} \parallel \overline{AB}$

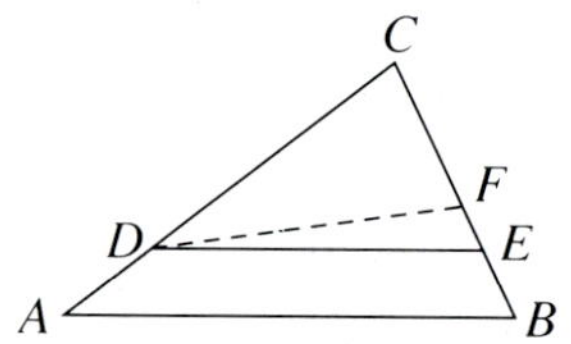

Proof Either (1) $\overline{DE} \parallel \overline{AB}$ or (2) there is another line $\overleftrightarrow{DF} \parallel \overline{AB}$, with $F \in \overline{BC}$.

If we assume (2), then $\overline{DF}$ divides the sides proportionally by Theorem 6-2, so $m\overline{AC}/m\overline{DC} = m\overline{BC}/m\overline{FC}$. But we are given $m\overline{AC}/m\overline{DC} = m\overline{BC}/m\overline{EC}$, so $m\overline{BC}/m\overline{EC} = m\overline{BC}/m\overline{FC}$ and $m\overline{EC} = m\overline{FC}$. This means $F = E$ (Segment Construction, Theorem 2-4), so $\overline{DF} = \overline{DE}$ and it must be true that $\overline{DE} \parallel \overline{AB}$.

Example

Given $\triangle ACD$, $\overline{DC} \parallel \overline{BE}$.
(a) Find s if $r = 5$, $u = 4$, $v = 3$.
(b) Find r if $m\overline{AC} = 8$, $m\overline{BC} = 3$, $m\overline{DE} = 6$.
(c) Find $m\overline{AC}$ if $s = 4$, $v = 2$, $m\overline{AD} = 10$.

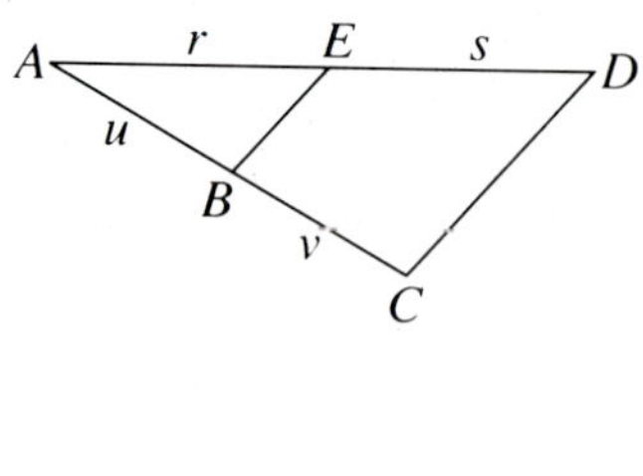

Solution

(a) $(u+v)/u = (r+s)/r$ by Theorem 6-2(a); $\frac{7}{4} = (5+s)/5$.
$20 + 4s = 35,\ 4s = 15,\ s = 3\frac{3}{4}$

(b) $(u+v)/v = (r+s)/s$ by Theorem 6-2(c), so $\dfrac{8}{3} = \dfrac{r+6}{6}$,
$3(r+6) = 8(6),\ 3r + 18 = 48,\ 3r = 30$ and $r = 10$.

(c) From Theorem 6-2(c) we obtain $m\overline{AC}/2 = \frac{10}{4}$; thus $m\overline{AC} = \frac{20}{4} = 5$.

Theorem 6-4 (*SSS Similarity Theorem*) If the corresponding sides of two triangles are proportional, the triangles are similar.

Given: $\dfrac{m\overline{AB}}{m\overline{DE}} = \dfrac{m\overline{BC}}{m\overline{EF}} = \dfrac{m\overline{CA}}{m\overline{FD}}$

Prove: $\triangle ABC \sim \triangle DEF$

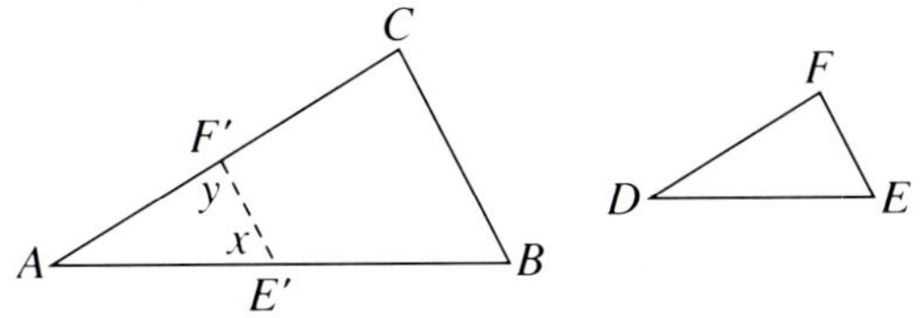

Proof

Statements	*Reasons*
1. Let E' and F' be points on $\overrightarrow{AB}$ and $\overrightarrow{AC}$ such that $m\overline{AE'} = m\overline{DE}$ and $m\overline{AF'} = m\overline{DF}$	1. Segment Construction Theorem (Theorem 2-4)
2. $\dfrac{m\overline{AB}}{m\overline{DE}} = \dfrac{m\overline{AC}}{m\overline{DF}}$	2. Given
3. $\dfrac{m\overline{AB}}{m\overline{AE'}} = \dfrac{m\overline{AC}}{m\overline{AF'}}$	3. Substitution Axiom
4. $\overline{E'F'} \parallel \overline{BC}$	4. If ℓ intersects two sides of a triangle and divides them proportionally, ℓ is parallel to the third side (Theorem 6-3)
5. $\therefore \angle x \cong \angle B$ and $\angle y \cong \angle C$	5. Corresponding angles are congruent (Theorem 5-15)
6. $\triangle ABC \sim \triangle AE'F'$	6. If two angles of a triangle are congruent to two corresponding angles of a second triangle, the triangles are similar. (Corollary 6-1(a))
7. $\dfrac{m\overline{BC}}{m\overline{E'F'}} = \dfrac{m\overline{AB}}{m\overline{AE'}}$	7. Definition of Similar Triangles (Def. 6-3)

Statements	*Reasons*
8. $\frac{m\overline{BC}}{m\overline{EF}} = \frac{m\overline{AB}}{m\overline{DE}} = \frac{m\overline{AB}}{m\overline{AE'}}$	8. Given and Substitution Axiom for Real Numbers (E-4)
9. $\therefore m\overline{E'F'} = m\overline{EF}$	9. Statements 7 and 8
10. $\triangle AE'F' \cong \triangle DEF$	10. SSS Postulate
11. $\angle A \cong \angle D$ and $\angle x \cong \angle E$	11. CPCTC
12. $\therefore \angle B \cong \angle E$	12. Angle congruence is transitive (Theorem 3-1(c))
13. $\triangle ABC \sim \triangle DEF$	13. Same as 6

Theorem 6-5 (*SAS Similarity Theorem*) If two triangles have two pairs of corresponding sides proportional and the angles included by them congruent, the triangles are similar.

Given: $\frac{m\overline{AB}}{m\overline{DE}} = \frac{m\overline{AC}}{m\overline{DF}}$, $\angle A \cong \angle D$
Prove: $\triangle ABC \sim \triangle DEF$

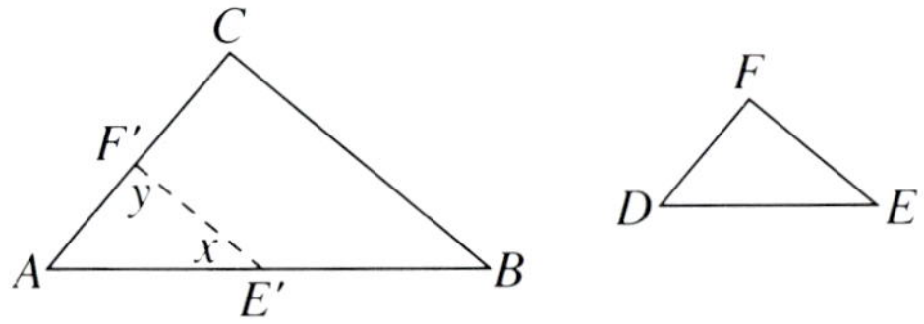

Proof Let E' and F' be points on $\overrightarrow{AB}$ and $\overrightarrow{AC}$ such that $m\overline{AE'} = m\overline{DE}$ and $m\overline{AF'} = m\overline{DF}$. Substituting we have $m\overline{AB}/m\overline{AE'} = m\overline{AC}/m\overline{AF'}$, and by Theorem 6-3 $\overline{E'F'} \parallel \overline{BC}$. Thus $\angle x \cong \angle B$ and $\angle y \cong \angle C$, since they are corresponding angles, and $\triangle AE'F' \cong \triangle DEF$ by the *SAS* Postulate. From this we obtain $\angle x \cong \angle E$ and $\angle y \cong \angle F$, so that $\angle B \cong \angle E$ and $\angle C \cong \angle F$, which means $\triangle ABC \sim \triangle DEF$ by the AAA Similarity Theorem.

A useful special case of Theorems 6-2 and 6-3 can be stated as follows.

Theorem 6-6 The segment joining the midpoints of two sides of a triangle is parallel to the third side and its length is equal to one-half the length of the third side.

Given: $\triangle ABC$ with points D and E bisecting $\overline{AC}$ and $\overline{BC}$, respectively
Prove: $\overline{DE} \parallel \overline{AB}$ and $m\overline{DE} = \frac{1}{2}m\overline{AB}$

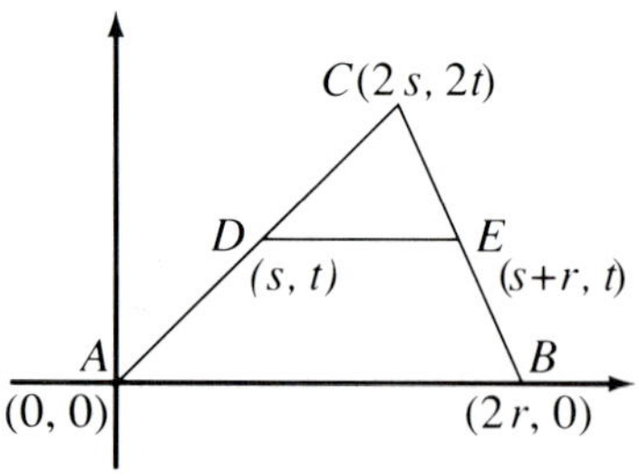

Proof Place a coordinate system on the triangle with the origin at A and $\overline{AB}$ on the positive x-axis. We can let $A(0, 0)$, $B(2r, 0)$ and $C(2s, 2t)$ be the coordinates of the vertices. Then D and E will have coordinates (s, t) and $(r + s, t)$, and

$$m_{\overline{AB}} = \frac{0}{2r} = 0$$

$$m_{\overline{DE}} = \frac{t - t}{r + s - s} = \frac{0}{r} = 0,$$

so the slopes are equal, and

$$\text{m}\overline{AB} = |2r - 0| = |2r|$$
$$\text{m}\overline{DE} = |r + s - s| = |r|,$$

which shows that $\text{m}\overline{DE} = \frac{1}{2}\text{m}\overline{AB}$.

Construction of Numbers

We have seen in Construction 11 that a segment of length p can be divided into n congruent parts, so we can construct a representation of any rational number of the form p/q. We can also add or subtract segments of lengths p and q to find $p + q$ and $p - q$. By means of similar triangles we can find products and quotients as follows. (Refer to Fig. 6-3.)

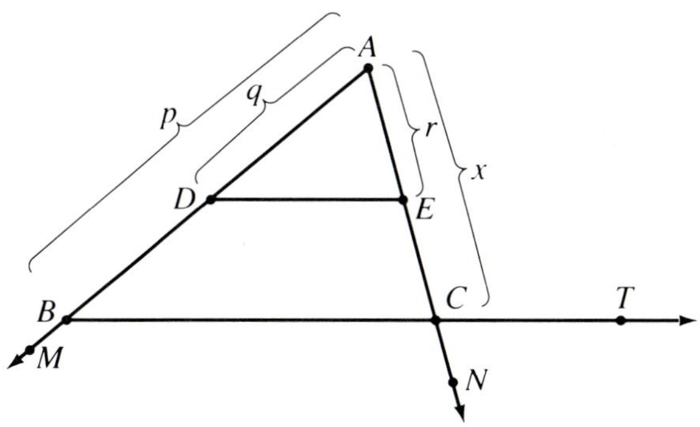

Figure 6-3

Draw any angle MAN. On $\overrightarrow{AM}$ construct $\overline{AB}$ with length p and $\overline{AD}$ with length q. On $\overrightarrow{AN}$ construct $\overline{AE}$ with length r. Now construct $\angle ABT \cong \angle ADE$, so that $\overrightarrow{BT}$ intersects $\overrightarrow{AN}$ at C. Then $\triangle ABC \sim \triangle ADE$, and $\text{m}\overline{AB}/\text{m}\overline{AD} = \text{m}\overline{AC}/\text{m}\overline{AE}$. If $\text{m}\overline{AC}$ x, this means $p/q = x/r$, or $x = pr/q$. When $q = 1$ unit, $x = pr$. If $r = 1$ unit, $x = p/q$, another construction for the quotient.

Exercise Set 6.3

1. If two triangles are congruent, are they necessarily similar? Why?
2. Are two equilateral triangles necessarily similar? Why?
3. Are two isosceles triangles similar? Why?

4. If the ratio of similitude of two triangles is $\frac{3}{4}$ and the sides of the smaller triangle are 7, 10, and 14, find the lengths of the sides of the larger triangle.

5. In the following figure, if the segments have lengths as given, is $\overline{DE} \parallel \overline{AB}$? Why?

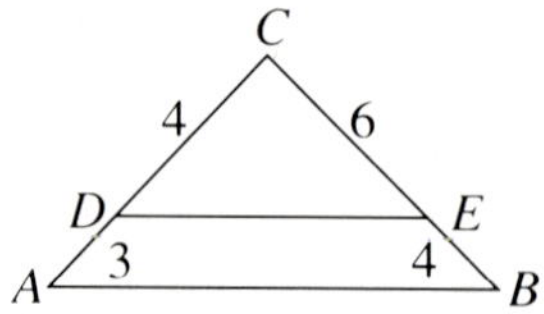

6. In the following figure, $\overline{UV} \parallel \overline{RS}$. Prove $\triangle RST \sim \triangle UVT$.

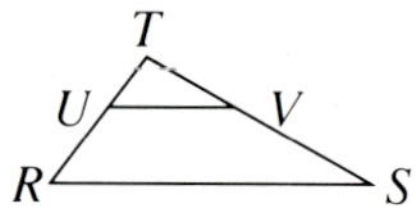

7. If triangles PQR and XYZ are similar, $m\angle P = 80$, $\angle R \cong \angle Z$, $m\angle Q = 30$, $m\overline{PQ} = 12$, and $m\overline{PQ}/m\overline{XY} = \frac{3}{2}$,

(a) what is the measure of $\angle Z$?

(b) what is the length of $\overline{XY}$?

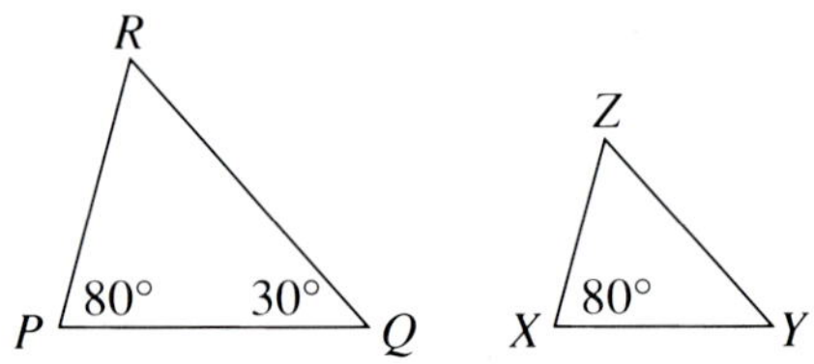

8. In the following figure $\overline{AC} \perp \overline{AE}$, $\overline{AC} \perp \overline{CD}$, $\overline{AC} \cap \overline{DE} = B$.

(a) Prove $\dfrac{m\overline{AE}}{m\overline{CD}} = \dfrac{m\overline{AB}}{m\overline{BC}}$

(b) Prove $\dfrac{m\overline{AB}}{m\overline{AE}} = \dfrac{m\overline{BC}}{m\overline{CD}}$

(c) Prove $\dfrac{m\overline{AE}}{m\overline{EB}} = \dfrac{m\overline{CD}}{m\overline{DB}}$

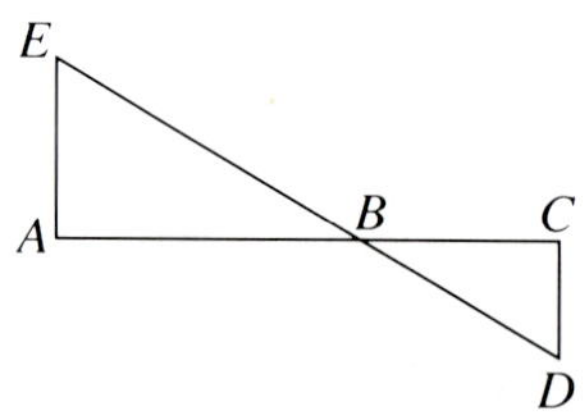

9. In the following figure $\overline{AC} \cap \overline{DE} = B$ and $\overline{AE} \parallel \overline{DC}$. Prove $\triangle ABE \sim \triangle CBD$

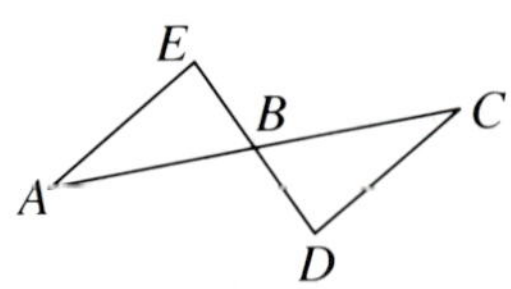

10. In the following figure $\overline{AE}$ and $\overline{BD}$ are altitudes of the triangle.
   (a) Prove $\triangle ACE \sim \triangle BCD$.
   (b) Prove $\triangle ADF \sim \triangle BEF$.
   (c) Prove $\dfrac{m\overline{AE}}{m\overline{DB}} = \dfrac{m\overline{AC}}{m\overline{BC}}$

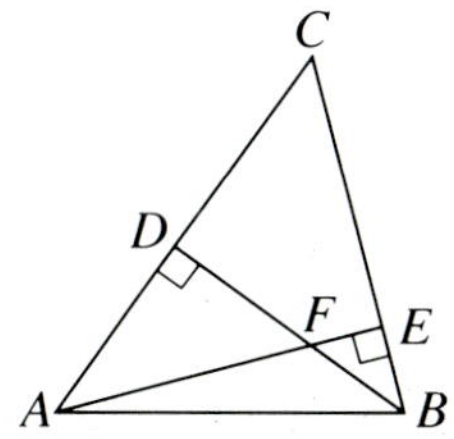

11. In the following figure, U is the midpoint of $\overline{RT}$ and V is the midpoint of $\overline{ST}$.
   (a) If $m\overline{RS} = 24$, then $m\overline{UV} =$ ________? Why?
   (b) If $m\overline{UV} = 15$, then $m\overline{RS} =$ ________?
   (c) $\angle TUV \cong \angle TRS$. Why?

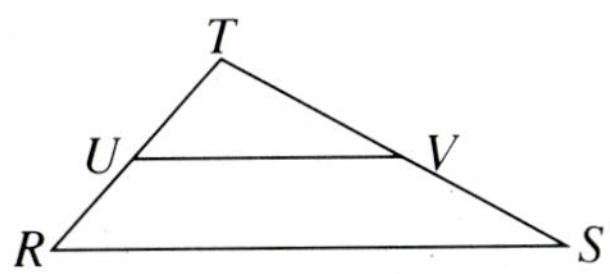

12. Refer to the following figure. From the given information, which theorem or corollary can best be used to show $\triangle MPQ \sim \triangle XYZ$?
   (a) $m\overline{MQ} = 12$, $m\overline{PQ} = 16$, $m\overline{MP} = 20$, $m\overline{XZ} = 9$, $m\overline{YZ} = 12$, $m\overline{XY} = 15$
   (b) $m\angle Q = m\angle Z = 80$, $m\angle M = m\angle X = 60$
   (c) $m\angle Q = m\angle Z = 80$, $m\overline{MQ} = 15$, $m\overline{PQ} = 18$, $m\overline{XZ} = 10$, $m\overline{YZ} = 12$

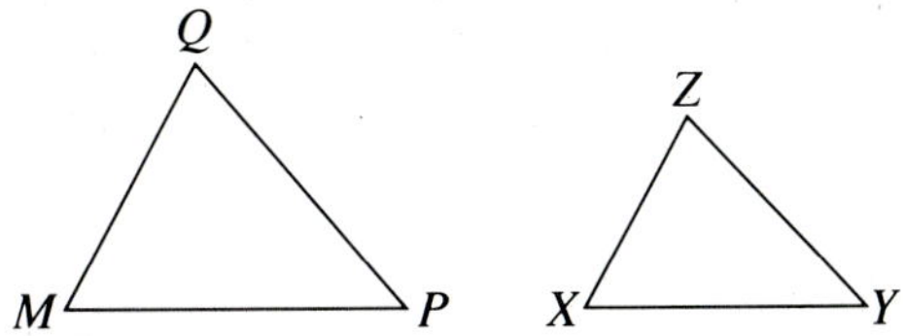

13. Prove that in similar triangles, the corresponding altitudes have the same ratios as corresponding sides.
14. Prove that in similar triangles, the corresponding medians have the same ratios as corresponding sides.
15. Prove that similarity of triangles is an equivalence relation.
16. If, in the following figure, $\triangle ABC$ is a right triangle and $\overline{DC}$ is the altitude from the right angle vertex to the hypotenuse, prove $\triangle ADC \sim \triangle CDB \sim \triangle ACB$.

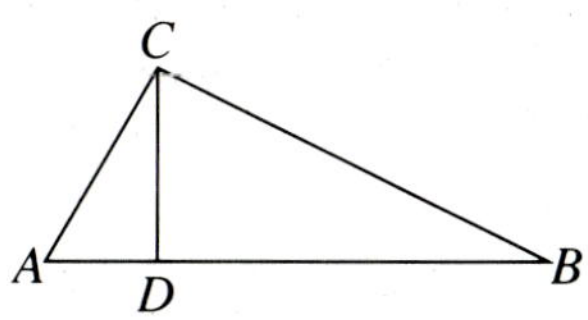

17. Prove that if the vertex angles of two isosceles triangles are congruent, the triangles are similar.

18. If a 20-ft lamppost casts a shadow of 15 ft, find the height of a tree whose shadow is 80 ft at the same time of day. (Assume the sun's rays to be parallel.)

19. If a man 6 ft 3 in. tall is standing next to a telephone pole that is 42 ft tall and the pole casts a shadow 37 ft long, how long is the man's shadow?

*20. If $\triangle RST \sim \triangle MNP$ and if the coordinates are $R(6, 3)$, $S(2, 5)$, $T(5, 8)$, $M(10, 1)$, $N(0, 6)$, and $P(x, y)$, find a set of values of x and y using algebra. (R, S, M, and N are collinear.)

6-2 RIGHT TRIANGLE SIMILARITIES

In a right triangle the altitude from the vertex of the right angle to the hypotenuse has many important properties.

Theorem 6-7 In a right triangle, the altitude to the hypotenuse forms two right triangles that are similar to each other and to the given triangle.

Given: $\triangle ACB$ is a right triangle with altitude $\overline{CD}$ from C to the hypotenuse $\overline{AB}$

Prove: $\triangle ACB \sim \triangle ADC \sim \triangle CDB$

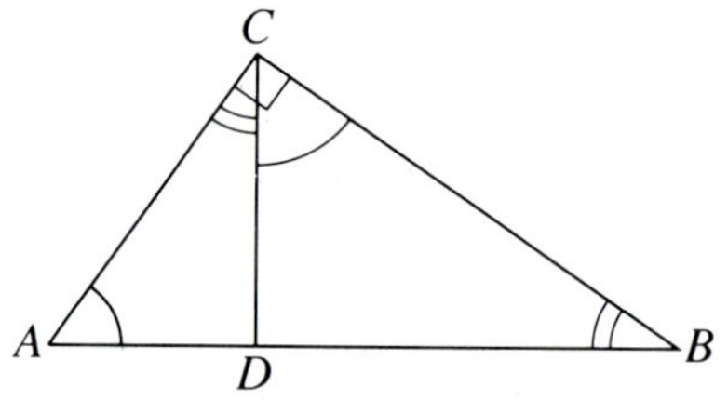

Proof The restatement of the theorem tells us the correspondence between the triangles that we wish to prove similar. By matching the right angles in the triangle, we can see that each of the smaller triangles has an acute angle matching an acute angle of the large triangle, so the remaining angles will then match.

Since $\angle A$ is common to right triangles ADC and ACB, they are similar by Corollary 6-1(b); likewise $\angle B$ is common to $\triangle ACB$ and $\triangle CDB$, so $\triangle ACB \sim \triangle CDB$. By the definition of similar triangles, we have $\angle A \cong \triangle DCB$; hence $\triangle ADC \sim \triangle CDB$ and we have the desired relation $\triangle ACB \sim \triangle ADC \sim \triangle CDB$.

Corollary 6-7(a) In a right triangle, the altitude from the right angle to the hypotenuse is the geometric mean between the two segments formed on the hypotenuse.

(Referring to Fig. 6-4, x is the geometric mean between a' and b', since $a'/x = x/b'$ or $x = \sqrt{a'b'}$. Sometimes x is called the *mean proportional* between a' and b'.)

* The more difficult problems we indicated by an asterisk.

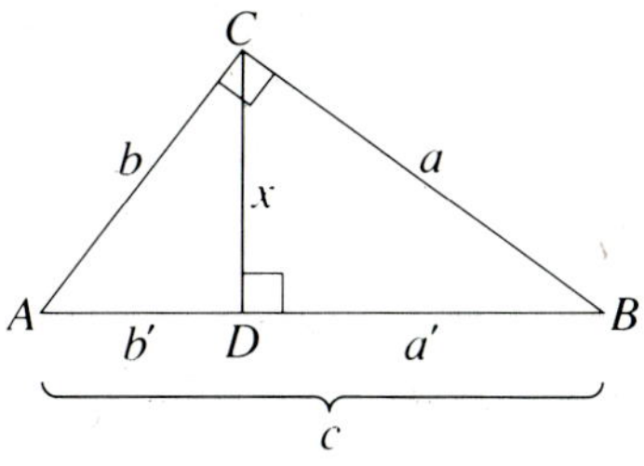

Figure 6-4

Corollary 6-7(b) The altitude to the hypotenuse of a right triangle divides the hypotenuse into two segments such that the length of a leg of a right triangle is the geometric mean between the length of the hypotenuse and the length of the adjacent segment of the hypotenuse.

(Again referring to Fig. 6-4, $c/b = b/b'$ or $b = \sqrt{c \cdot b'}$, and $c/a = a/a'$ or $a = \sqrt{c \cdot a'}$.)

Examples

1. In a right triangle ABC with hypotenuse $\overline{AB}$ and $\overline{CD} \perp \overline{AB}$, if $m\overline{AD} = 6$ and $m\overline{DB} = 8$, find $m\overline{CD}$.

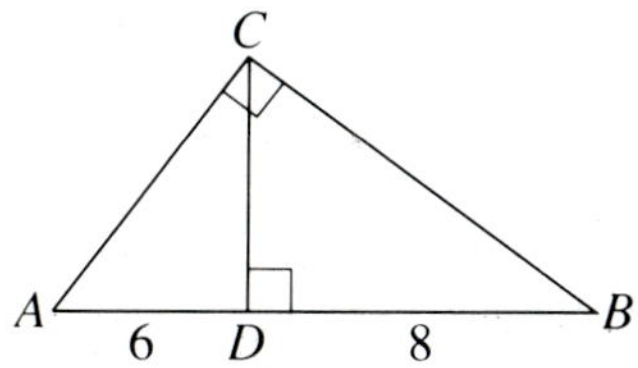

Solution Using Corollary 6-7(a), we find $m\overline{CD} = \sqrt{6 \cdot 8} = \sqrt{48} = \sqrt{16} \cdot \sqrt{3} = 4\sqrt{3}$. (Radical expressions should be written in simplest form.)

2. Using the figure in the previous example, this time with $m\overline{AC} = 8$ and $m\overline{BD} = 12$, find $m\overline{AD}$.

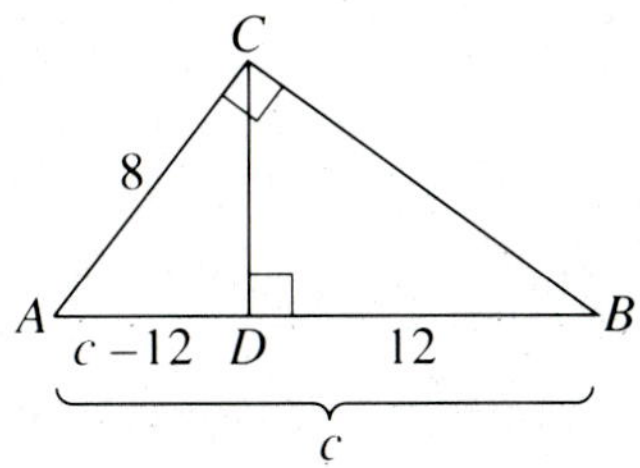

Solution According to Corollary 6-7(a), we have $m\overline{AD}/m\overline{AC} = m\overline{AC}/m\overline{AB}$. If $m\overline{AB} = c$, $m\overline{AD} = c - 12$, so $(c - 12)/8 = 8/c$, and $c(c - 12) = 64$ or $c^2 - 12c - 64 = 0$. This can be factored as $(c - 16)(c + 4) = 0$; thus $c - 16 = 0$ or $c + 4 = 0$, and $c = 16$ or $c = -4$. Since all lengths are positive, $c \neq -4$ and $c = 16 = m\overline{AB}$. Thus $m\overline{AD} = 16 - 12 = 4$.

We recognize the equation $c^2 - 12c - 64 = 0$ as a quadratic equation. Equations of this form, that is, $ax^2 + bx + c = 0$ with a, b, and c real numbers ($a \neq 0$), may sometimes be solved as above by factoring or always by the *quadratic formula*, as follows.

For all quadratic equations

$$ax^2 + bx + c = 0 (a \neq 0)$$

$$x = \frac{-b + \sqrt{b^2 - 4ac}}{2a} \quad \text{or} \quad x = \frac{-b - \sqrt{b^2 - 4ac}}{2a}$$

If we apply the formula to the equation of the preceding example, we have

$$x^2 - 12x - 64 = 0, \text{ so } a = 1, b = -12 \text{ and } c = -64$$

$$\sqrt{b^2 - 4ac} = \sqrt{(-12)^2 - 4(1)(-64)} = \sqrt{400} = 20$$

so

$$x = \frac{-(-12) + 20}{2(1)} = \frac{12 + 20}{2} = \frac{32}{2} = 16$$

or

$$x = \frac{-(-12) - 20}{2} = \frac{12 - 20}{2} = \frac{-8}{2} = -4$$

(Recall that -4 is not a feasible solution.)

Exercise Set 6.4

1. Solve the following quadratic equations either by factoring, when possible, or by the quadratic formula.
   (a) $x^2 - 2x = 0$
   (b) $x^2 + 5x + 6 = 0$
   (c) $x^2 + 2x - 15 = 0$
   (d) $2x^2 - x = 1$
   (e) $8x^2 - 6x + 1 = 0$
   (f) $2x^2 + 3x - 2 = 0$
   (g) $x^2 + 10x + 21 = 0$
   (h) $\frac{x}{2} + \frac{12}{x} = x - 1$ (*Hint:* clear the fractions)
   (i) $\frac{x^2}{6} + 1 = \frac{7}{6}$
   (j) $\frac{5x - 7}{9} + \frac{14}{2x - 3} = x - 1$
2. In the following figure,
   (a) if $m\overline{BD} = 3$ and $m\overline{AB} = 12$, find $m\overline{BC}$
   (b) if $m\overline{BC} = 4$ and $m\overline{AD} = 6$, find $m\overline{BD}$
   (c) if $m\overline{AC} = 12$ and $m\overline{BD} = 10$, find $m\overline{AB}$.

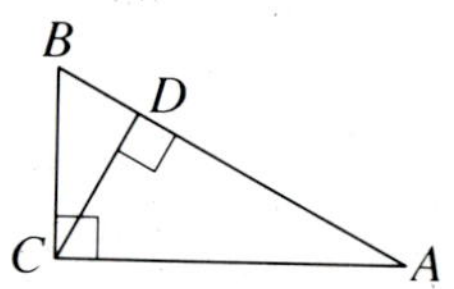

3. Find the unknown lengths in the following figure.

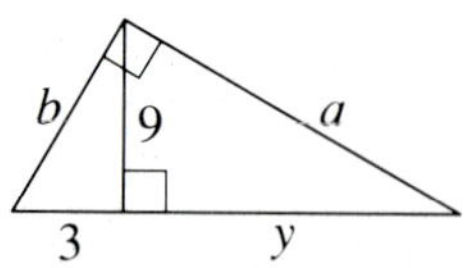

4. Repeat Exercise 3.

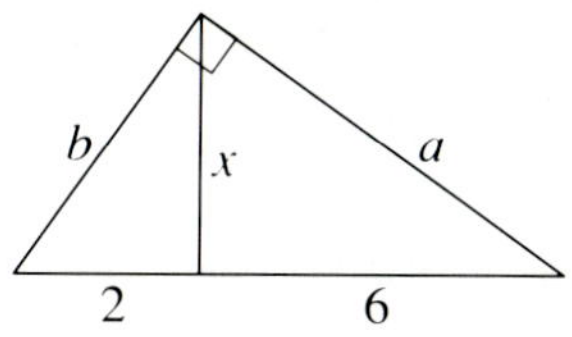

5. Repeat Exercise 3.

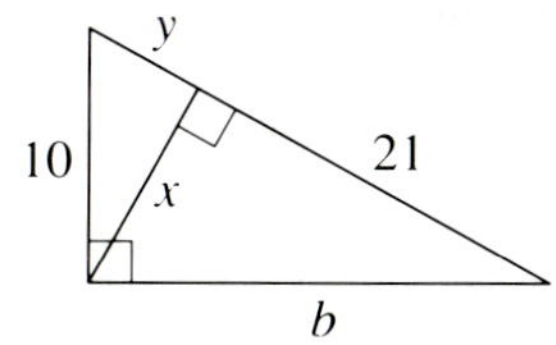

6. In the following figure, $\overline{BC} \parallel \overline{ED}$.
(a) Express d in terms of a, b, and c.
(b) Find the length b if $d = 4$, $a = 6$ and $c = 8$.

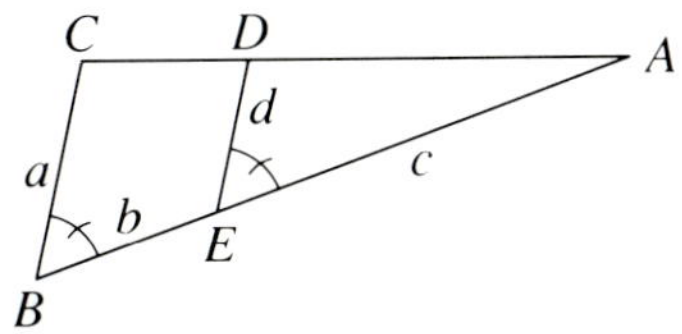

7. **(a)** If $m = 27$ and $n = 3$, find a, p, and q.
(b) If $p = 15$ and $n = 9$, find m and q.

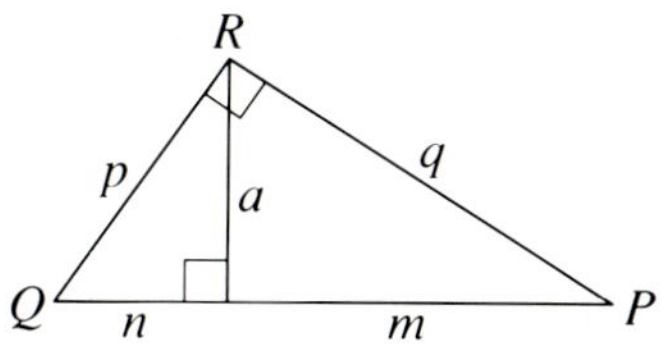

8. Prove Corollary 6-7(a).

6-3 THE THEOREM OF PYTHAGORAS

We are now ready to prove the theorem referred to in the first chapter as one of the most famous and useful in mathematics. It was said to have been first proved by the Pythagoreans, a colony of Greeks who lived on the southern coast of Italy around 500 B.C. Although their main interest was music, their studies included the mathematics of the day, which involved mostly geometry. Some sources say that theirs was a secret society, and when one of the members told an outsider about the theorem, he was taken out to sea and cast overboard. The proof of the theorem given here is relatively simple. The proof attributed to Euclid requires many more steps.

PYTHAGORAS (?-ca. 497 B.C.) There are more than three hundred known proofs of the Pythagorean Theorem. Pythagoras' teacher, Thales, may have actually developed the first proof of this theorem. (*New York Public Library Picture Collection*)

Theorem 6-8 (*Pythagorean Theorem*) The square of the length of the hypotenuse of a right triangle is equal to the sum of the squares of the lengths of the legs.

Given: Right $\triangle ABC$ with legs of lengths a and b, and hypotenuse of length c
Prove: $c^2 = a^2 + b^2$

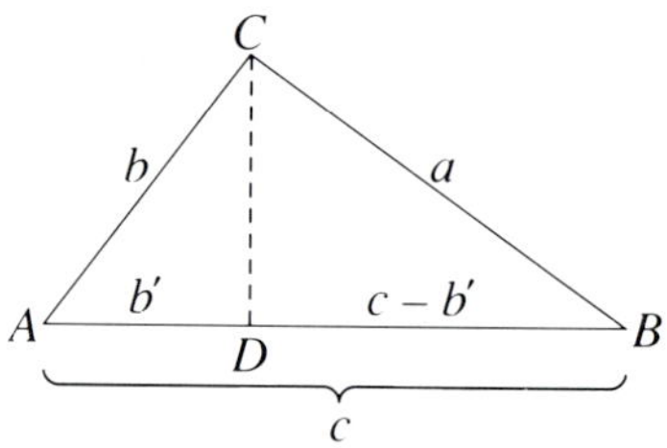

Proof

Statements	*Reasons*
1. If $\overline{CD}$ is the altitude to the hypotenuse, $b'/b = b/c$ and $(c - b')/a = a/c$	1. Corollary 6-7(b)
2. $b'c = b^2$ and $c^2 - b'c = a^2$	2. Property of Proportions
3. $c^2 - b^2 = a^2$	3. Substitution Axiom
4. $c^2 = a^2 + b^2$	4. Addition Law for Equals

Corollary 6-8(a) The square of the length of a side of a right triangle is equal to the square of the length of the hypotenuse less the square of the length of the other side.

Given: Right triangle ABC with C the right angle
Prove: $a^2 = c^2 - b^2$ and $b^2 = c^2 - a^2$

(The proofs of this corollary and the ones that follow are left to the reader.)

Corollary 6-8(b) If the hypotenuse and leg of one right triangle are congruent, respectively, to the hypotenuse and leg of another right triangle, the triangles are congruent.

Corollary 6-8(c) A point in the interior of an angle that is equidistant from the sides of the angle lies on the bisector of the angle.

The converse of Theorem 6-8 also holds and is also extremely useful.

Theorem 6-9 If a triangle ABC has sides of length a, b, and c, respectively, opposite angles A, B, and C, such that $c^2 = a^2 + b^2$, then the triangle is a right triangle and C is the right angle.

Given: $\triangle ABC$ with $a^2 + b^2 = c^2$
Prove: $\angle C$ is a right angle

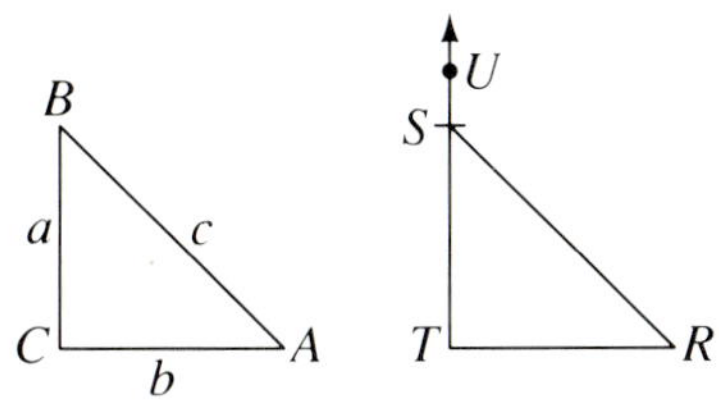

Proof

Statements	*Reasons*
1. There is a segment $\overline{TR} \cong \overline{CA}$ such that $m\overline{TR} = b$	1. Segment Construction Theorem (Theorem 2-4)
2. At T on $\overline{TR}$ there is a ray $\overrightarrow{TU}$ such that $\angle RTS$ is a right angle	2. Angle Construction Postulate (Post. 3-2)
3. On $\overrightarrow{TU}$ there is a point S such that $m\overline{TS} = a$	3. Same as Reason 1
4. RST is a right triangle	4. Definition of right triangle (Def. 4-3)
5. $[m\overline{RS}]^2 = [m\overline{ST}]^2 + [m\overline{RT}]^2 = a^2 + b^2$	5. Pythagorean Theorem (Theorem 6-8) and Substitution Axiom
6. $c^2 = a^2 + b^2$, so $[m\overline{RS}]^2 = c^2$	6. Given; Substitution Axiom
7. $m\overline{RS} = c$	7. Theorem I-8, Appendix I
8. $\triangle ABC \cong \triangle RST$	8. SSS (Post. 4-3)
9. $\angle C \cong \angle T$	9. CPCTC
10. $\therefore C$ is a right angle	10. An angle congruent to a right angle is a right angle (Corollary 3-6(a))

Examples

1. If $\triangle ABC$ is a triangle with right angle at C,
(a) Find the length of the hypotenuse if $a = 13$ and $b = 4$.
(b) Find the length of the other leg if $a = 5$ and $c = 13$.

Solution (a) $c^2 = a^2 + b^2$, so $c^2 = 13^2 + 4^2 = 169 + 16 = 185$ and $c = \sqrt{185}$.
(b) From Corollary 6-8(a), $b^2 = c^2 - a^2$, so $b^2 = 13^2 - 5^2 = 169 - 25 = 144$ and $b = \sqrt{144} = 12$.

2. If the hypotenuse of a right triangle has a length of 4 and the length of one leg is half the length of the hypotenuse, how long is each leg?

Solution One leg measures $\frac{1}{2}(4) = 2$. If the length of the other is x, then $x^2 = 4^2 - 2^2 = 16 - 4 = 12$, so $x = \sqrt{12} = \sqrt{4 \cdot 3} = 2\sqrt{3}$.

3. An isosceles triangle has congruent sides of length 11 and a base of length 12. What is the length of the altitude, correct to the nearest hundredth?

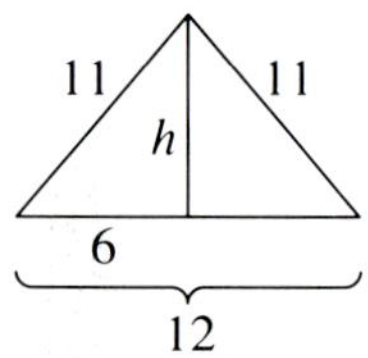

Solution The altitude bisects the base, so $h^2 = 11^2 - 6^2 = 121 - 36 = 85$, or $h = \sqrt{85}$. Since the value of h is required correct to the nearest hundredth, we refer to Table II in Appendix IV and find $h = \sqrt{85} \approx 9.22$. (The symbol $\approx$ means "is approximately equal to".) ◁

We are now able to prove the distance formula, which was introduced in Chapter 3.

Theorem 6-10 The distance from $R(x_1, y_1)$ to $S(x_2, y_2)$ is the real number

$$d(\overline{RS}) = \sqrt{(x_2 - x_1)^2 + (y_2 - y_1)^2}$$

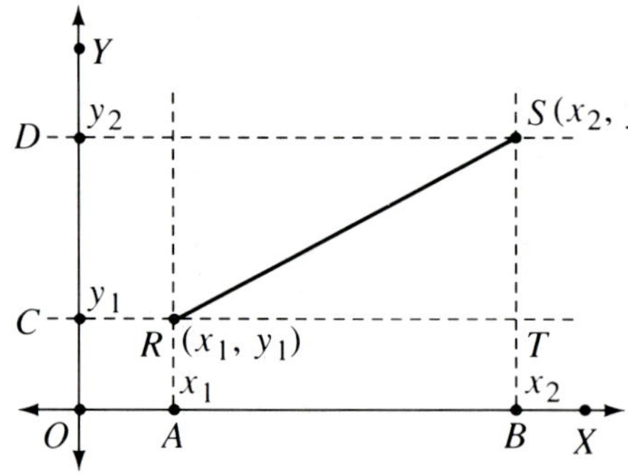

Proof If $\overleftrightarrow{RS}$ is neither vertical nor horizontal, we may draw a line through R parallel to $\overleftrightarrow{OX}$ and a line through S parallel to $\overleftrightarrow{OY}$. These lines will intersect at some point T so that $\overline{RT}$ and $\overline{ST}$ are perpendicular, and $\triangle RTS$ is a right triangle with hypotenuse $\overline{RS}$. Now draw lines through R and S parallel to $\overleftrightarrow{OY}$ and $\overleftrightarrow{OX}$. The segments cut off will be congruent (Theorem 5-19), so $m\overline{AB} = m\overline{RT} = |x_2 - x_1|$ and $m\overline{CD} = m\overline{ST} = |y_2 - y_1|$. Since

$$[m\overline{RS}]^2 = [m\overline{RT}]^2 + [m\overline{ST}]^2$$

then

$$[m\overline{RS}]^2 = |x_2 - x_1|^2 + |y_2 - y_1|^2$$

or

$$m\overline{RS} = \sqrt{(x_2 - x_1)^2 + (y_2 - y_1)^2}$$

If $\overline{RS}$ is horizontal or vertical, we do not need the Pythagorean Theorem, but we shall consider the case where $\overline{RS}$ is horizontal. If $\overline{RA}$ and $\overline{SB}$ are drawn parallel to $\overleftrightarrow{OY}$, then $m\overline{AB} = m\overline{RS} = |x_2 - x_1|$. Since $y_2 = y_1$, $y_2 - y_1 = 0$, and $m\overline{RS} = |x_2 - x_1| = \sqrt{(x_2 - x_1)^2 + (y_2 - y_1)^2}$.

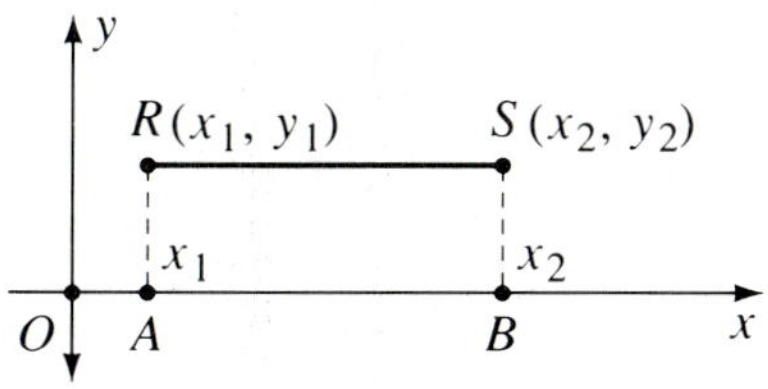

Exercise Set 6.5

1. Using the converse of the Pythagorean Theorem, find which of the following number triples could be the lengths of sides of right triangles.

	a	b	c
(a)	8	15	17
(b)	7	9	14
(c)	7	24	25
(d)	10	24	26
(e)	16	20	34

2. Using the Pythagorean Theorem, if each of the following are lengths of sides of a right triangle, find the length of the third side if a and b are legs and c is the hypotenuse.

	a	b	c
(a)	6	8	?
(b)	5	?	13
(c)	3	5	?
(d)	1	?	2
(e)	1	2	?
(f)	?	4	8
(g)	?	4	9
(h)	1	3	?

3. If x is a positive real number, for what values of x will 5, $2x$ and $(2x + 1)$ be sides of a right triangle?

4. For what values of u and v will

$$(u^2 + v^2)^2 = (2uv)^2 + (u^2 - v^2)^2?$$

5. If the hypotenuse of a right triangle is 10 and its legs are of equal length, what are their lengths?

6. A part of an electric typewriter has dimensions as indicated on the following figure. What is the distance from A to B (to the nearest hundredth)?

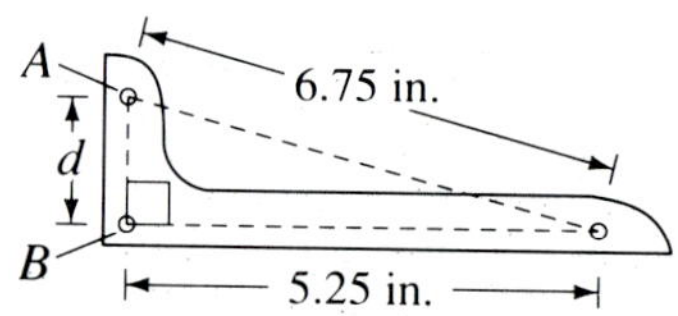

7. A plumber needs to run a pipe from point R to point S underneath a house. What is m$\overline{RS}$, if R is 6 ft from the corner at T and S is 9 ft (to the nearest hundredth)?

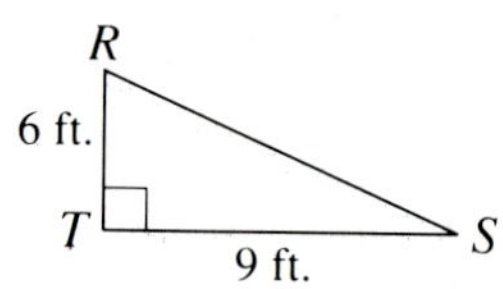

8. A 10-ft brace is to be attached $2\frac{1}{2}$ ft below the bed of a trailer. How far to the left of point X will it meet the trailer bed? (Nearest tenth.)

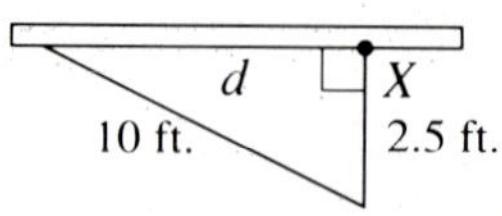

9. An isosceles right triangle has a leg of length 3. What is the length of the hypotenuse?
10. Prove that if two altitudes of a triangle are congruent, the triangle is isosceles.
11. **(a)** Prove Corollary 6-8(a).
    (b) Prove Corollary 6-8(b).
    (c) Prove Corollary 6-8(c).

6-4 SPECIAL RIGHT TRIANGLES

The 30°, 60° Right Triangle

Theorem 6-11 In a right triangle with acute angles of measure 30° and 60° and hypotenuse length c, the length of the side opposite the 30° angle is $\frac{1}{2}c$ and the length of the side opposite the 60° angle is $(\frac{1}{2}c)\sqrt{3}$.

Given: Right triangle ABC with $m\angle A = 60$ and $m\angle B = 30$
Prove: $b = \frac{1}{2}c$ and $a = (\frac{1}{2}c)\sqrt{3}$

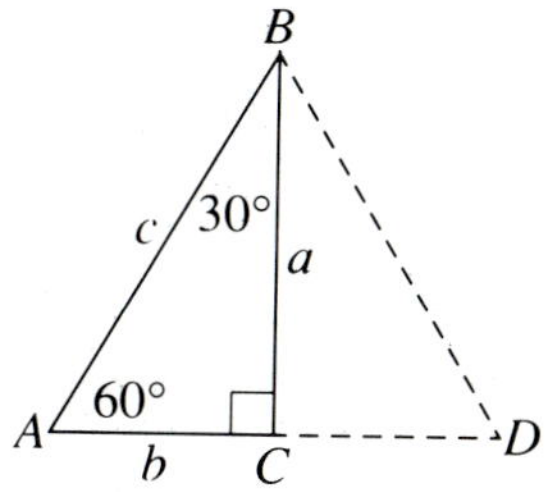

Proof

Statements	*Reasons*
1. Extend $\overrightarrow{AC}$ to D so that $m\overline{CD} = m\overline{AC}$	1. Segment Construction Theorem (Theorem 2-4)
2. BCD is a triangle	2. Definition of Triangle (Def. 4-1)
3. $\angle BCD$ is a right angle	3. Theorem 3-11
4. $\triangle BCA \cong \triangle BCD$	4. SAS (Post. 4-1)
5. $\therefore m\angle D = 60$, $m\angle CBD = 30$, and $m\overline{BD} = c$	5. CPCTC
6. $\triangle ABD$ is equiangular, so it is equilateral and $m\overline{AD} = 2b = c$	6. Corollary 4-6
7. $b = \frac{1}{2}c$	7. Multiplication Law of Equals
8. $a^2 = c^2 - b^2 = c^2 - (\frac{1}{2}c)^2$	8. Corollary 6-8(a) and Substitution Axiom
9. $a^2 = c^2 - \frac{1}{4}c^2 = \frac{3}{4}c^2$; $\therefore a = \sqrt{\frac{3}{4}c^2} = \sqrt{\frac{1}{4}c^2 \cdot 3} = (\frac{1}{2}c)\sqrt{3}$	9. Substitution; Properties of Radical Expressions

Example

Given the following triangle.

(a) Show that $b = a/\sqrt{3} = a\sqrt{3}/3$.
(b) Find a and b if $c = 4$.
(c) Find a and b if $c = 6$.
(d) Find a and c if $b = 6$.
(e) Find b and c if $a = 4$.

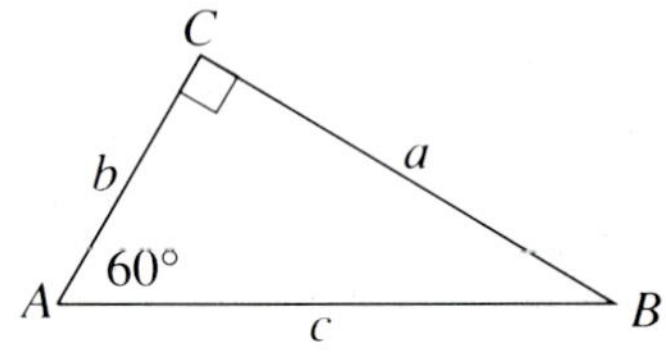

Solutions

(a) Since $m\angle A = 60$ and $m\angle C = 90$, $m\angle B = 30$ because the acute angles of a right triangle are complementary. Thus by Theorem 6-11 $b = \frac{1}{2}c$ and $a = (\frac{1}{2}c)\sqrt{3}$, so $a/\sqrt{3} = \frac{1}{2}c$. Thus $b = a/\sqrt{3}$ by Axiom E-3, so $b = (a/\sqrt{3})(\sqrt{3}/\sqrt{3}) = a\sqrt{3}/3$.

(b) Again using Theorem 6-11, $b = \frac{1}{2}c = \frac{1}{2}(4) = 2$ and $a = (\frac{1}{2}c)\sqrt{3} = (\frac{1}{2} \cdot 4)\sqrt{3} = 2\sqrt{3}$.

(c) $b = \frac{1}{2}(6) = 3$; $a = (\frac{1}{2} \cdot 6)\sqrt{3} = 3\sqrt{3}$.

(d) Since $b = \frac{1}{2}c$, $6 = \frac{1}{2}c$ and $c = 12$, then $a = (\frac{1}{2} \cdot 12)\sqrt{3} = 6\sqrt{3}$.

(e) If $a = 4$, $4 = (\frac{1}{2}c)\sqrt{3}$ or $8 = \sqrt{3}c$, so $c = 8/\sqrt{3} = 8\sqrt{3}/3$; then $b = \frac{1}{2}(8\sqrt{3}/3) = 4\sqrt{3}/3$. (Or use the result of part (a))

The 45°, 45° Right Triangle

If a right triangle is isosceles, then its legs have equal lengths, so the acute angles have equal measures. The acute angles are also complementary, so the measure of each is 45. If the legs are a and b and the hypotenuse is c, then $c^2 = a^2 + b^2$ and since $a = b$, $c^2 = a^2 + a^2$, so $c^2 = 2a^2$ or $c = \sqrt{2} \cdot a$.

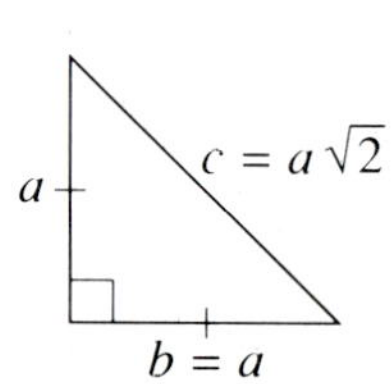

Figure 6-5

Thus we have the following theorem.

Theorem 6-12 In an isosceles right triangle the length of the hypotenuse is equal to the length of a leg multiplied by $\sqrt{2}$.

Exercise Set 6.6

1. Find the lengths of the altitude and base of an isosceles triangle if one of its congruent sides measures 12 and a base angle is
   (a) 30° **(b)** 60° **(c)** 45°
2. In the following table, find the lengths of two sides in terms of the given side.

c (hypotenuse)	a (opp. 30°∠)	b (opp. 60°∠)
c		
	a	
		b

(a)

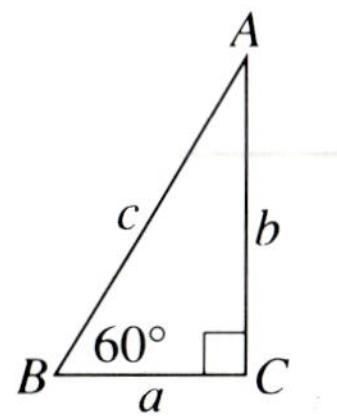

c	a	b
c		
	a	
		b

(b)

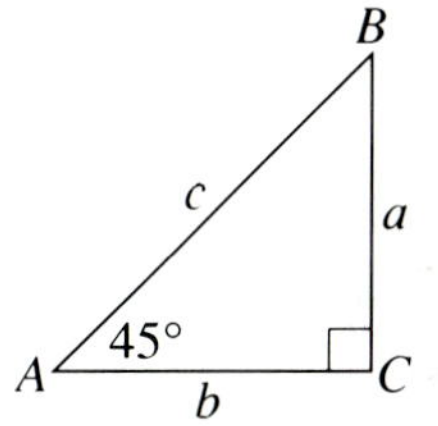

3. In each figure, use the given information to find the lengths and angle measures indicated.

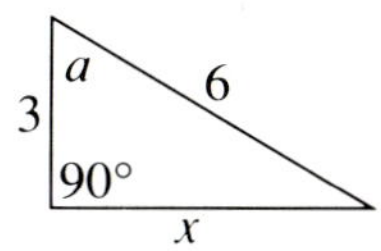

$a =$

$x =$

(a)

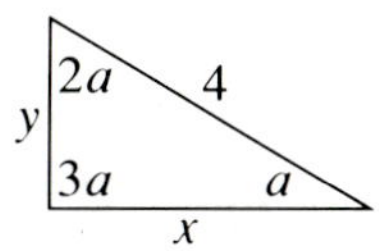

$a =$

$x =$

$y =$

(b)

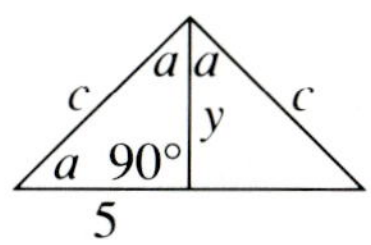

$a =$

$y =$

$c =$

(c)

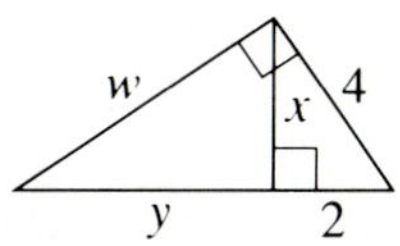

$x =$

$y =$

$w =$

(d)

4. If the length of the altitude of an equilateral triangle is 8, find the length of the sides.
5. If ABC is a right triangle, use the given information to find the lengths that are missing in the following table:

	A	a	b	c	m	n	h_c
(a)			4	5			
(b)		5	12				
(c)		7		25			
(d)		10					6
(e)				18	2		
(f)	30°			8			
(g)	45°			8			
(h)	60°			8			

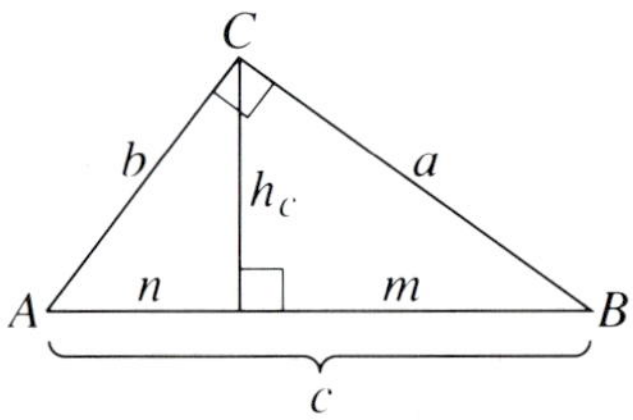

6. Prove that the length of the altitude of the hypotenuse of a right triangle is equal to the product of the lengths of the legs divided by the length of the hypotenuse. (In the figure for Exercise 5, $h_c = ab/c$.)
7. Find the length of the altitude to one of the congruent sides in each of the triangles of Exercise 1.
*8. If the sides of a triangle measure 7, 8, and 9, find the length of the altitude to the side of length 8. (*Hint:* Is this a right triangle? Draw the figure.)
9. Construct a 30°, 60° right triangle.
10. Construct a 45°, 45° right triangle.
11. (A puzzle-like problem.) Our church tower is square and has a pinnacle on each corner. Driving one day along a straight road in the neighborhood, I noticed that two pinnacles, the northeast and the southeast, were in a line. I set the trip mileage to zero and drove on. I noted that after 1.5 mi the northwest and southeast pinnacles were in a line, and at 3.5 mi the southwest and southeast pinnacles were in line. How far was I from the church?

* The more difficult problems we indicated by an asterisk.

6-5 SLOPE OF PERPENDICULAR LINES*

In Exercise 13 of Exercise Set 3.14, page 78, we found that if the product of the slopes of two lines is -1, the lines appear to be perpendicular. It is now possible to prove that this is the case. First we prove the converse.

Theorem 6-13 If two nonvertical lines are perpendicular, the product of their slopes is -1.

Given: $\ell_1 \perp \ell_2$

Prove: $m_{\ell_1} \cdot m_{\ell_2} = -1$

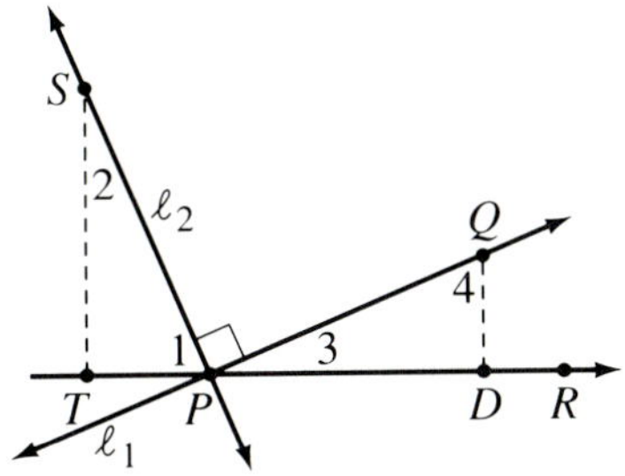

Proof Let P be the point of intersection of ℓ_1 and ℓ_2. Draw a horizontal line PR through P. From $Q \in \ell_1$ and $S \in \ell_2$ draw lines perpendicular to $\overleftrightarrow{PR}$, intersecting $\overleftrightarrow{PR}$ at T and D. Then STP and PDQ are right triangles and $m\angle 1 + m\angle 2 = 90$. Also $m\angle 1 + m\angle SPQ + m\angle 3 = 180$, so $m\angle 1 + m\angle 3 = 90$. Thus $m\angle 1 + m\angle 2 = m\angle 1 + m\angle 3$, and $m\angle 2 = m\angle 3$. Therefore $\triangle STP \sim \triangle PDQ$ and the corresponding sides are proportional: $m\overline{TP}/m\overline{DQ} = m\overline{ST}/m\overline{PD}$, from which $m\overline{TP}/m\overline{ST} = m\overline{DQ}/\overline{PD}$. Applying the formula for slope of a line, $m_{\ell_1} = m\overline{DQ}/m\overline{PD}$ and $m_{\ell_2} = -m\overline{ST}/m\overline{TP}$, and $m\overline{TP}/m\overline{ST} = -1/m_{\ell_2}$. From this, $m_{\ell_1} = -1/m_{\ell_2}$, $m_{\ell_1} \cdot m_{\ell_2} = -1$.

Theorem 6-14 Two nonvertical lines are perpendicular if the product of their slopes is -1.

Given: $m_{\ell_1} \cdot m_{\ell_2} = -1$

Prove: $\ell_1 \perp \ell_2$

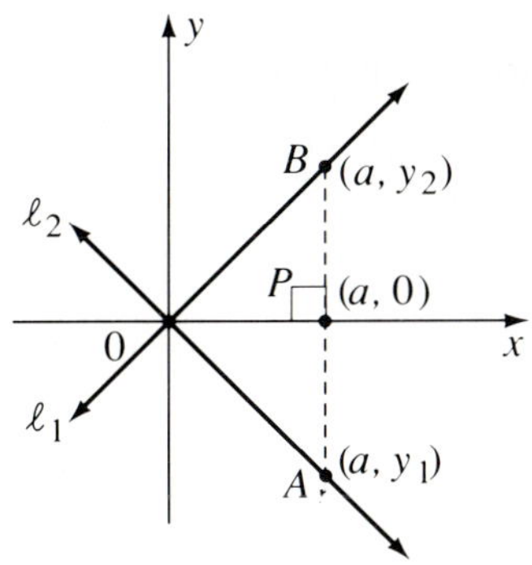

* This section may be omitted.

Proof Place a coordinate system on the figure so that the intersection of the lines is at the origin and neither line lies on the y-axis. Locate a point $P(a, 0)$ on the x-axis and draw a perpendicular intersecting ℓ_1 at $B(a, y_2)$ and $A(a, y_1)$, as shown in the figure.

Statements	*Reasons*
1. $m_{\ell_1} = \frac{y_2}{a}$ and $m_{\ell_2} = \frac{y_1}{a}$	1. Definition of slope (Def. 3-13)
2. $[m\overline{AO}]^2 = a^2 + y_1^2$ $[m\overline{BO}]^2 = a^2 + y_2^2$ $[m\overline{AB}]^2 = (y_2 - y_1)^2$	2. Distance Formula (Theorem 6-10)
3. $\triangle ABO$ is a right triangle and $\ell_1 \perp \ell_2$ if $[m\overline{AB}]^2 = [m\overline{AO}]^2 + [m\overline{BO}]^2$	3. Converse of Pythagorean Theorem (Theorem 6-9)
4. $[m\overline{AB}]^2 = [m\overline{AO}]^2 + [m\overline{BO}]^2$ if $(y_2 - y_1)^2 = a^2 + y_1^2 + a^2 + y_2^2$, $y_2^2 - 2y_2y_1 + y_1^2 = 2a^2 + y_1^2 + y_2^2$, $- 2y_2y_1 = 2a^2$, $y_2y_1 = -a^2$	4. Substitution Axiom
5. Since $m_{\ell_1} \cdot m_{\ell_2} = -1$, $\frac{y_2}{a} \cdot \frac{y_1}{a} = -1$ or $y_2 \cdot y_1 = -a^2$	5. Given; Multiplication Law of Equals

Hence $\triangle ABC$ is a right triangle and $\ell_1 \perp \ell_2$.

Exercise Set 6.7

1. Given $A(0, 2)$, $B(5, 0)$, $S(5, 7)$, $T(1, b)$, and $\overline{AB} \perp \overline{ST}$; find the value of b.
2. Plot each set of points and determine whether $\overline{AC} \perp \overline{BC}$ by means of Theorem 6-14:
   (a) $A(5, 9)$, $B(1, 2)$, $C(7, 5)$
   (b) $A(-1, 4)$, $B(5, 1)$, $C(0, -2)$
   (c) $A(-2, 2)$, $B(6, 3)$, $C(4, 6)$
3. Show that the triangles containing right angles in Exercise 2 are right triangles by means of the Distance Formula and Theorem 6-9.
4. Prove analytically that the median to the base of an isosceles triangle is perpendicular to the base.
5. Prove analytically that altitudes to the sides of an equilateral triangle are congruent.

6-6 NUMERICAL TRIGONOMETRY*

Many of the applications of geometry to the solution of problems in engineering, physics, navigation, and astronomy depend upon indirect measurements. A navigator, for example, may locate his position by measuring the angle between two

* This section may be omitted.

objects a known distance apart. The engineer tunnels through mountains along a path that has been computed in the field or offices. The branch of mathematics which involves this sort of computation is known as *trigonometry*, a word meaning "triangle measurement."

Although the study of trigonometry as a field of mathematics can involve a great deal of time and effort, the geometry we have considered thus far together with the most basic ideas of trigonometry will permit a wide range of practical applications from the very basic level to what many people would regard as extremely complex. We hope the reader will be able to appreciate the power of this simple mathematical tool.

Right Triangle Ratios

In the last section we proved that if an acute angle of one right triangle is congruent to an acute angle of another right triangle, then they are similar.

If the three right triangles shown in Fig. 6-6 have $\angle A \cong \angle D \cong \angle G$, then we know they are similar, and from the definition we have

$$\frac{a}{d} = \frac{c}{f} \quad \text{or} \quad \frac{a}{c} = \frac{d}{f} \quad \text{and} \quad \frac{d}{g} = \frac{f}{i} \quad \text{or} \quad \frac{d}{f} = \frac{g}{i}$$

so

$$\frac{a}{c} = \frac{d}{f} = \frac{g}{i}$$

In like manner we may obtain

$$\frac{b}{c} = \frac{e}{f} = \frac{h}{i} \quad \text{and} \quad \frac{a}{b} = \frac{d}{e} = \frac{g}{h}$$

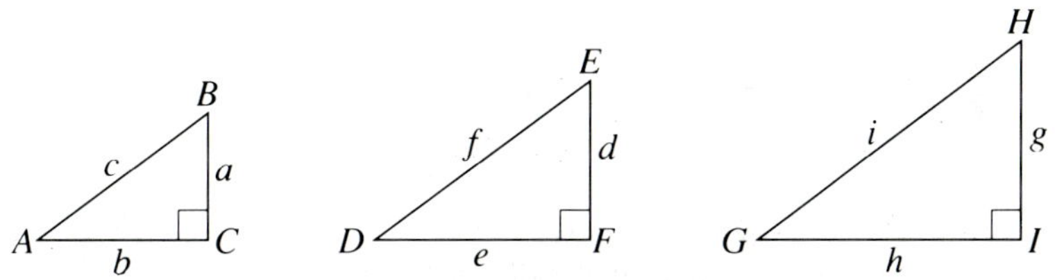

Figure 6-6

These relationships can also be demonstrated by a figure as in Fig. 6-7 on the following page, where we see that triangles ABC, $AB'C'$ and $AB''C''$ are similar.

Here we see that for a given acute angle A of a right triangle, the ratios of any two sides of the similar triangles remain the same. The ratios depend on the measure of the angle, *not* on the size of the triangle. These particular ratios are basic to our work in trigonometry and each has a special name. They are called the sine, cosine, and tangent of angle A and are defined as shown at the top of the next page.

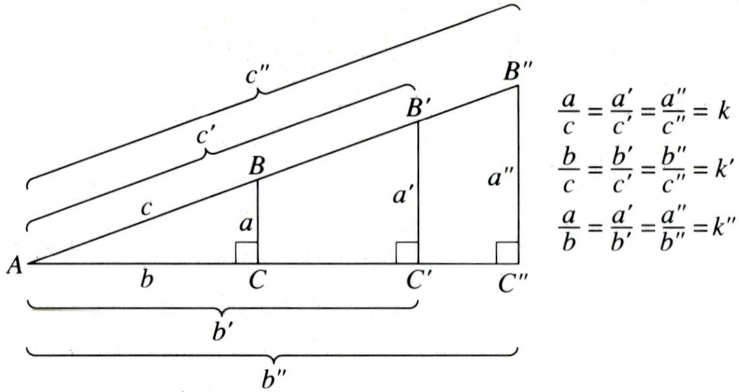

Figure 6-7

Definition 6-5 In the right triangle ABC with right angle at C, c the length of the hypotenuse, a the length of the side opposite $\angle A$, and b the length of the side adjacent to $\angle A$,

$$\textbf{sine } \angle A = \textbf{sin } A = \frac{a}{c} = \frac{\text{length of side opposite } \angle A}{\text{length of hypotenuse}}$$

$$\textbf{cosine } \angle A = \textbf{cos } A = \frac{b}{c} = \frac{\text{length of side adjacent } \angle A}{\text{length of hypotenuse}}$$

$$\textbf{tangent } \angle A = \textbf{tan } A = \frac{a}{b} = \frac{\text{length of side opposite } \angle A}{\text{length of side adjacent}}$$

See Fig. 6-7.

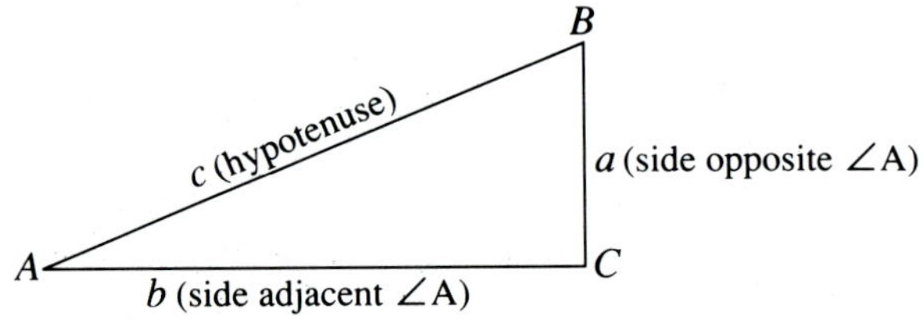

Figure 6-8

In like manner we can define the ratios for angle B; but note that b is the length of the side opposite and a becomes the length of the adjacent side, although the hypotenuse, c, remains the same. Abbreviating and referring only to the lengths of the sides, we have

$$\sin B = \frac{b}{c}, \qquad \cos B = \frac{a}{c}, \qquad \tan B = \frac{b}{a}$$

In order to use these efficiently it will be necessary for the reader to memorize the definitions and be able to recognize and read them from a right triangle in any position.

Example

In each of the triangles, write the values of the three trigonometric ratios for each acute angle.

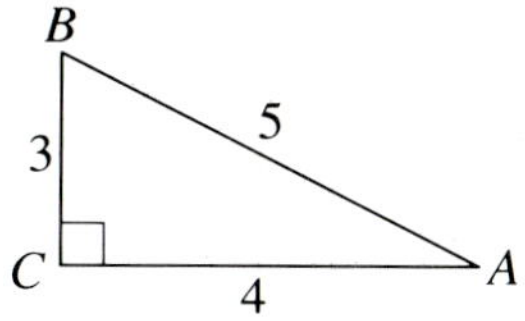

(a)

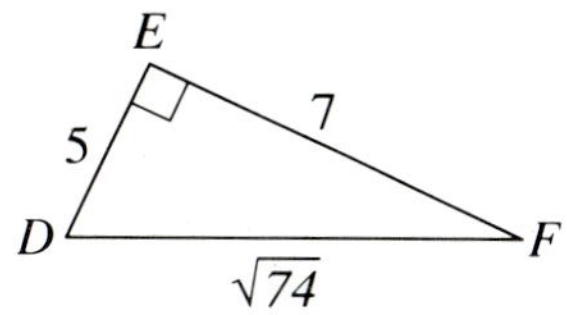

(b)

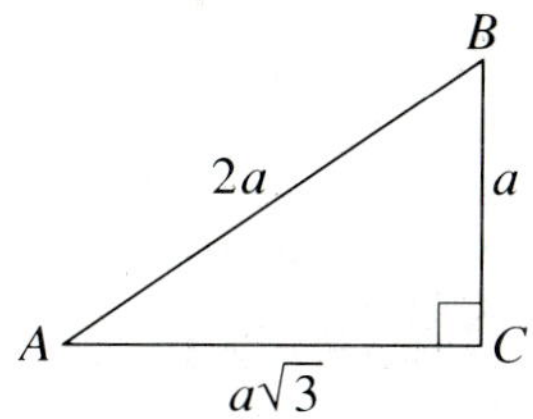

(c)

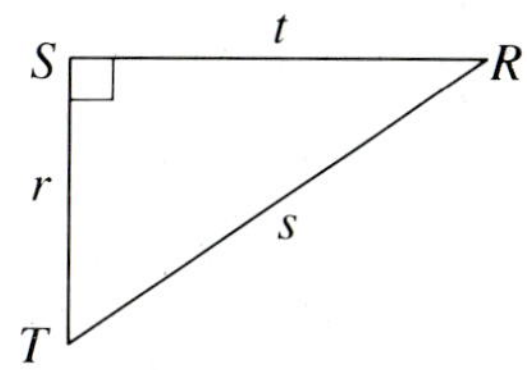

(d)

Solution Applying the definition, after identifying the hypotenuse, the side opposite and side adjacent, we have

(a) $\sin A = \frac{3}{5}$ $\qquad$ $\sin B = \frac{4}{5}$

$\cos A = \frac{4}{5}$ $\qquad$ $\cos B = \frac{3}{5}$

$\tan A = \frac{3}{4}$ $\qquad$ $\tan B = \frac{4}{3}$

(b) $\sin D = \dfrac{7}{\sqrt{74}} = \dfrac{7\sqrt{74}}{74}$ $\qquad$ $\sin F = \dfrac{5}{\sqrt{74}} = \dfrac{5\sqrt{74}}{74}$

$\cos D = \dfrac{5}{\sqrt{74}} = \dfrac{5\sqrt{74}}{74}$ $\qquad$ $\cos F = \dfrac{7}{\sqrt{74}} = \dfrac{7\sqrt{74}}{74}$

$\tan D = \dfrac{7}{5}$ $\qquad$ $\tan F = \dfrac{5}{7}$

(c) $\sin A = \dfrac{a}{2a} = \dfrac{1}{2}$ $\qquad$ $\sin B = \dfrac{a\sqrt{3}}{2a} = \dfrac{\sqrt{3}}{2}$

$\cos A = \dfrac{a\sqrt{3}}{2a} = \dfrac{\sqrt{3}}{2}$ $\qquad$ $\cos B = \dfrac{a}{2a} = \dfrac{1}{2}$

$\tan A = \dfrac{a}{a\sqrt{3}} = \dfrac{\sqrt{3}}{3}$ $\qquad$ $\tan B = \dfrac{a\sqrt{3}}{a} = \sqrt{3}$

(d) $\sin R = \dfrac{r}{s}$ $\qquad$ $\sin T = \dfrac{t}{s}$

$\cos R = \dfrac{t}{s}$ $\qquad$ $\cos T = \dfrac{r}{s}$

$\tan R = \dfrac{r}{t}$ $\qquad$ $\tan T = \dfrac{t}{r}$

Example

Find the trigonometric ratios for 30° and 45° angles.

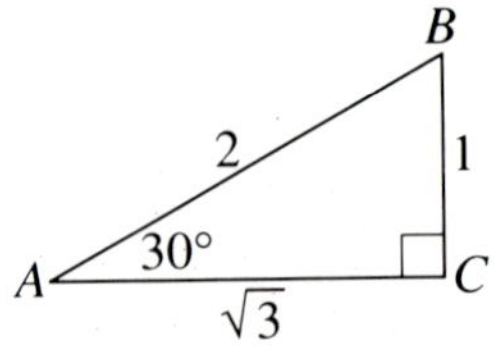

Solution If $c = 2$ in the 30°, 60° right triangle, we have $a = 1$ and $b = (\frac{1}{2} \cdot 2)\sqrt{3} = \sqrt{3}$ by Theorem 6-11. Since all 30°, 60° right triangles are similar, the ratios developed for this triangle will hold for all 30°, 60° triangles. Thus

$$\sin 30° = \frac{a}{c} = \frac{1}{2}, \qquad \cos 30° = \frac{b}{c} = \frac{\sqrt{3}}{2}$$

$$\tan 30° = \frac{a}{b} = \frac{1}{\sqrt{3}} \cdot \frac{\sqrt{3}}{\sqrt{3}} = \frac{\sqrt{3}}{3}$$

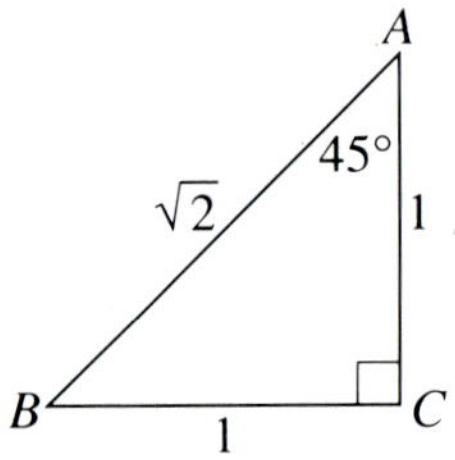

A 45°, 45° right triangle is isosceles, and if we let the legs have length 1, the hypotenuse has length $\sqrt{2}$ by Theorem 6-12. Then

$$\sin 45° = \frac{a}{c} = \frac{1}{\sqrt{2}} = \frac{\sqrt{2}}{2}, \qquad \cos 45° = \frac{b}{c} = \frac{1}{\sqrt{2}} = \frac{\sqrt{2}}{2}$$

and

$$\tan 45° = \frac{a}{b} = \frac{1}{1} = 1.$$

Since all 45°, 45° right triangles are similar, these ratios will be the same for all 45° angles.

Exercise Set 6.8

1. Find the sine, cosine and tangent of angles X and Y in the given triangles.

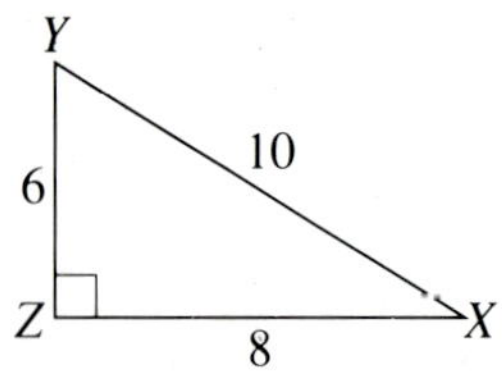

(a)

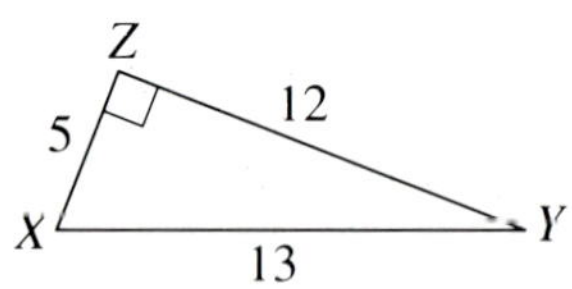

(b)

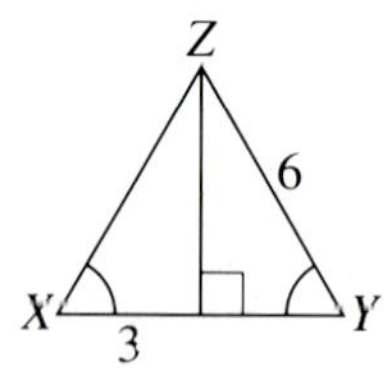

(c)

2. Using the following figure, find

$\sin A$	$\sin \angle BCD$
$\cos A$	$\cos \angle BCD$
$\tan A$	$\tan \angle BCD$

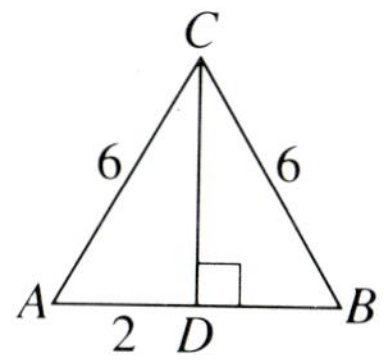

3. Using the following figure find:

$\sin 60°$

$\cos 60°$

$\tan 60°$

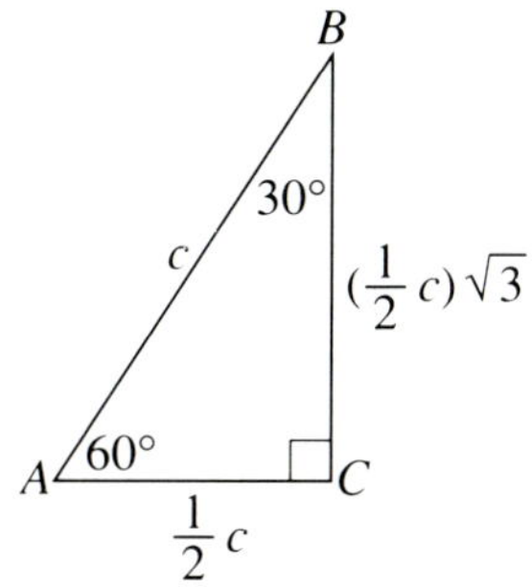

4. What ratio of angle A is indicated by each of the following?

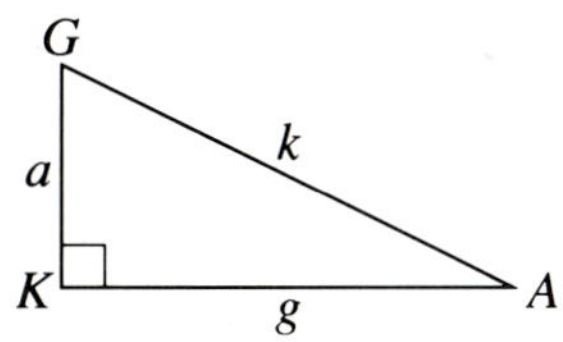

(a)

$\frac{a}{g} = ?$ $\frac{a}{k} = ?$

$\frac{g}{k} = ?$

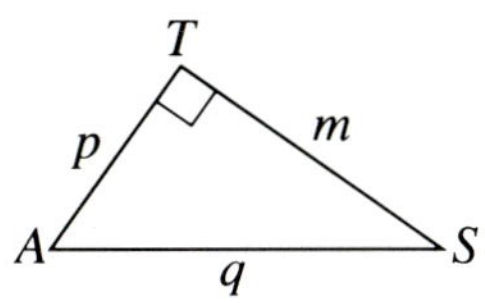

(b)

$\frac{m}{q} = ?$ $\frac{p}{q} = ?$

$\frac{m}{p} = ?$

Use of Tables of Trigonometric Ratios

Up to this point we have found the values of trigonometric ratios for the angles 30°, 45°, and 60° using the special relationships that exist between their sides. We could find *approximate* values of the ratios for other angles by measurement, but the results would tend to be too inaccurate for practical use. Formulas exist by means of which these ratios can be computed to any degree of accuracy desired, but we shall make use of tables or calculators. In Appendix IV you will find a table of the ratios for angles from 0° to 90° accurate to four decimal places. Part of such a table is shown here in Fig. 6-9.

Angle	sin	cos	tan	Angle	sin	cos	tan
16°	0.2756	0.9613	0.2867	**61°**	0.8746	0.4848	1.8040
17°	0.2924	0.9563	0.3057	**62°**	0.8829	0.4695	1.8807
18°	0.3090	0.9511	0.3249	**63°**	0.8910	0.4540	1.9626
19°	0.3256	0.9455	0.3443	**64°**	0.8988	0.4384	2.0503
20°	0.3420	0.9397	0.3640	**65°**	0.9063	0.4226	2.1445
21°	0.3584	0.9336	0.3839	**66°**	0.9135	0.4067	2.2460
22°	0.3746	0.9272	0.4040	**67°**	0.9205	0.3907	2.3559
23°	0.3907	0.9205	0.4245	**68°**	0.9272	0.3746	2.4751
24°	0.4067	0.9135	0.4452	**69°**	0.9336	0.3584	2.6051
25°	0.4226	0.9063	0.4663	**70°**	0.9397	0.3420	2.7475
26°	0.4384	0.8988	0.4877	**71°**	0.9455	0.3256	2.9042
27°	0.4540	0.8910	0.5095	**72°**	0.9511	0.3090	3.0777
28°	0.4695	0.8829	0.5317	**73°**	0.9563	0.2924	3.2709
29°	0.4848	0.8746	0.5543	**74°**	0.9613	0.2756	3.4874
30°	0.5000	0.8660	0.5774	**75°**	0.9659	0.2588	3.7321

FIGURE 6-9

Examples

1. Find the ratios indicated:

$$\sin 22^\circ \approx ?, \qquad \cos 27^\circ \approx ?, \qquad \tan 19^\circ \approx ?$$

Solution Referring to the row of the given angle and under the column headed by the appropriate ratio, we find sin 22° ≈ 0.3746, cos 27° ≈ 0.8910, and tan 19° ≈ 0.3443. (≈ means "approximately equal to")

2. Find the angle if the ratio is given:

$$\sin B \approx 0.9455, \qquad \cos R \approx 0.4540, \qquad \tan S \approx 0.5317$$

Solution Now we look under the column headed by sin, then down the column until we find 0.9455. This value is opposite 71° in the column headed angle, so $\angle B \approx 71°$. In like fashion we find $m\angle R \approx 63$ and $m\angle S \approx 28$. ◁

Use of Calculators for Finding Trigonometric Ratios

Students who have scientific calculators should be able to find the values of trigonometric ratios and solve problems quite simply with their calculators. A scientific calculator has keys with sine, cosine, and tangent, as well as y^x, $\sqrt{\ }$, and others. Your calculator must be in degree mode, which sometimes shows up as *deg* on the calculator display. There are other units used for measuring angles in trigonometry, the *radian* and the *grad,* which are indicated by *rad* or *grad* on the display. Be certain that your calculator is in the degree mode. One way to check is to enter 30 and then depress sin; the display should read 0.5. If it does not, then your calculator is not in degree mode. Consult your operator's manual for the method of changing modes. Finding sines, cosines, and tangents of angles is quite simple. Just enter the degree measure of the angle and then depress the trigonometric function desired. The value will immediately appear to as many decimal digits as your calculator can display. Your calculator will accept angles in decimal parts of a degree and probably with minutes and seconds as well, but for our purposes all angles will be rounded to the nearest degree.

Examples

1. To find sin 40° enter 40 and press [sin]. Your calculator will display 0.6427876. . . .

2. To find tan 40° enter 40 and press [tan]. Your calculator will display 0.8390996. . . .

3. To find cos 83° enter 83 and press [cos]. Your calculator will display 0.1218693. . . . ◁

Finding angle measures when given the trigonometric ratios is a bit more complicated. If tan $A = 0.7$, for example, you must enter 0.7 and then use a second function, or *inverse,* key to find the angle. The second function, or inverse, will usually be found over the trig function key and will be labeled "$\tan^{-1}$" or "arctan."

Examples

1. To find $m\angle A$ if tan $A = 0.7$, enter .7 and press [2nd] (or inverse), [tan]. Your calculator will display 34.9920. . . , so the angle measures 35°, correct to the nearest degree.

2. To find $m\angle B$ if sin $B = 0.4443$, enter .4443 and press [2nd], [sin]. Your calculator will display 26.37856. . . , so the angle measures 26°, correct to the nearest degree.

3. Given tan $B = \frac{2}{5}$ find

(a) ratios for sin B and cos B without using tables

(b) $m\angle B$ to the nearest degree

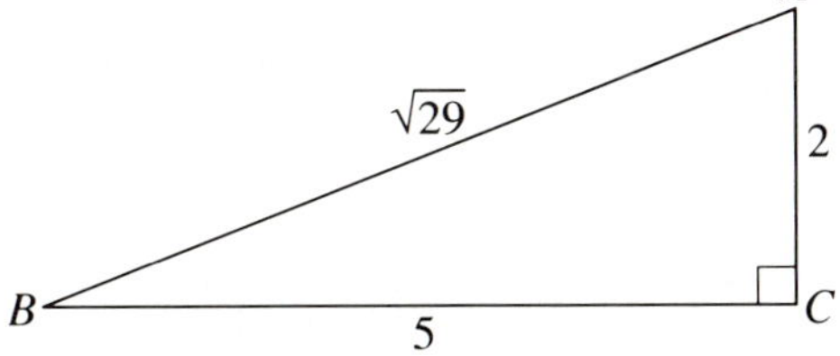

Solution Draw a right triangle and label the legs 2 and 5. Since $\tan B =$ (side opp.)/(side adj.) $= \frac{2}{5}$, the Pythagorean Theorem gives us $c^2 = 2^2 + 5^2 = 4 + 25 = 29$, or $c = \sqrt{29}$. Thus

(a) $\sin B = \dfrac{2}{\sqrt{29}} = \dfrac{2\sqrt{29}}{29}$ and $\cos B = \dfrac{5}{\sqrt{29}} = \dfrac{5\sqrt{29}}{29}$

(b) $\tan B = \frac{2}{5} = 0.400$, so $m\angle B \approx 22$

We choose the degree measure for which the ratio is closest. In Fig. 6-8 we find that $\tan 21° \approx 0.3839$ and $\tan 22° \approx 0.4040$. Since $0.4000 - 0.3839 = 0.0161$, whereas $0.4040 - 0.4000 = 0.0040$, $m\angle B$ must be closer to 22 than to 21. (When using the calculator, we obtain 21.8014. . .°, so we simply round to 22°.)

4. The base angles of an isosceles triangle are 28° and the length of the base is 10. Find the length of the altitude and the lengths of the congruent sides (to the nearest hundredth).

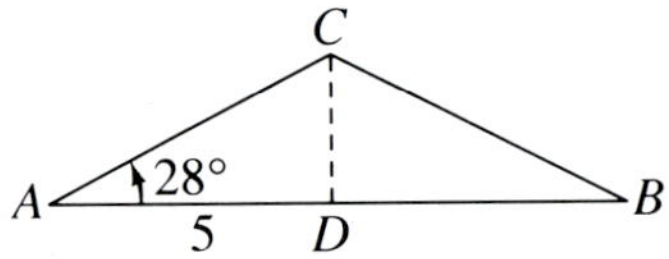

Solution The altitude to the base of an isosceles triangle bisects the base; hence

$$\cos 28° = \frac{5}{m\overline{AC}}; \qquad 0.8829 \approx \frac{5}{m\overline{AC}}$$

If

$$m\overline{AC}(0.8829) \approx 5, \qquad m\overline{AC} \approx \frac{5}{0.8829} \approx 5.66$$

$$\frac{m\overline{CD}}{5} = \tan 28° \approx 0.5317 \quad \text{so} \quad m\overline{CD} \approx 5(.5317) \approx 2.66$$

Exercise Set 6.9

1. Using a calculator or Table I in Appendix IV, find the values of the three trigonometric ratios for
   (a) 20° **(b)** 37° **(c)** 64°
2. Referring to Table I in Appendix IV, we see that as $m\angle A$ increases from 0 to 90, $\sin A$ increases from 0 to 1. Describe what happens to $\cos A$ and $\tan A$ as $m\angle A$ increases from 0 to 90.
3. Find the number of degrees in angle A for each of the following:
   (a) $\sin A = 0.6293$ **(b)** $\cos A = 0.3090$
   (c) $\tan A = 0.5095$ **(d)** $\cos A = 0.8572$
   (e) $\sin A = 0.9903$
4. Find (to the nearest degree) the measure of the angles of a triangle with sides of length
   (a) 3, 4, and 5 **(b)** 5, 12, and 13 ***(c)** 14, 14, and 20

5. One angle of a right triangle is 22° and the hypotenuse measures 12 units. Find the measure of the third angle (to the nearest degree) and the lengths of the legs (to the nearest tenth).

*6. The sides of a triangle measure 5, 8, and 10. Find
(a) the length of the altitude to the longest side (to the nearest tenth)
(b) the measure of the smallest angle (to the nearest degree)

Solution of Right Triangles

To "solve a triangle" is to find the angles and lengths of sides not already known. In solving a right triangle, it is suggested that the following steps be followed.

1. *Draw the triangle* as accurately as possible. The diagram will give you an idea of the relation between the sides and the angles.
2. *List the known parts* and their values; then *list the parts to be found.*
3. To find an unknown part, write a *trigonometric equation* that contains the unknown part and two given parts.
4. To find a second unknown part, write a trigonometric equation (try *not* to use a part previously computed) that involves the part to be found and two known parts.

Examples

1. In a right triangle ABC, $a = 5$, $m\angle A = 65$. Solve the triangle.
(a) Draw the figure.

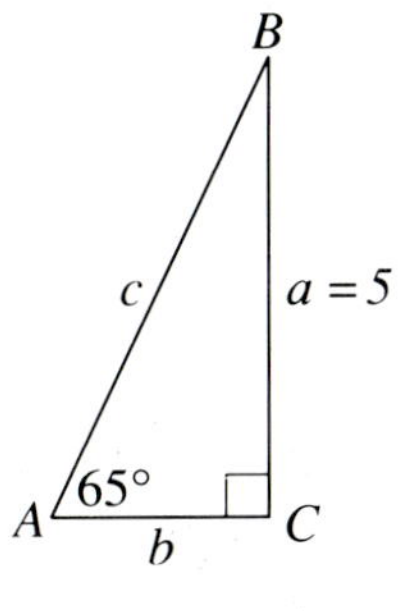

(b) $a = 5$, $m\angle A = 65$. Find c, b, $m\angle B$.
(c) To find $m\angle B$: $m\angle A + m\angle B = 90$; $m\angle B = 90 - 65 = 25$.
(d) To find c: $\sin 65° = a/c = 5/c$

$$\therefore c = \frac{5}{\sin 65°} \approx \frac{5}{0.9063} \approx 5.52$$

(e) To find b: $\tan A = a/b$, $b = a/\tan A = 5/\tan 65° \approx 5/2.1445$

$$\therefore b \approx 2.33$$

2. In a right $\triangle ABC$, $c = 8.55$ and $a = 5.6$. Find $m\angle A$, $m\angle B$, and b.
(a) Draw the figure and label the given parts.

* The more difficult problems we indicated by an asterisk.

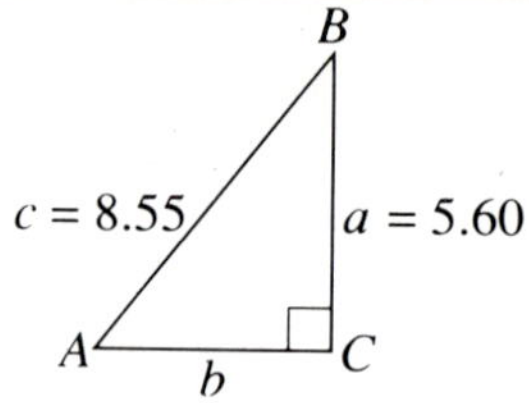

(b) To find m$\angle A$:

$$\sin A = \frac{a}{c} = \frac{5.60}{8.55}, \qquad \sin A \approx 0.6549$$

$$\therefore \angle A \approx 41°$$

(c) To solve for $\angle B$:

$$m\angle B = 90 - m\angle A = 90 - 41 = 49$$

(d) To solve for b:

$$\cos A = \frac{b}{c}; \qquad b = c \cos A$$

$$\therefore b \approx 8.55(0.7547) \approx 6.45$$

3. In right $\triangle ABC$, if $\angle A = 35°$, and $a = 10$, find b.

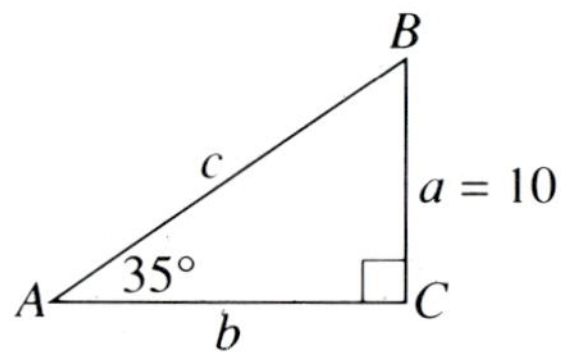

(a) $\tan 35° = \frac{a}{b}$; $b = \frac{a}{\tan 35°} \approx \frac{10}{0.7002} \approx 14.28.$

(b) m$\angle A = 35$, m$\angle B = 55$ and $\tan B = b/a$, $b = a\,(\tan B)$. Thus $b \approx 10(1.4281) \approx 14.28$. Note that computation is simpler in the second solution. ◁

In Fig. 6-10 we picture a situation that is sufficiently common for us to attach special names to the angles involved. The angle from the lower horizontal upward to the object on top is called the *angle of elevation,* that is, x is the angle of elevation. The angle from the upper horizontal looking downward is called the *angle of depression,* that is, y in the figure is the angle of depression.

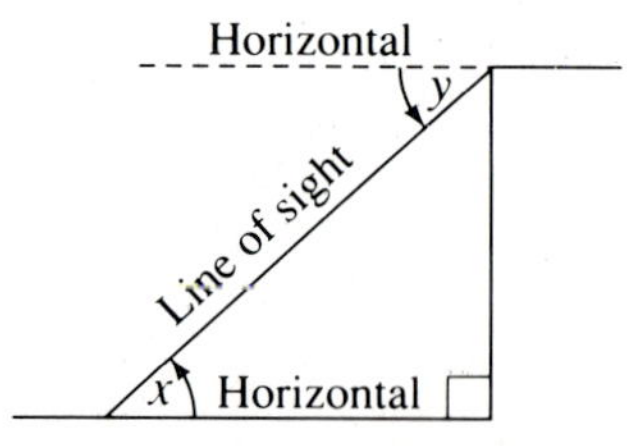

Figure 6-10

Example

1. You are standing 110 ft from the foot of a tower and the angle of elevation to the top is 72°. What is the height of the tower (to the nearest foot)?

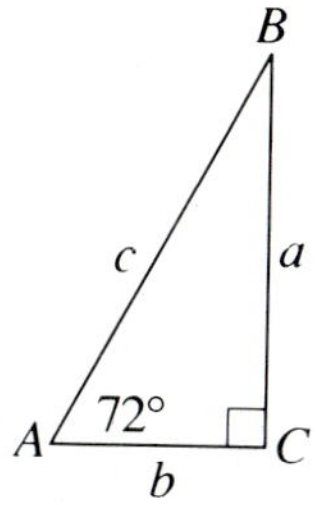

Solution $\angle A = 72°$, $b = 110$ ft, $\tan A = a/b$; $\therefore a = b(\tan A)$ and $a \approx 110(3.0777) \approx 339$ ft.

2. The operator of a drawbridge 80 ft above the water sees a boat approaching. If the angle of depression of the boat is 4°, how far is the boat from the bridge (to the nearest foot)?

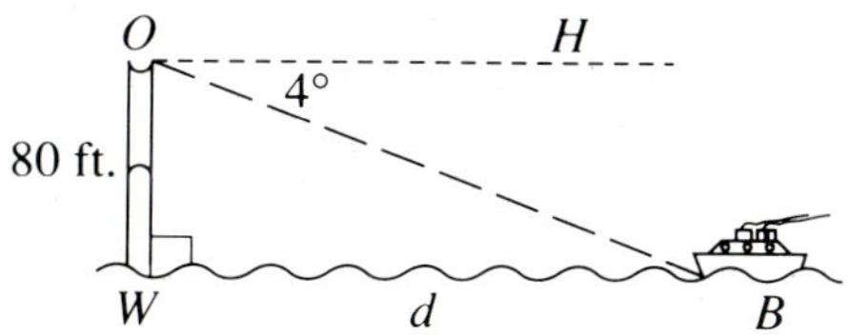

Solution If $m\angle BOH$ is 4, then $m\angle OBW = 4$, or $m\angle WOB = 86$, since the angles are complementary. Then $\tan 86° = d/80$, so $d = 80 \tan 86° \approx 80(14.3007) \approx 1144$ ft. ◁

Exercise Set 6.10

In Problems 1–5, refer to the following figure. Find measures of angles to the nearest degree, sides to the nearest tenth.

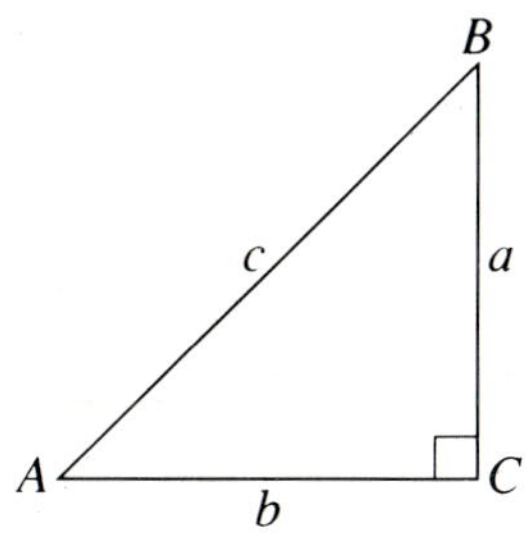

1. If $m\angle A = 45$ and $a = 5$, find b and c.
2. If $m\angle B = 30$ and $a = 3$, find b and c.
3. If $\angle B = 47°$ and $b = 8$, find a.

4. If $a = 5$, $b = 7$, find m$\angle A$ and side c.
5. If $a = 8.4$ and $c = 12.7$, find m$\angle A$, side b.
6. If when standing on the top of a building 120 ft above the ground, you sight an automobile when you look downward with an angle of depression of 15°, how far away is the automobile from the foot of the building (to the nearest foot).
7. Find the measures of the base angles of an isosceles triangle with base of length 9 if one of the congruent sides measures 14 (to the nearest degree).
8. If the slope of a canyon floor is 7°, how far back up the canyon will a dam of height 125 ft force the water (to the nearest ten feet)?
9. Refer to Exercise 7 on page 168. Find the measure of angle RST to the nearest degree.
10. An observer at point O wants to know the height of the clouds at night, so she measures the angle of elevation to the point of reflection R from a searchlight at S, which is 1500 ft from O. If m$\angle ROS = 33$, what is the height of the clouds (to the nearest ten feet)?

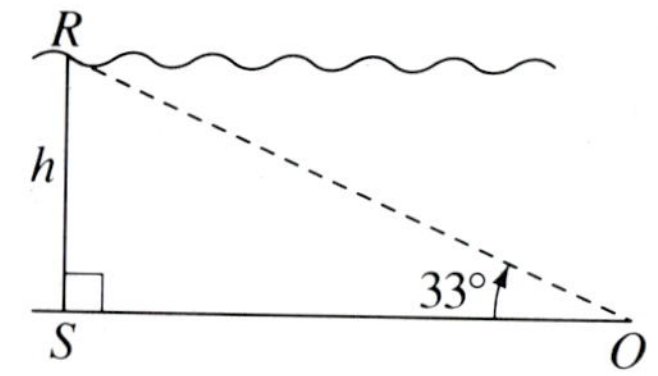

11. **(a)** Find m$\angle AOB$ and m$\overline{OA}$ in the following figure.
    (b) If m$\angle COB = 140$ and m$\overline{OC} = 9.5$, find the coordinates of point C (to the nearest tenth).

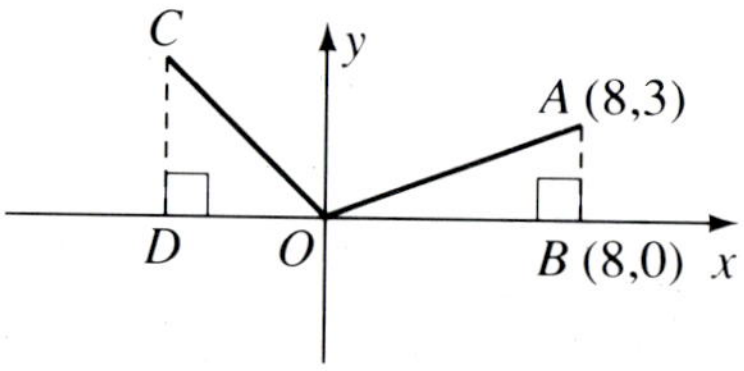

12. A helicopter leaves a ship and flies on a course of 125°, while the ship continues on a course of 90°. After flying 25 mi to point A, what is the distance from A to point B, which is directly ahead of the ship?

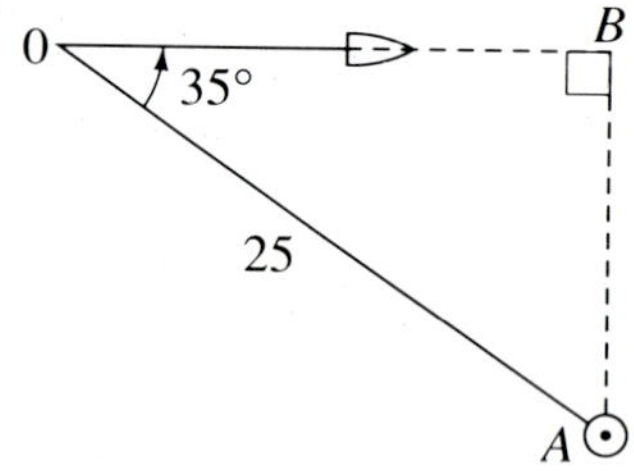

13. A rescue squad must dig a tunnel in order to rescue a child trapped in an abandoned well. If the depth of the well is 20 ft and the tunnel is started 25 ft from the top of the well, at what angle from the surface must the tunnel be dug?

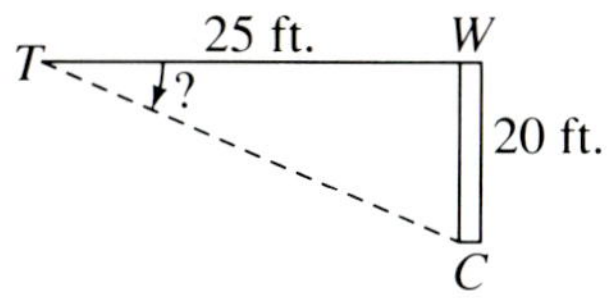

14. During a certain portion of liftoff, astronauts are unable to move their heads up or down, but can move their eyes from straight ahead through an angle of depression ($\angle HEL$) of 42°. If the distance to the control panel is 14.0 in., over what vertical distance are they able to see?

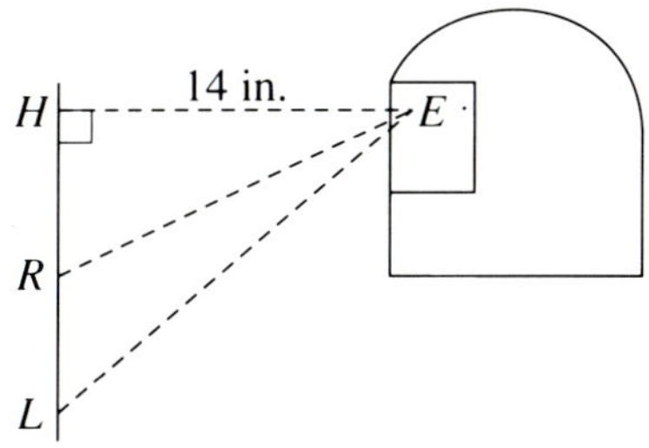

15. Refer to Exercise 14. If the maximum effective range of sight (distance $\overline{ER}$) at this time is determined to be 16.0 in., find $m\angle HER$, then $m\overline{HR}$.

16. An inductive circuit, an electronic circuit containing resistance, an inductor, and impedance to current flow, can be analyzed using a right triangle diagram, as shown, where R is the resistance of the circuit in ohms, X_L the inductive reactance in ohms, Z the impedance in ohms and θ is the phase angle.

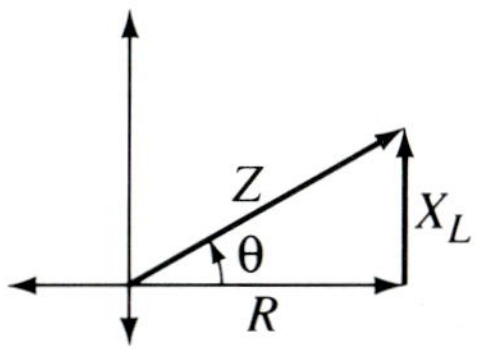

(a) If, in the given circuit, $R = 150$ ohms (Ω) and $X_L = 80\ \Omega$, find Z in ohms and θ in degrees.

(b) If $\theta = 42°$ and $Z = 50\ \Omega$, find R and X_L.

Note: In all exercises sets throughout the text, a problem requiring the use of trigonometry will be denoted by ▶ preceding the problem number.

REVIEW EXERCISES

Solve for N in Exercises 1–6.

1. $\dfrac{91}{65} = \dfrac{133}{N}$

2. $\dfrac{39}{N} = \dfrac{N}{351}$

3. $\frac{54}{3} = \frac{3}{N}$

4. $\frac{N + 1}{6} = \frac{20}{8}$

5. $\frac{N - 2}{3} = \frac{N + 1}{5}$

6. $\frac{N + 3}{N} = \frac{N - 1}{N - 3}$

7. If the lengths of the sides of a triangle are 6, 7, and 8, respectively, find the lengths of the corresponding sides of a similar triangle if its shortest side measures 40.

8. If the altitude of an equilateral triangle measures h, find the lengths of the sides.

9. How long a ladder will be required to reach a window 24 ft high if the lower end of the ladder must be 10 ft from the wall?

10. The perimeters of two similar triangles are 200 ft and 300 ft, respectively. If a side of the first measures 80 ft, find the length of the corresponding side of the second triangle.

11. A tree casts a shadow 90 ft long when a 6-ft post casts a shadow of 4 ft. How tall is the tree?

▶ 12. A tree casts a shadow 18 ft long. How tall is the tree if the angle made by the shadow and the top of the tree is 50°?

13. If the legs of a right triangle measure 8 and 12, find the lengths of the projections of the legs upon the hypotenuse, and the length of the altitude to the hypotenuse.

14. If the sides of a triangle measure 6, 12, and $6\sqrt{3}$, find (a) the length of the altitude to the longest side, (b) the length of the median to the side of length 12, and (c) the angle formed by the altitude and the median of parts (a) and (b).

15. Prove that the bisector of an angle of a triangle divides the opposite side into segments which are proportional to the adjacent sides. (*Hint:* Through A, draw $\overrightarrow{AF} \parallel \overline{CD}$ and extend $\overrightarrow{BD}$ intersecting $\overrightarrow{AF}$ at E.)

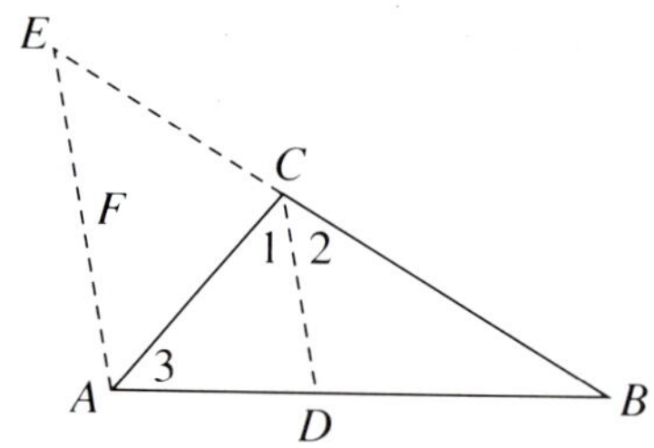

16. If the base and altitude of a triangle measure 7 ft and 5 ft, respectively, and the length of the altitude of a similar triangle is 8 ft 6 in., find the length of the corresponding base.

17. In triangle RST, m$\overline{RT}$ = 15, $\overline{UV} \parallel \overline{ST}$, m$\overline{RU}$ = 3, m$\overline{US}$ = 8, and m$\overline{ST}$ = 18. Find the lengths of $\overline{RV}$ and $\overline{UV}$.

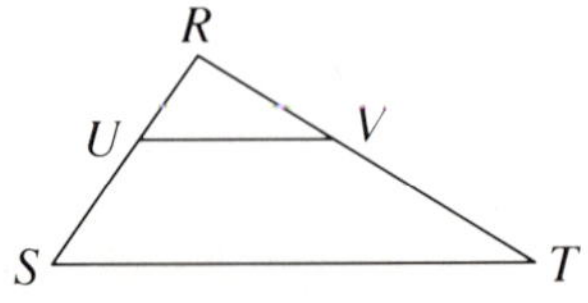

▶ **18.** Find the lengths of sides $\overline{XZ}$ and $\overline{YZ}$ in the following triangle (to the nearest tenth).

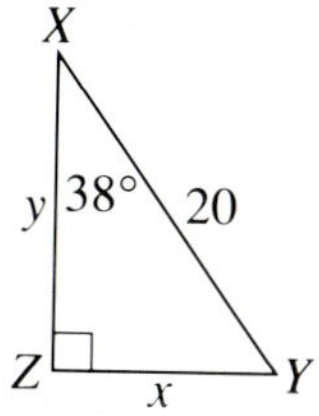

▶ **19.** Find the measure of the smallest angle of a triangle with sides which measure 2, 4, and $2\sqrt{5}$.

20. Which angle of triangle PQR is a right angle if $P(-5, -1)$, $Q(4, 2)$ and $R(2, 8)$? Prove your result.

Cumulative Review—Chapters 2–6

Supply the word or phrase that will make the following correct.

1. The measure of an *acute angle* is a, such that ________.

2. Two angles are said to be *adjacent* if they have a common ________, the same ________ and the intersection of their interiors is ________.

3. Two angles are said to be ________ if the sum of their measures is 180.

4. If $m\angle A = x$ and $m\angle B = x$, then $\angle A$ and $\angle B$ are said to be ________.

5. If $m\angle R = x$, $m\angle S = y$ and $x + y = 90$, then $\angle R$ and $\angle S$ are said to be ________.

6. If $\angle A \cong \angle B$ and $\angle B \cong \angle C$, then $\angle$________ $\cong \angle$________ by the ________ property for congruent angles.

7. The angles formed by two intersecting lines are either ________ or ________.

8. In the statement "If P, then Q," P is called the ________ and Q the ________.

9. A proof is made up of ________ and ________, the ________, and a sequence of ________ with supporting ________.

10. A triangle whose angles are congruent is said to be ________.

11. A triangle with no sides congruent is called a(an) ________.

12. A triangle having an obtuse angle is a(an) ________.

13. ________ are lines that form congruent adjacent angles.

14. Lines that are drawn and used but not given in the hypothesis are called ________.

15. Lines in a plane whose intersection is the empty set are said to be ________.

16. Each exterior angle of an equilateral triangle has measure ________.

17. If $a/b = c/d$, then $c/a =$ ________.

18. If $a/b = c/d$, then $a/c =$ ________.

19. If $xy = zw$, then $x/w =$ ________.

20. If $3a = 4b$, then $a/b =$ ________.

21. If on a road map the scale reads 1 in. = 10 mi, 27 mi corresponds to ________ in.

▶ **22.** The sine of ________° is $\sqrt{3}/2$.

▶ **23.** Tan 60° = ________.

▶ **24.** Cos 45° = ________.

25. A triangle whose sides are proportional to 3, 4, and 5 is a ________ triangle.

Are the following statements true or false?

1. If two triangles are congruent, the corresponding sides are congruent.

2. A triangle may be both scalene and obtuse.

3. A right triangle may be equilateral.

4. A triangle may be obtuse and isosceles.

5. A median of a triangle is a segment from a vertex perpendicular to a side.

6. An acute triangle may be isosceles.

7. If the corresponding angles of two triangles are congruent, the triangles are congruent.

8. If on $\overline{AB}$ A-P-B such that $\overline{AP} \cong \overline{PB}$ and there is a point Q, not on $\overline{AB}$, such that $\overline{AQ} \cong \overline{BQ}$, then $\overline{AB} \perp \overline{PQ}$.

Problems

1. If A, B, and C are collinear, m$\angle DBF = 32$, m$\angle FBC = 58$ and $\angle ABD \cong \angle EBF$, find m$\angle ABF$ and m$\angle DBC$.

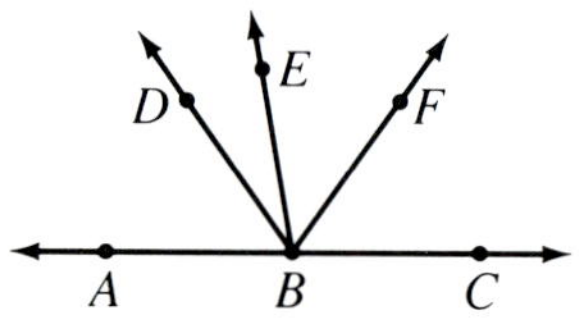

2. In the figure for Problem 1, the point E is in the interior of what angles?

3. In the figure for Problem 1, the point E is in the exterior of what angles?

4. In the figure for Problem 1, list the angles that contain the point F in their interior.

5. Find the measure of the complement of each of the following angles:

(a) 24° **(b)** 60°
(c) 17.4° **(d)** x°
(e) $(90 - x)$° **(f)** $(40 - x)$°

6. If the measure of an angle is 25 more than that of its supplement, what is the measure of each angle?

7. The measure of an angle when increased by 10 is three times the measure of its complement. Find the measure of each angle.

8. In the following figure, if m$\angle AFE$ is 115 and m$\angle EGF$ is 50, find m$\angle AFB$, m$\angle EFG$, m$\angle FGC$, m$\angle CGD$, m$\angle BFG$.

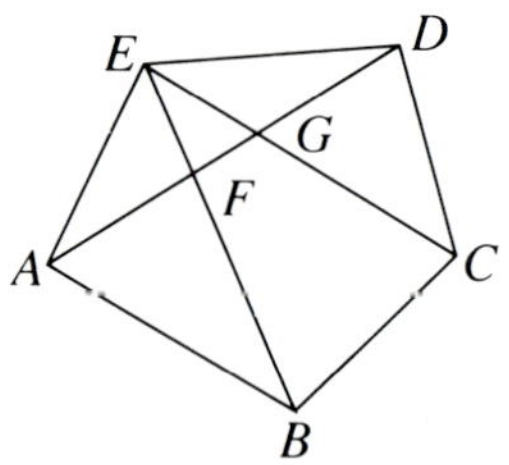

9. Using the figure from Problem 8, list
  (a) the adjacent angles at vertex A; vertex E
  (b) the vertical angles at F
  (c) the supplementary angles at G

10. Identify the following triangles.

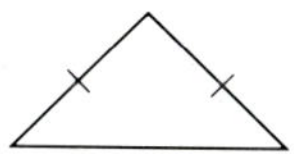
(a)

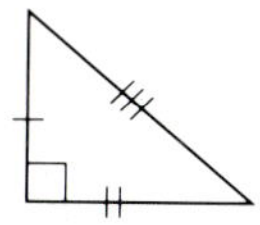
(b)

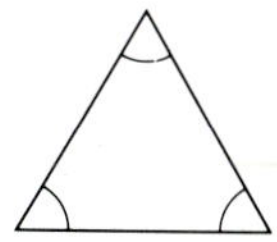
(c)

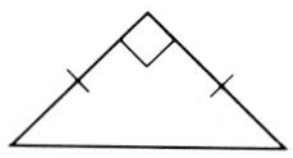
(d)

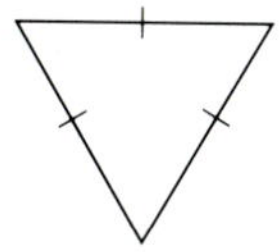
(e)

11. Draw a triangle for each set of following conditions:
  (a) No sides congruent and one right angle
  (b) No sides congruent and one obtuse angle
  (c) No sides congruent and all acute angles
  (d) Two sides congruent and one right angle
  (e) Two sides congruent and one obtuse angle
  (f) Two sides congruent and one acute angle
  (g) Three sides congruent

12. If the measure of one angle of a triangle is 90 and the measure of a second angle is x, find the measure of the third angle.

13. Locate the following sets of points and draw the segments connecting them. Determine the lengths of the segments, then classify the triangles.
  (a) $A(2, 1)$, $B(3, 4)$, $C(-1, 2)$
  (b) $A(0, -1)$, $B(-3, -5)$, $C(0, -3)$
  (c) $A(4, -4)$, $B(-4, 0)$, $C(0, 4)$

14. The following triangle is isosceles ($\overline{AB} \cong \overline{BD}$). If $\angle DBC$ is 140°, find the measure of $\angle A$.

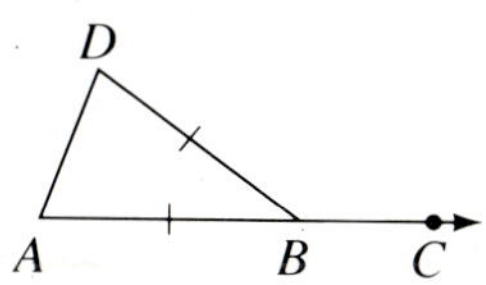

15. If two distinct isosceles triangles have the same base, prove that the line connecting their vertices bisects the base.

16. A triangle has sides of length 14, 20, and 24. Find the perimeter of the triangle formed by connecting the midpoints of the sides.

17. Prove if $\overline{AC} \cong \overline{BC}$, $\overline{DE} \parallel \overline{AB}$, $\overline{DE} \cong \overline{DF}$, then $m\overline{AB}/m\overline{EF} = m\overline{AC}/m\overline{DE}$.

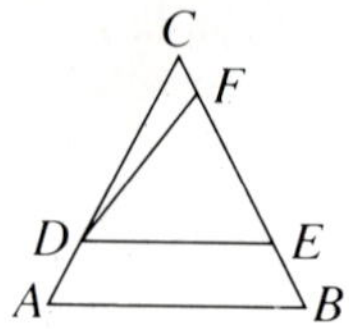

18. Prove Theorem 6-3 analytically.

19. $\triangle ABC$ is a right triangle and $\overline{DC}$ an altitude to the hypotenuse.

(a) If $a = 20$, $b = 15$ and $h = 12$, find $m\overline{AB}$, m, and n.

(b) If $m = 4$ and $n = 9$, find h.

(c) If $b = 6\sqrt{2}$ and $m = 6$, find $m\overline{AB}$.

(d) If $a = 20$ and $m = 9$, find n.

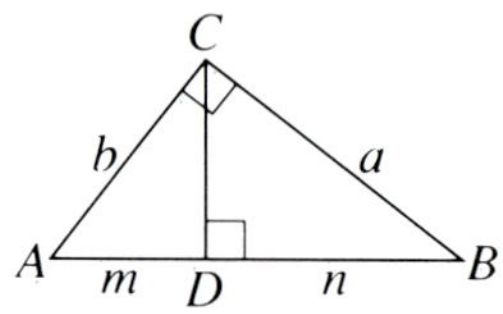

▶ **20.** If $\cos(30° + 60°) = \cos 90°$, does $\cos 30° + \cos 60° = \cos 90°$? Prove your answer.

▶ **21.** Prove that $\tan A = (\sin A)/(\cos A)$.

▶ **22.** **(a)** Find the exact value of $(\sin 30°)^2 + (\cos 30°)^2$. (*Hint:* See the example on page 178.)

(b) Find the exact value of $(\sin 45°)^2 + (\cos 45°)^2$.

(c) Use a calculator or refer to the tables to find the approximate value of $(\sin 27°)^2 + (\cos 27°)^2$.

(d) Given right triangle ABC with C the vertex of the right angle. Prove that $(\sin A)^2 + (\cos A)^2 = 1$. (*Hint:* Recall that $a^2 + b^2 = c^2$, $\sin A = a/c$, and $\cos A = b/c$.)

7

Polygons and Polygonal Regions

Land surveying, as well as architecture, the construction of machinery no less than mathematical physics, are continually calculating relations of space of the most varied kind by geometrical principles; they expect that the success of their constructions and experiments shall agree with these calculations; and no case is known in which this expectation has been falsified, provided the calculations were made correctly and with sufficient data.

HERMAN VON HELMHOLZ (1821–1894)

7-1 CLASSIFICATION OF POLYGONS

In Chapter 4 we found that three noncollinear points determine a triangle. Now let us consider the figures that might be obtained by connecting four coplanar points, no three of which are collinear, with segments. If we join them so that a point is connected with only two other points, we see that there are three different figures that may be formed (Fig. 7-1).

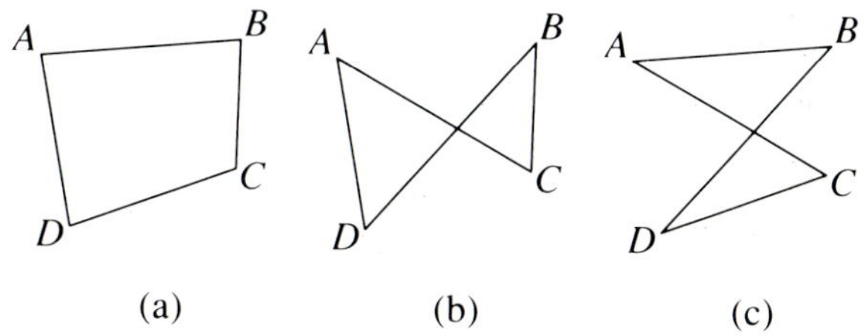

Figure 7-1

If there are 5 coplanar points joined under the same conditions, the figures formed may look like those in Fig. 7-2. Thus we find that many different forms appear when no restrictions are placed on the *order* in which the points are joined. In Fig. 7-1 we see that (a) appears to be a single surface, whereas (b) and (c) appear to be made up of triangles. Again in Fig. 7-2, (a)–(c) are distinct, whereas (d), (e), and (f) are composites. We will classify the distinct figures as *polygons*.

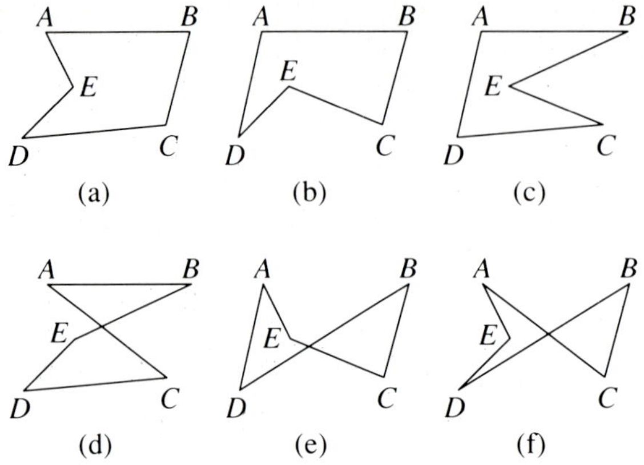

Figure 7-2

Definition 7-1 If $A_1, A_2, A_3, \ldots, A_n$ $(n \geq 3)$ is a set of distinct points in a plane and the segments $\overline{A_1A_2}, \overline{A_2A_3}, \overline{A_3A_4}, \ldots, \overline{A_{n-1}A_n}, \overline{A_nA_1}$ have the following properties:

(a) no two segments intersect except at their endpoints, and

(b) no two segments with a common endpoint are collinear, then the union of these n sets is a **polygon.** Each point $A_1, A_2, A_3, \ldots, A_n$ is a **vertex** of the polygon and each segment $\overline{A_1A_2}$, etc., is a **side** of the polygon.

Definition 7-2 Polygons having the number of sides indicated are named the following:

4 sides—**quadrilateral**
5 sides—**pentagon**
6 sides—**hexagon**
7 sides—**heptagon**
8 sides—**octagon**
9 sides—**nonagon**
10 sides—**decagon**
12 sides—**dodecagon**
n sides—***n*-gon**

A polygon is classified as *convex* if its interior is a convex set. A polygon whose interior is not a convex set is a *concave* polygon. The first three polygons in Fig. 7-3 are convex, and the hexagon in that figure is concave.

In the future, when we refer to a polygon we will mean convex polygon unless otherwise stated at the time.

Definition 7-3 In a polygon two sides are said to be **consecutive sides** if they have a common vertex, and two angles are **consecutive angles** if they have a common side. A **diagonal** is a segment joining two nonconsecutive vertices of a polygon.

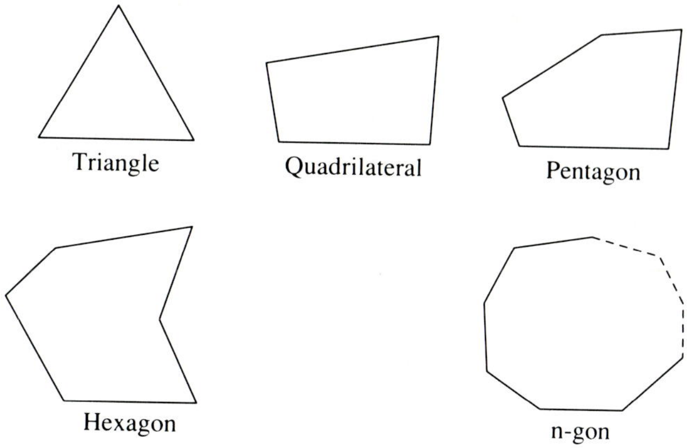

Figure 7-3

When we wish to identify a polygon we use the letters of the vertices of the angles listed in clockwise or counterclockwise order, so that *ABCDE* or *CDEAB* are various ways of identifying the polygon shown in Fig. 7-4.

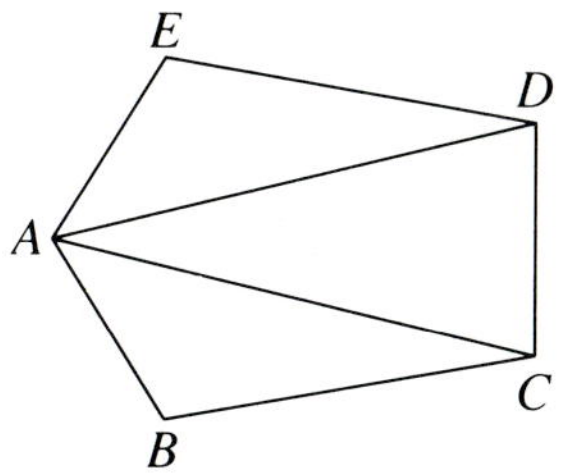

Figure 7-4

We also observe in Fig. 7-4 that $\overline{AB}$ and $\overline{BC}$ are consecutive sides, $\angle A$ and $\angle B$ are consecutive angles, and $\overline{AC}$ and $\overline{AD}$ are diagonals.

The terms equilateral and equiangular also apply to polygons, as they did to triangles; i.e., an *equilateral* polygon is a polygon with all its sides congruent and an *equiangular* polygon has all of its angles congruent.

Example

(a) Name the vertices and sides of the following figures.

(b) In figure (a) how many diagonals can be drawn? Name them. Name and draw all diagonals in figure (b).

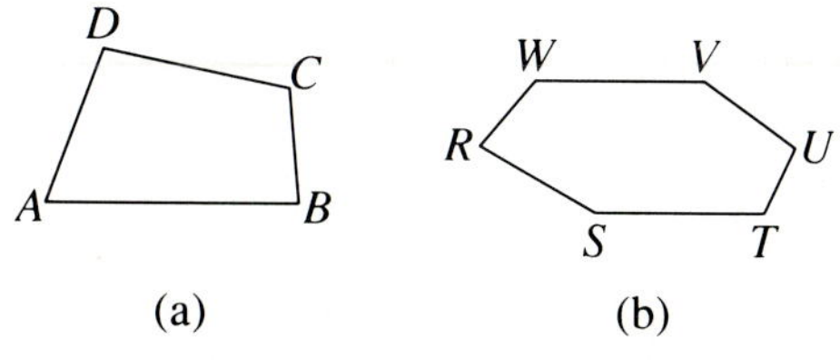

Solution

(a) In figure (a) the vertices are A, B, C, D; the sides are $\overline{AB}$, $\overline{BC}$, $\overline{CD}$, $\overline{DA}$. In figure (b) the vertices are R, S, T, U, V, W, and the sides are $\overline{RS}$, $\overline{ST}$, $\overline{TU}$, $\overline{UV}$, $\overline{VW}$, and $\overline{WR}$.

(b) In figure (a) diagonal $\overline{AC}$ may be drawn from A and diagonal $\overline{BD}$ from B. No new diagonals may be drawn from C and D. Thus there are two. In figure (b) we may draw $\overline{RV}$, $\overline{RU}$, $\overline{RT}$, $\overline{SW}$, $\overline{SV}$, $\overline{SU}$, $\overline{TW}$, $\overline{TV}$, $\overline{UW}$. ◁

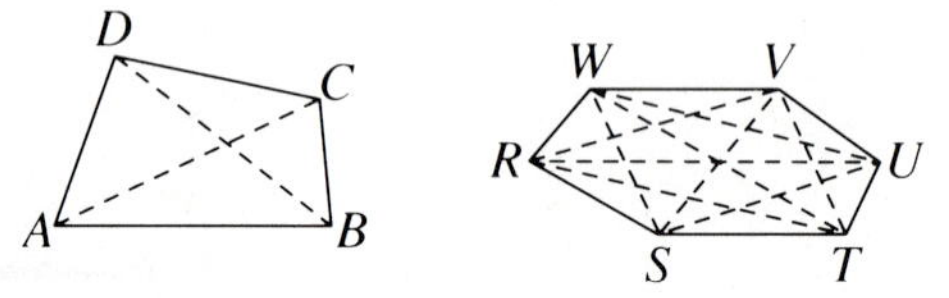

(a) (b)

Exercise Set 7.1

1. Is the following polygon convex or concave?
   (a) How many vertices does it have?
   (b) How many sides?
   (c) Name the sides of the polygon.
   (d) Name three pairs of consecutive sides.
   (e) Name two pairs of nonconsecutive sides.
   (f) Draw and name the diagonals from the point E.

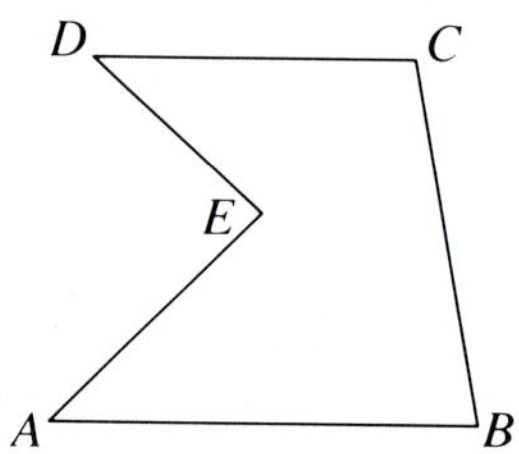

2. **(a)** Is the following figure a quadrilateral? Why?
   (b) Is the figure a convex set? Why?
   (c) Draw and list *all* the diagonals. How many did you find?

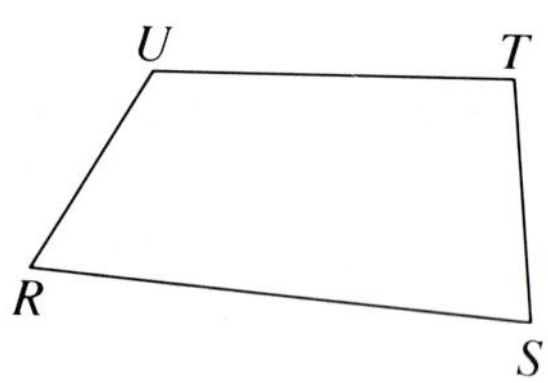

3. How many diagonals may a polygon have if the number of sides is
   (a) 3 **(b)** 4
   (c) 5 **(d)** 6
   (e) 9 **(f)** n
   (g) 67?

From early Greek times to the present day it has been considered by many that the most beautiful dimensions for a rectangle are those involving the Golden Section, or divine proportion, exemplified by the Parthenon. The length l and the width w of such a rectangle satisfy the proportion. $\dfrac{l + w}{l} = \dfrac{l}{w}$.

4. Draw a convex quadrilateral $ABCD$ and all possible diagonals from vertex A. How many triangles are formed? What is the sum of the measures of the angles of a triangle? What is the sum of the measures of the angles of a quadrilateral?
5. **(a)** Repeat Exercise 4 for a convex pentagon, a convex hexagon, and a convex octagon.
   (b) Write a formula relating the number of sides (n) of a convex polygon and the number of triangles (t) formed when all possible diagonals are drawn from one vertex.

7-2 QUADRILATERALS

The quadrilateral in its varied forms and sizes is seen throughout the world in nature, particularly where people have built structures for living, working or playing. Some of the important properties of quadrilaterals are given in the following pages.

Definition 7-4 The **opposite sides** of a quadrilateral are two sides that do not intersect. The **opposite angles** of a quadrilateral are two nonconsecutive angles.

In Fig. 7-5 the quadrilateral $ABCD$ has consecutive sides $\overline{AB}$ and $\overline{BC}$, $\overline{BC}$ and $\overline{CD}$, $\overline{CD}$ and $\overline{DA}$, and $\overline{DA}$ and $\overline{AB}$. Its opposite sides are $\overline{AD}$ and $\overline{BC}$, $\overline{AB}$ and $\overline{CD}$, while A and C, B and D are opposite angles.

Definition 7-5 A **trapezoid** is a quadrilateral with one and only one set of opposite sides parallel. A trapezoid with nonparallel sides congruent is an **isosceles trapezoid.**

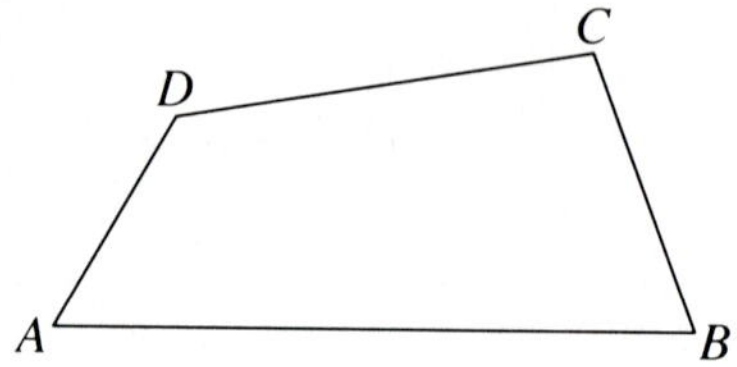

Figure 7-5

The parallel sides are said to be the **bases.** The **base angles** are a pair of angles which have the endpoints of a base as their vertices (Fig. 7-6).

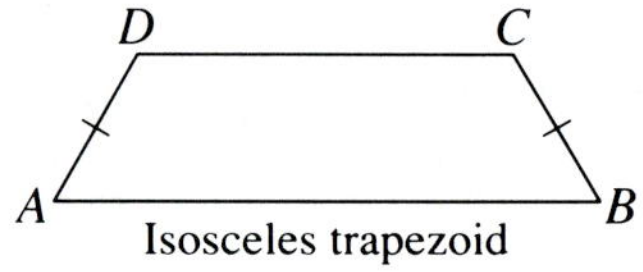

Figure 7-6

Definition 7-6 A **parallelogram** is a quadrilateral with both sets of opposite sides parallel (Fig. 7-7).

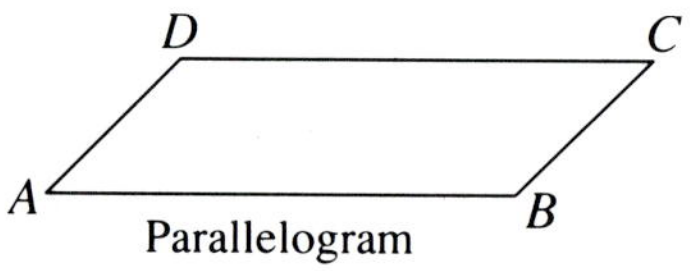

Figure 7-7

Definition 7-7 A **rhombus** is a parallelogram all of whose sides are congruent (Fig. 7-8).

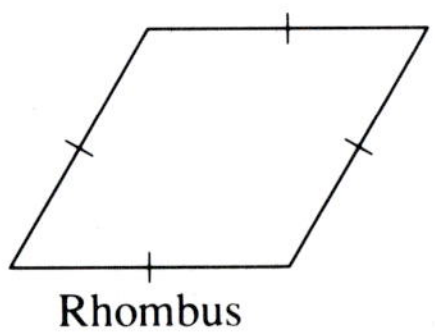

Figure 7-8

Definition 7-8 A **rectangle** is a parallelogram all of whose angles are right angles (Fig. 7-9).

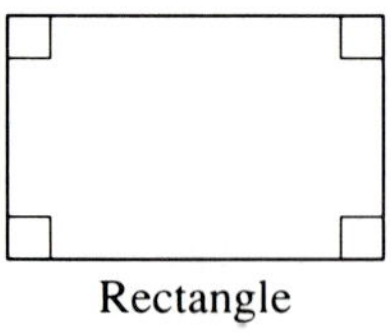

Figure 7-9

Definition 7-9 A **square** is a rectangle all of whose sides are congruent (Fig. 7-10).

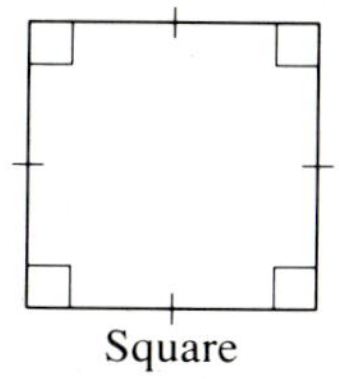

Figure 7-10

Now that we have defined some of the special types of quadrilaterals, we can prove some theorems which bring forth some of their properties or which tell us how to identify a quadrilateral as one of these special types. The proofs will be left as exercises.

Theorem 7-1 A diagonal of a parallelogram separates it into two congruent triangles (See Fig. 7-11.)

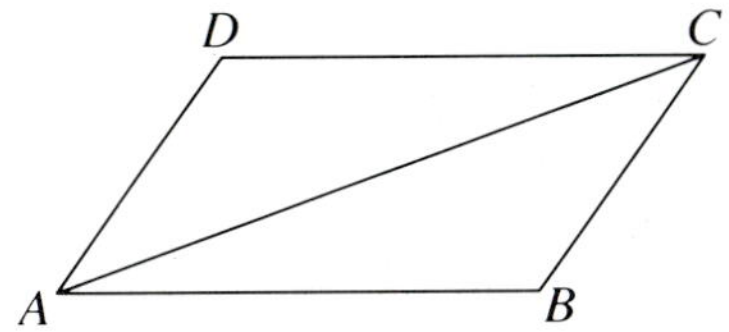

Figure 7-11

Corollary 7-1(a) The opposite sides of a parallelogram are congruent.

Corollary 7-1(b) The opposite angles of a parallelogram are congruent.

Theorem 7-2 The diagonals of a rectangle are congruent.

Theorem 7-3 In a parallelogram any two consecutive angles are supplementary.

Some of the theorems and corollaries that follow are very easy to prove when a coordinate system is used. Although we usually will show the axes in the horizontal-vertical position, as in Fig. 7-12(a), we may place them anywhere in the plane, as in Fig. 7-12(b).

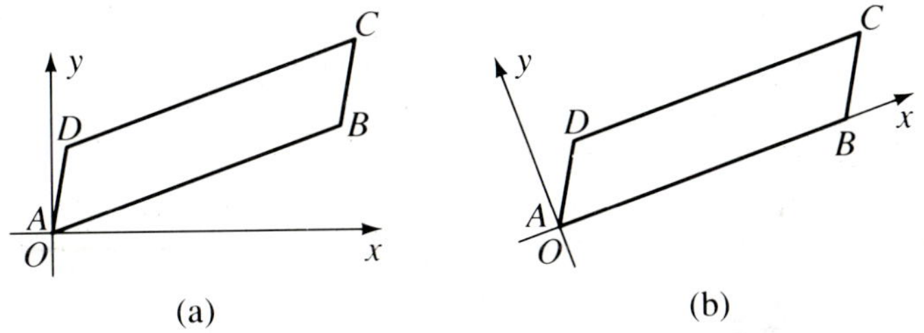

Figure 7-12

Our aim will be to place the axes in such a way that the coordinates of key points are as simple as possible, but we must also take into account certain special properties of the figure involved, such as parallel sides.

Example

Assign coordinates to the vertices of a parallelogram $ABCD$ if A is at the origin and $\overline{AB}$ is part of the x-axis.

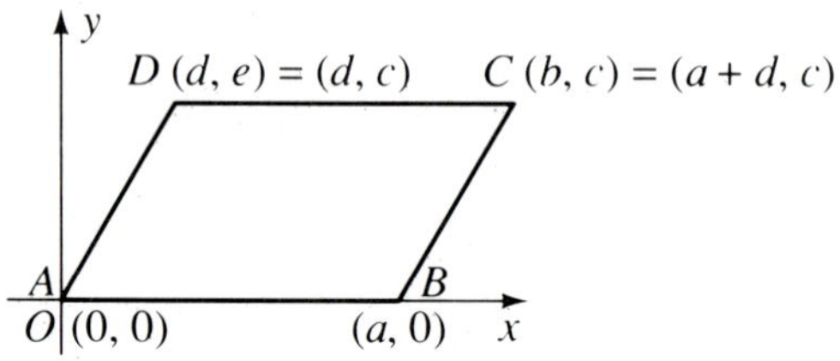

Solution The coordinates of A are $(0, 0)$ and to B we can assign $(a, 0)$. Let the coordinates of C be (b, c) and of $D(d, e)$. Since $ABCD$ is a parallelogram, however, the opposite sides are parallel. Thus $\overline{AB} \parallel \overline{CD}$ and the slope of each is 0, which means $m_{\overline{CD}} = (c - e)/(b - d) = 0$, or $c - e = 0$, so $c = e$. Also $\overline{AD} \parallel \overline{BC}$, so $c/d = c/(b - a)$, and $d = b - a$ or $b = a + d$.

Thus the coordinates of C are $(a + d, c)$ and those of D are (d, c). We might note that as a consequence of the sides being parallel and their slopes equal, as shown above, we can prove Corollary 7-1(a), since $m\overline{AD} = \sqrt{d^2 + c^2}$, $m\overline{BC} = \sqrt{[(a + d) - a]^2 + c^2} = \sqrt{d^2 + c^2}$ and $m\overline{AB} = |a|$, $m\overline{CD} = \sqrt{[(a + d) - d]^2 + (c - c)^2} = \sqrt{a^2} = |a|$. Using the notation of this example it is easy to prove the following theorem. ◁

Theorem 7-4 The diagonals of a parallelogram bisect each other.

Exercise Set 7.2

1. What is the figure formed when two congruent isosceles triangles are drawn with their vertices on opposite sides of the same base? Why?
2. In the parallelogram $ABCD$, $m\angle A = x - 30$ and $m\angle D = 2x + 15$. Find $m\angle A$, $m\angle B$, $m\angle C$, $m\angle D$.

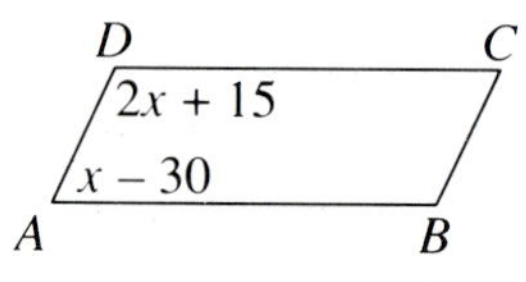

3. Prove that the segments joining the midpoints of adjacent sides of a quadrilateral are congruent to and parallel to the segment joining the midpoints of the other two sides.
4. If the following are vertices of a quadrilateral $ABCD$, identify each as a parallelogram, rectangle, rhombus, square, trapezoid, or simply a quadrilateral. (Be prepared to prove your answer by referring to Theorem 5-13, Theorem 6-10, or Theorem 6-14.)

	A	B	C	D
(a)	(3, 1)	(8, 5)	(9, 11)	(4, 7)
(b)	(3, 3)	(7, 1)	(7, 4)	(5, 5)
(c)	(−1, −3)	(5, 3)	(5, 6)	(−4, 0)
(d)	(−1, −3)	(1, 2)	(−4, 4)	(−6, −1)
(e)	(3, −2)	(5, 3)	(0, 6)	(−2, −2)

5. **(a)** Prove Theorem 7-1.
   (b) Prove Corollary 7-1(a).
   (c) Prove Corollary 7-1(b).
6. Prove Theorem 7-2 by means of coordinates.
7. Prove Theorem 7-2 synthetically.
8. Prove Theorem 7-3.
9. Prove Theorem 7-4. (*Hint:* Label the vertices as in the example on page 200. Draw the diagonals. Find the midpoints of $\overline{AC}$ and $\overline{BC}$.)
10. In the $\triangle ABC$, if $\overline{AB} \parallel \overline{DE}$, then
    (a) $ABED$ is a ________.
    (b) If $\overline{AC} \cong \overline{BC}$, then $ABED$ is ________.

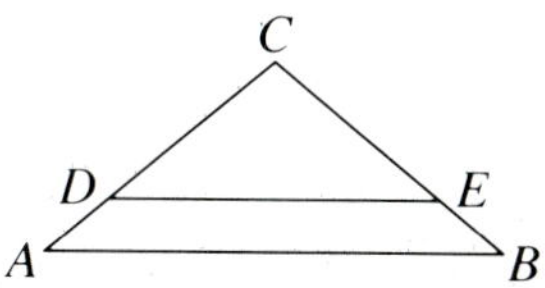

11. Prove that the base angles of an isosceles trapezoid are congruent.
12. If $ABCD$ is an isosceles trapezoid with $\overline{AB} \parallel \overline{CD}$, find x and y.

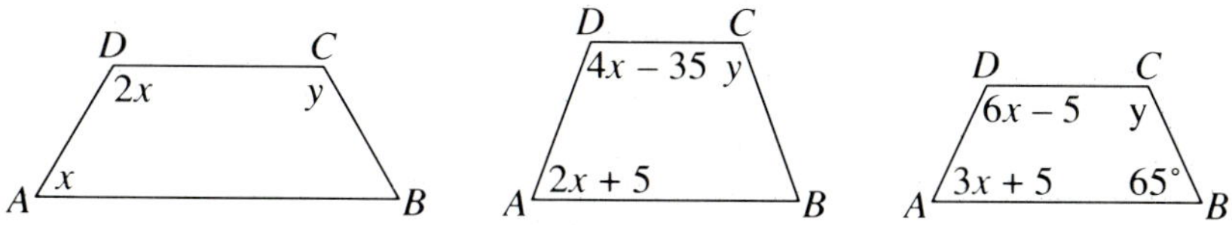

(a) (b) (c)

13. Prove that the median of a trapezoid is parallel to the bases and its measure is one half the sum of their measures. (A *median* of a trapezoid is a segment that joins the midpoints of the nonparallel sides.)
    Given: Trapezoid $ABCE$ with E and F midpoints of $\overline{AC}$ and $\overline{DB}$, respectively.
    Prove: $\overline{EF} \parallel \overline{AB}$ and $\overline{EF} \parallel \overline{CD}$, $m\overline{EF} = \frac{1}{2}[m\overline{AB} + m\overline{CD}]$

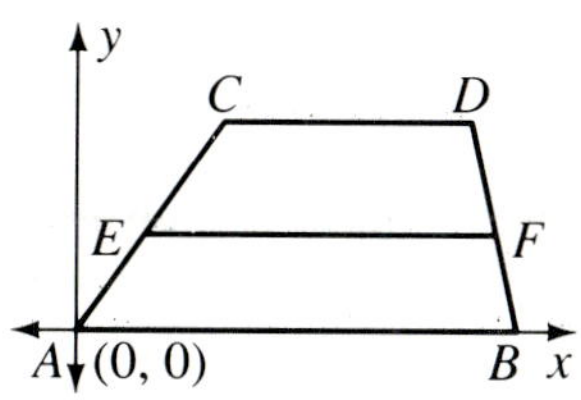

14. The coordinates of the vertices of a rectangle are $A(0, 0)$, $B(6, 0)$, $C(6, 4)$, and $D(0, 4)$. Draw diagonal $\overline{AC}$ and find $m\angle BAC$.
15. Disprove the following statement: If a diagonal of a quadrilateral determines two congruent triangles, the quadrilateral is a parallelogram (Converse of Theorem 7-1).
16. Prove that the segments joining the midpoints of the opposite sides of a quadrilateral bisect each other (use coordinates).

The theorems and corollaries proved up to now have concerned given parallelograms. Next we have a set which enable us to show that certain quadrilaterals *are* parallelograms.

Theorem 7-5 If a quadrilateral has both sets of opposite sides congruent, it is a parallelogram. (Note that this is the converse of Corollary 7-1(a)).

Theorem 7-6 If a quadrilateral has one set of opposite sides congruent and parallel, it is a parallelogram.

Theorem 7-7 If the diagonals of a quadrilateral bisect each other, it is a parallelogram.

Theorem 7-8 If a parallelogram has one right angle, then it is a rectangle.

Corollary 7-8 If the diagonals of a parallelogram are congruent, it is a rectangle.

The Euclidean proof of the next theorem is easy and will be left as an exercise, but the proof using coordinates is interesting and we outline it here.

Theorem 7-9 The diagonals of a rhombus are perpendicular to each other.

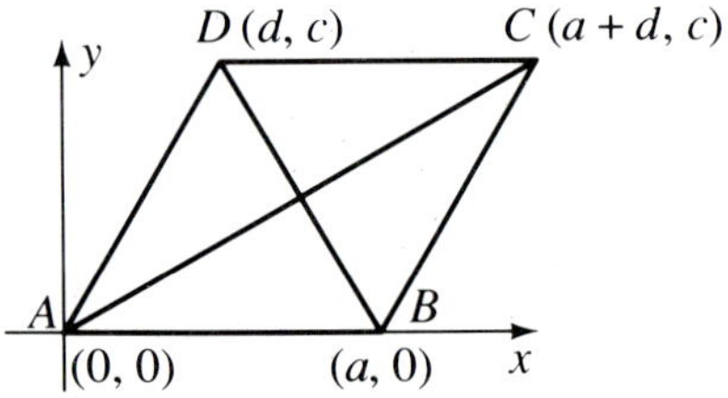

Proof We can label the vertices as in the example on page 200, since a rhombus is a parallelogram. We need to show that $\overline{AC} \perp \overline{BD}$, which will be true if $m_{\overline{AC}} \cdot m_{\overline{BD}} = -1$. (Why?)

$$m_{\overrightarrow{AC}} \cdot m_{\overrightarrow{BD}} = \frac{c}{a+d} \cdot \frac{c}{d-a} = \frac{c^2}{d^2 - a^2}$$

Now $c^2/(d^2 - a^2)$ will equal -1 if $c^2 = a^2 - d^2$. Does it? (We have not used the fact that this is a *rhombus* yet. What makes a parallelogram a rhombus? Complete the proof.)

Theorem 7-10 If the diagonals of a quadrilateral bisect each other and are perpendicular, then the quadrilateral is a rhombus.

Examples

1. In the following rhombus, $m\angle 1 = 40$ and $m\angle 4 = 40$.

(a) Find the measure of each of the numbered angles.

▶ (b) If $m\angle 1 = 40$ and $m\overline{AD} = 10$, find $m\overline{AC}$ and $m\overline{BD}$.

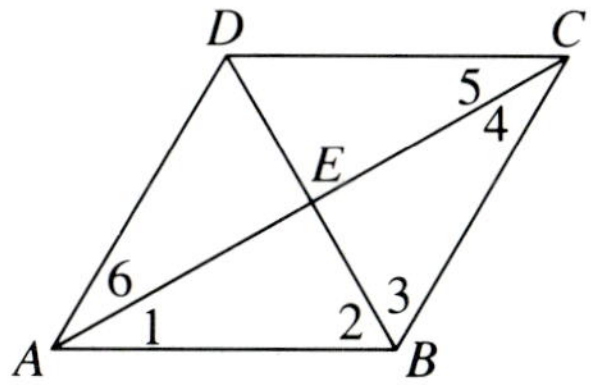

Solution

(a) A rhombus is a parallelogram, so $\overline{AB} \parallel \overline{CD}$ and $\overline{AD} \parallel \overline{BC}$, which implies $\angle 4 \cong \angle 6$ (alternate-interior angles). Thus $m\angle 1 = m\angle 5 = 40$ and $m\angle 6 = 40$. By the addition of angular measures. Post. 3-3, $m\angle 1 + m\angle 6 = m\angle DAB = 80$. Since $\angle ABC$ and $\angle DAB$ are supplementary (Theorem 5-16), this means $m\angle ABC = 100$. The diagonals are perpendicular, so AEB is a right triangle with $\angle 1$ the complement of $\angle 2$, giving $m\angle 2 = 50$. Finally $m\angle 3 = m\angle ABC - m\angle 2 = 100 - 50 = 50$.

▶ (b) Since $\triangle AEB$ is a right triangle, $\sin 40° = m\overline{BE}/10$ and $m\overline{BE} = 10 \sin 40° \approx 6.428$. Also $\cos 40° = m\overline{AE}/10$ and $m\overline{AE} = 10 \cos 40° \approx 7.660$. Thus, $m\overline{AC} = 2m\overline{AE} \approx 15.320$ and $m\overline{BD} = 2m\overline{BE} \approx 12.856$. ◁

2. Suppose $\angle 1 \cong \angle 2$ and $\overline{AB} \cong \overline{CD}$ in $ABCD$.

(a) What is the relation between $\angle 3$ and $\angle 4$?

(b) What is the relation between $\overline{AD}$ and $\overline{BC}$?

(c) What kind of a figure is $ABCD$?

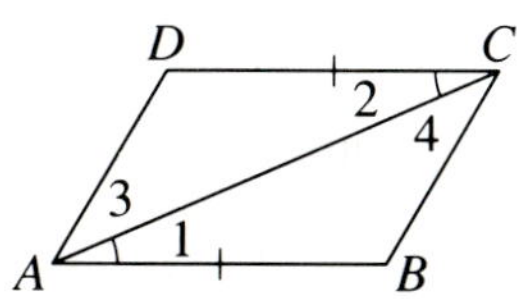

Solution

(a) With $\angle 1 \cong \angle 2$ and $\overline{AB} \cong \overline{CD}$ and $\overline{AC} \cong \overline{AC}$, $\triangle ABC \cong \triangle CDA$. Thus $\angle 3 \cong \angle 4$ by CPCTC.

(b) $\overline{AD} \cong \overline{BC}$ for the same reason.

(c) Since both sets of opposite sides are congruent, the figure is a parallelogram. ◁

Exercise Set 7.3

1. If $\triangle ABC$ is an isosceles triangle with midpoints D, E, and F as indicated, prove that $AFED$ and $BFDE$ are parallelograms.

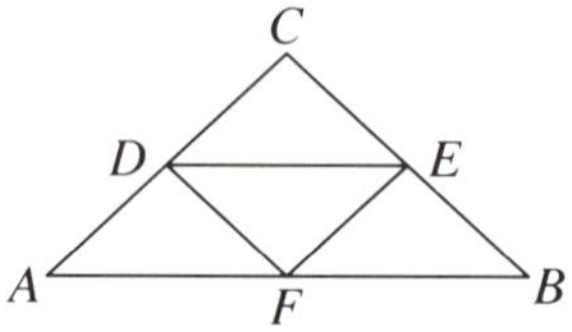

2. Prove Theorem 7-5 (*Hint:* Draw a diagonal).
3. Prove Theorem 7-6 (*Hint:* Draw a diagonal).
4. Prove Theorem 7-7.
5. Prove Theorem 7-8.
6. Prove Theorem 7-9 without the use of coordinates.
7. Prove Theorem 7-10.
8. Prove that the segments joining the midpoints of the sides of a quadrilateral, taken in order, form a parallelogram.
9. Given a quadrilateral with vertices $A(2, 2)$, $B(10, 2)$, $C(8,10)$, and $D(2, 8)$.
   (a) Find the midpoints of the segments and label them R, S, T, U, respectively.
   (b) Prove that $RSTU$ is a parallelogram.
   (c) Prove that the diagonals of $RSTU$ bisect each other and are perpendicular.
   (d) Prove that $RSTU$ is a square.
10. ▶ If $RSTU$ is a parallelogram with $R(0, 0)$ and $S(6, 0)$, diagonal $\overline{RT}$ of length 15 and $m\angle SRT = 40$, find the coordinates of U and T (to the nearest tenth).

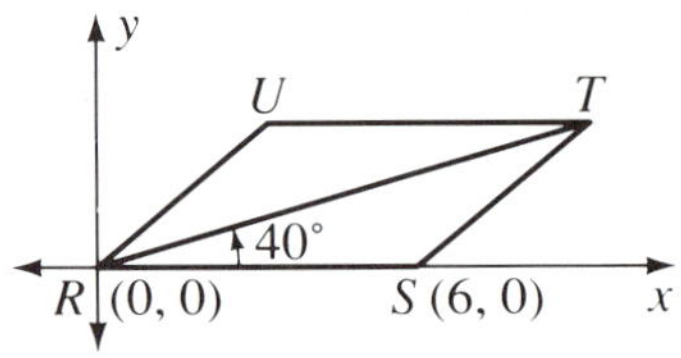

11. Prove Corollary 7-8 by means of coordinates.
12. Which quadrilaterals in the set {parallelogram, rectangle, rhombus, square} always have the characteristics given below?
    (a) Both pairs of opposite angles are respectively congruent.
    (b) The diagonals are congruent.
    (c) One pair of opposite sides are congruent and parallel.
    (d) Each pair of consecutive angles supplementary.
    (e) Consecutive sides congruent.
    (f) Diagonals bisect each other and are perpendicular.
13. Construct a rhombus that has diagonals with lengths equal to $m\overline{AB}$ and $m\overline{AC}$ in the following figure.

Definition 7-10 The **perimeter** of a polygon is the sum of the measures of its sides.

The perimeter of a triangle ABC is $m\overline{AB} + m\overline{BC} + m\overline{CA}$. The perimeter of a quadrilateral $ABCD$ is $m\overline{AB} + m\overline{BC} + m\overline{CD} + m\overline{DA}$. Since the opposite sides of a rectangle are congruent, the perimeter is $l + w + l + w$, or $P = 2l + 2w$ if P represents perimeter, l the length, and w the width.

When a diagonal of quadrilateral is drawn, the figure is separated into two triangles (Fig. 7-13). Since the sum of the measures of the angles of each triangle is 180, the sum of the measures of the angles of two triangles is 360. Therefore, we may conclude the following.

Theorem 7-11 The sum of the measures of the angles of a quadrilateral is 360.

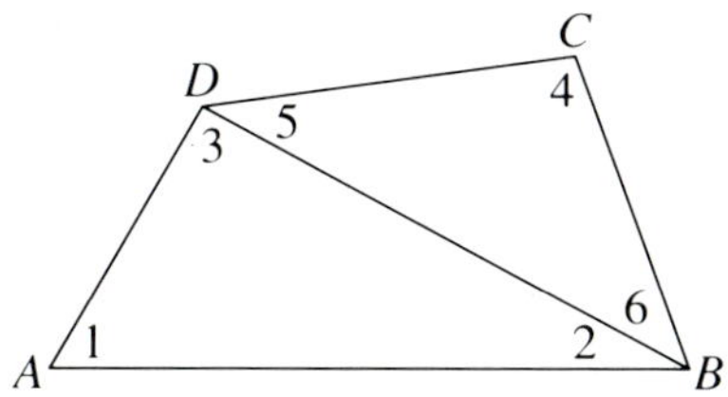

Figure 7-13

Exercise Set 7.4

1. Show that the sum of the measures of the angles of an n-gon is $(n - 2)180$ for
   (a) $n = 3$ **(b)** $n = 5$
   (c) $n = 6$ **(d)** $n = 7$
   (e) $n = 10$
2. **(a)** If t is the length of the side of an equilateral triangle, what is its perimeter?
   (b) If s is the length of the side of a square, what is its perimeter?
   (c) If r is the length of the diagonal of a square, then what is its perimeter?
3. Find the perimeter of a polygon with vertices at
   (a) $A(-3, 2)$, $B(9, -3)$, $C(-3, -3)$
   (b) $A(2, 1)$, $B(5, 5)$, $C(10, -7)$, $D(2, -1)$
4. **(a)** Prove that the sum of the measures of the exterior angles of a triangle (formed by extending each side in succession) is 360.
   (b) Prove that the sum of the measures of the exterior angles of a quadrilateral is 360. (Given quadrilateral $ABCD$ with exterior angles of measure a, b, c, and d, respectively, prove that $a + b + c + d = 360$.)

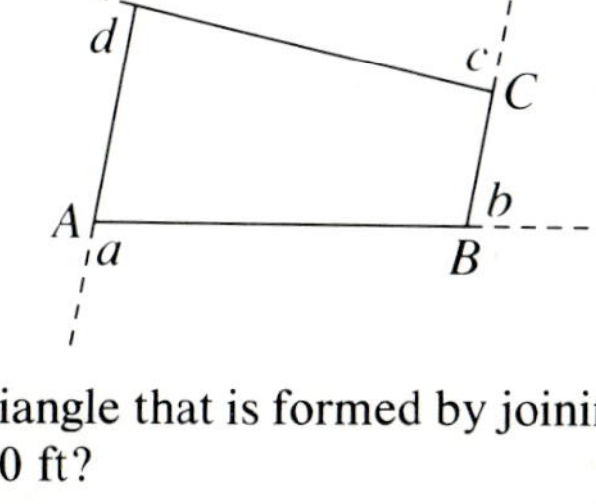

5. What is the perimeter of a triangle that is formed by joining the midpoints of the sides of a triangle with perimeter 250 ft?

7-3 POLYGONAL REGIONS AND AREAS

Thus far our discussion has been limited almost exclusively to sets of points contained in lines, rays, or segments. Although we have defined triangles and other polygons, we have developed properties of their sides and angles, paying little attention to the portion of the plane "inside" these figures. The world is greatly concerned with measures of the regions bounded by these figures, however, whether it is from the standpoint of painting a wall or of mapping a boundary between nations.

Definition 7-11 A **triangular region** is the union of the triangle and its interior. A **polygonal region** is the union of a finite number of coplanar triangular regions that intersect in at most a finite number of segments and points (i.e., no two triangular regions *overlap*).

Shown in Fig. 7-14 are some examples of ways in which polygons may be divided into triangular regions.

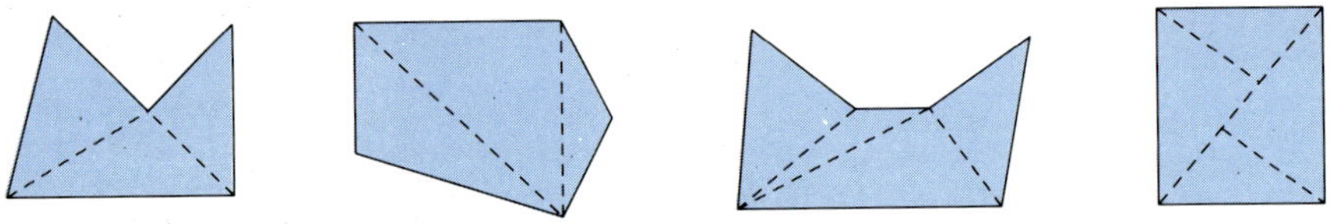

Figure 7-14

With segments and angles we associated real numbers that were the measures of the length and size; likewise we need a measure for the region of a polygon.

Postulate 7-1 To every polygonal region there corresponds a unique positive real number.

Definition 7-12 The number K corresponding to each polygonal region is called its **area.**

Postulate 7-2 An area of a polygonal region R is the sum of the areas of the distinct triangular regions $R_1, R_2, \ldots, R_n$ whose union is R (Fig. 7-15).

$R_1 \cup R_2 = R$
Area R_1 + area R_2 = area R
or $K_1 + K_2 = K$

$R_1 \cup R_2 \cup R_3 \cup R_4 \cup R_5 = R$
Area R_1 + area $R_2 + \cdots +$ area R_5 = area R
or $K_1 + K_2 + \cdots + K_5 = K$

Figure 7-15

Postulate 7-3 Congruent triangles determine regions having equal areas.

As we continue to develop ideas concerning areas, we will be using the terms *base* and *altitudes* of polygons, primarily for triangles, rectangles, parallelograms and trapezoids. Consider the triangles shown in Fig. 7-16. Here we see that the altitude of a triangle is the perpendicular segment from a vertex to the line containing opposite side, as defined in Chapter 4.

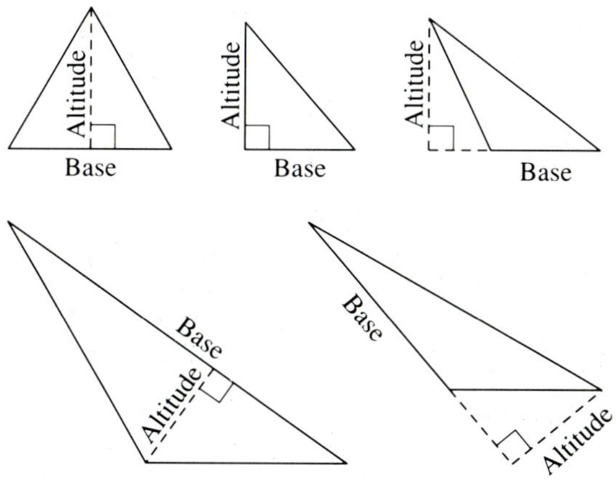

Figure 7-16

The unit of area measure will be called the *square unit*. If the measures of base and altitude are in inches the area will be in square inches, if in feet, then square feet will be the unit of area measure, etc. This derives from the usual introduction involving separation of a rectangular region into an integral number of squares, as in Fig. 7-17. It is impossible to separate some regions into squares of the same size, however, so we base our method of area computation on a postulate involving a right triangle.

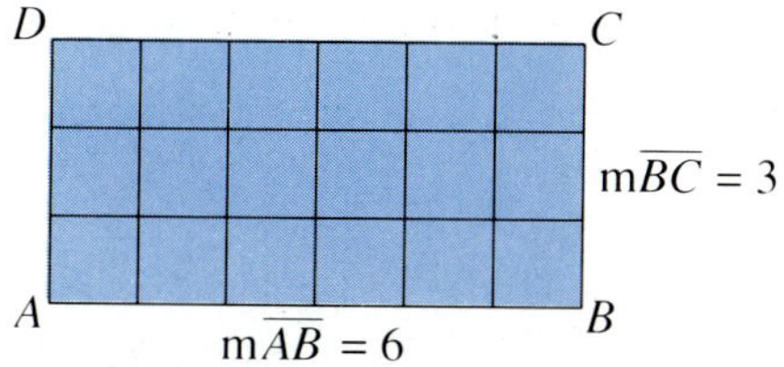

Figure 7-17

Postulate 7-4 The area (K) of a right triangular region is one-half the product of the lengths of the legs.

In a right triangle the legs form a base and altitude, so the postulate might be restated in terms of the triangle in Fig. 7-18: The area (K) of a right triangular region is $\frac{1}{2}ab$, if a is the length of the altitude and b is the length of the base. ($K_{ABC} = \frac{1}{2}ab$)

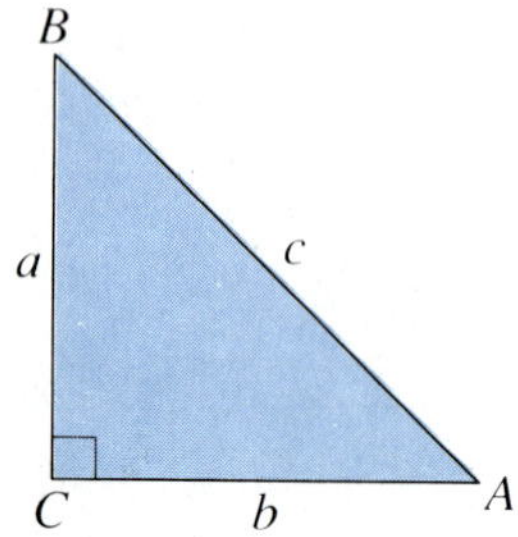

Figure 7-18

Henceforth, we will use "area of a triangle" in place of area of a triangular region, "area of a rectangle" in place of area of a rectangular region, etc.

Theorem 7-12 The area (K) of a triangle is equal to one-half the product of the length of a base and the length of the altitude to that base.

Given: $\triangle ABC$ with $\overline{BD}$ the altitude to base $\overline{AC}$, $m\overline{AC} = b$ and $m\overline{BD} = h$

Prove: $K = \frac{1}{2}bh$

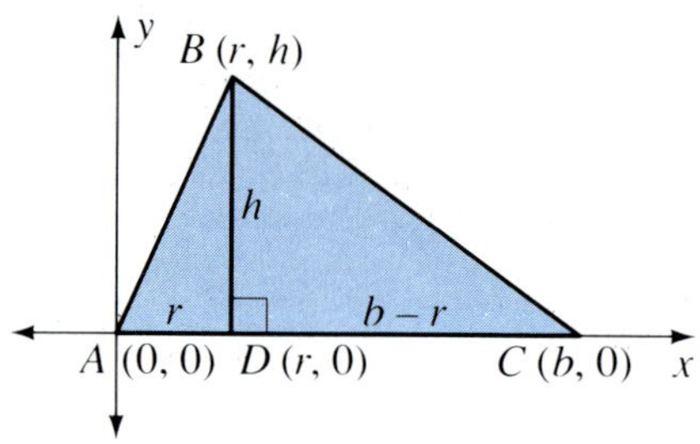

Proof

Statements	*Reasons*
1. Let $\overline{AC}$ determine a horizontal axis with $A(0, 0)$ and $\overline{BD} \perp \overline{AC}$, so that $D(r, 0)$, $C(b, 0)$ and $B(r, h)$	1. Definition of altitude (Def. 5-3)
2. $\triangle ADB$ and $\triangle CDB$ are right triangles	2. Definition of perpendicular (Def. 3-10) and definition of right triangle (Def. 4-3)
3. $K_{ADB} = \frac{1}{2}rh$ and $K_{CDB} = \frac{1}{2}(b - r)h$	3. Post. 7-4 (Since $b > r$, $b - r > 0$)
4. $K_{ABC} - K_{ADB} + K_{CDB}$	4. Post. 7-2
5. $K = \frac{1}{2}rh + \frac{1}{2}(b - r)h$ $= \frac{1}{2}h[r + (b - r)] = \frac{1}{2}h \cdot b$ $= \frac{1}{2}bh$	5. Axioms for real numbers

Corollary 7-12(a) Triangles that have bases of equal length and altitudes of equal length have equal areas (Fig. 7-19).

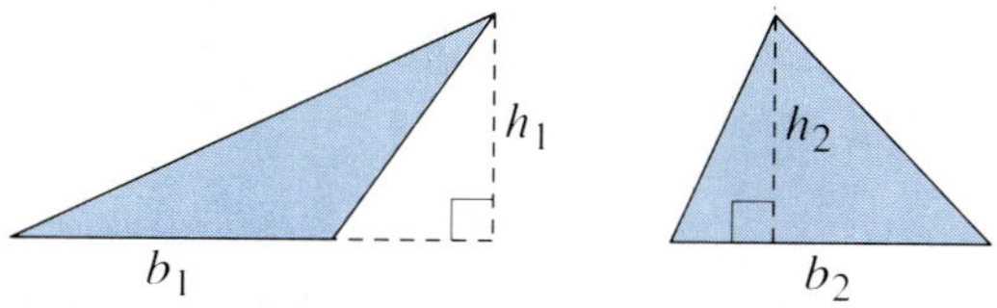

Figure 7-19

Corollary 7-12(b) The areas of two triangles having bases of equal length have the same ratio as the lengths of their altitudes.

Corollary 7-12(c) The areas of two triangles having altitudes of equal length have the same ratio as the lengths of their bases.

Corollary 7-12(d) The areas of two triangles have the same ratio as the products of the lengths of their bases and altitudes.

Examples

1. In a right triangle with the right angle at C, $m\overline{AC} = 5$, $m\overline{BC} = 12$ and $m\overline{AB} = 13$, find

(a) the area of triangle ABC

(b) the length of the altitude to the hypotenuse

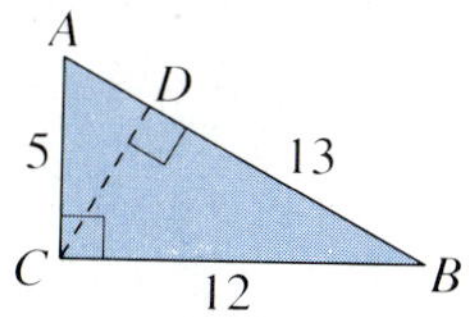

Solution

(a) $K_\triangle = \frac{1}{2}a \cdot b = \frac{1}{2}m\overline{BC} \cdot m\overline{AC} = \frac{1}{2}(12)(5) = 30.$

(b) To find $m\overline{CD}$ we can use the fact that $K_\triangle = \frac{1}{2}m\overline{AB} \cdot m\overline{CD}$; hence $30 = \frac{1}{2}(13) \cdot m\overline{CD}$, and $m\overline{CD} = \frac{60}{13}$, or $4\frac{8}{13}$.

2. If $\triangle ABC$ has a base of length 8 and an area of 36 square units, and $\triangle RST$ has a base of 8 and altitude to that base of 12, find

(a) the length of the corresponding altitude of $\triangle ABC$

(b) the ratio of their areas

Solution

(a) Let K_1 and h_1 be the measures of the area and the altitude of $\triangle ABC$. Since $K = \frac{1}{2}bh$, $36 = \frac{1}{2}(8)h_1$, and $h_1 = 9$.

(b) Let K_2 and h_2 be the measures of the area and the altitude of $\triangle RST$. Then $\frac{K_1}{K_2} = \frac{h_1}{h_2} = \frac{9}{12} = \frac{3}{4}$ by Corollary 7-12(b). ◁

Exercise Set 7.5

1. Show that each of the following figures is a polygonal region by indicating a separation into triangular regions such that Definition 7-11 is satisfied.

(a) (b) (b)

2. Find the area of triangle RST when
   (a) $m\overline{RS} = 9$, $m\overline{TV} = 6$
   (b) $m\overline{TS} = 12$, $m\overline{UR} = 5$
   (c) $m\overline{RV} = 3$, $m\overline{VS} = 12$, $m\angle T = 90$
   (d) $m\overline{TS} = 21$, $m\overline{TR} = 14$ and $\overline{TR} \perp \overline{ST}$
   (e) $m\overline{TU} = 8$, $m\overline{US} = 20$, $m\angle S = 45$
   (f) $m\overline{TS} = 18$, $m\overline{RS} = 20$, $m\angle S = 30$
   (g) $m\overline{RV} = 5$, $m\overline{TV} = 10$, $m\angle T = 90$

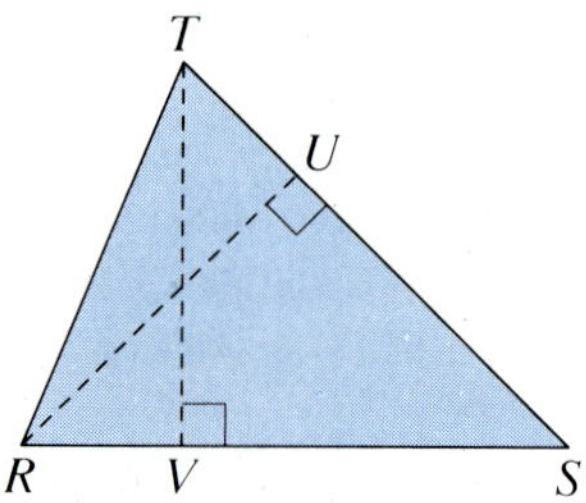

3. In the right triangle ABC, $\overline{BC} \perp \overline{AC}$ and $\overline{CD} \perp \overline{AB}$. If $m\overline{AC} = 21$, $m\overline{AB} = 29$, and the area is 210 square units, find $m\overline{BC}$ and $m\overline{CD}$.

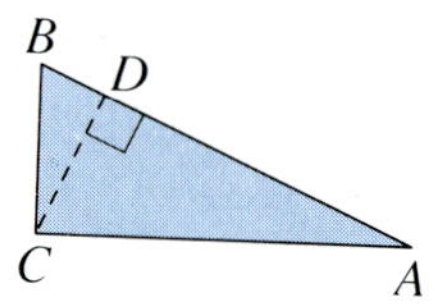

4. **(a)** Prove Corollary 7-12(a) **(b)** Prove Corollary 7-12(b)
   (c) Prove Corollary 7-12(c). **(d)** Prove Corollary 7-12(d).
5. Find the length of the base of a triangle with an area of 468 square units and an altitude of 39 units.
6. In the following triangles $\overline{RS} \perp \overline{TU}$ and $\overline{XY} \perp \overline{ZW}$. If $m\overline{TU} = m\overline{ZW}$, $m\overline{RS}/m\overline{XY} = \frac{5}{3}$, and the area of $\triangle RST$ is 75, what is the area of $\triangle XYZ$?

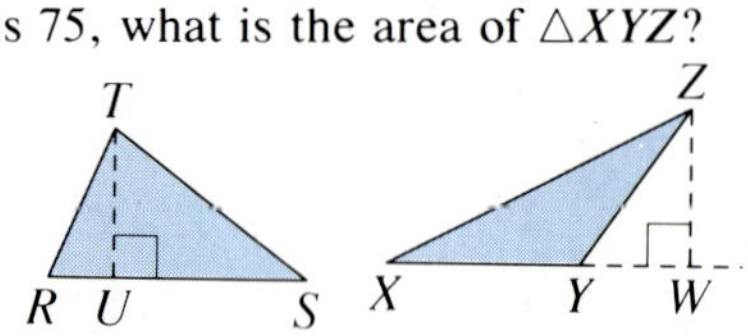

7. Referring to the figures in Exercise 6, with $\overline{TU} \perp \overline{RS}$, $\overline{WZ} \perp \overline{XY}$, $m\overline{RS} = m\overline{XY}$, the area of $\triangle RST = 54$, and the area of $\triangle XYZ = 84$, what is the ratio of their altitudes?

8. If $ABCD$ is a parallelogram and E is the midpoint of $\overline{AB}$, prove that $\triangle AED$ and $\triangle BEC$ have equal areas.

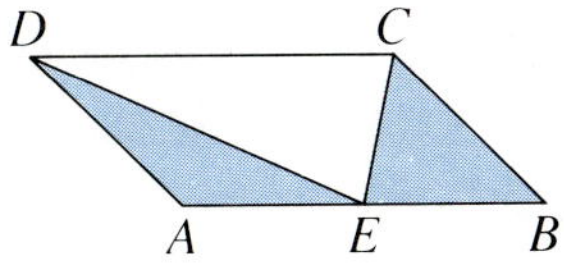

9. Prove that a median of a triangle separates the triangle into two triangles, each with an area one-half the area of the original triangle.

10. If $\overline{CD}$ and $\overline{AE}$ are medians of triangle ABC intersecting at F, prove that the areas of triangles ADF and CEF are equal. (Use the statement of the preceding problem.)

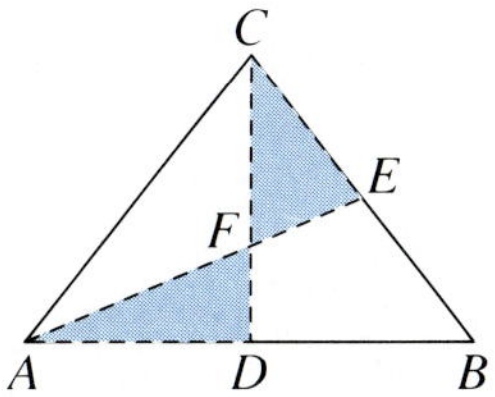

11. If the area of a right triangle is 72 square units and the length of the altitude to the hypotenuse is 9, find the length of the hypotenuse.

▶ **12.** In $\triangle ABC$, $m\angle A = 35$, $m\angle C = 90$, $m\overline{AC} = 12$. Find $m\overline{BC}$ and the area of the triangle.

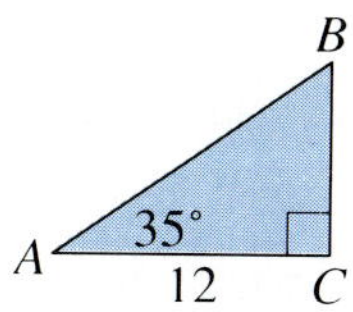

▶ **13.** In $\triangle RST$, $m\overline{RS} = 15$, $m\overline{RT} = 10$ and $m\angle R = 20$. Also $\overline{TU} \perp \overline{RS}$ and $m\overline{TU} = h$. Find h and the area of $\triangle RST$.

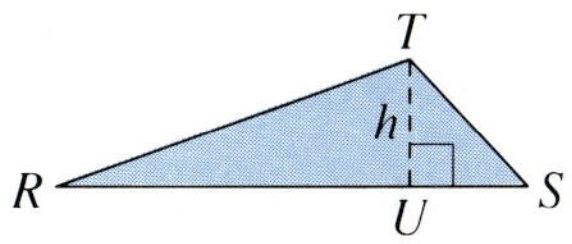

▶ **14.** In $\triangle XYZ$, $m\angle Y = 38$, $m\overline{XY} = 12$, $m\overline{YZ} = 10$. Find the area of $\triangle XYZ$.

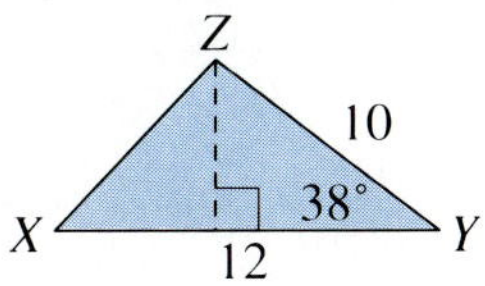

▶ **15.** Prove that the area of any acute triangle equals half the product of the lengths of two sides times the sine of their included angle. (In the following figure, $K = \frac{1}{2}bc \sin A$.)

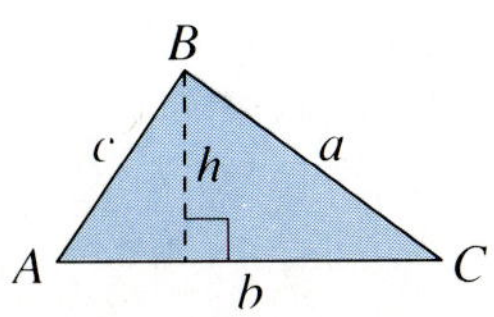

16. Find the area of a triangle with sides of length 8 in. and 15 in. and an included angle of 80° (to the nearest tenth).

17. Find the area of an isosceles triangle if two sides measure 15 in. and a base angle measures 59 (to the nearest tenth).

Area of Quadrilaterals

Since polygonal regions are defined in terms of triangular regions and the area of a polygonal region is the sum of a set of triangular regions, it will be easy to develop formulas for areas of quadrilaterals.

Definition 7-13 An **altitude of a parallelogram** or **trapezoid** is a segment having its endpoints in the lines containing two parallel sides and perpendicular to one of them (Fig. 7-20).

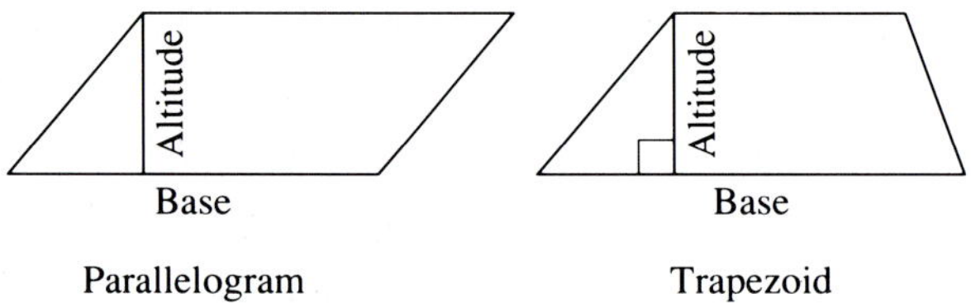

Figure 7-20

Definition 7-14 The **bases of a trapezoid** are the two parallel sides. The **base of a parallelogram** may be any one of the sides.

Theorem 7-13 The area (K) of a parallelogram is equal to the product of the lengths of a base (b) and an altitude (a) to that base. ($K = a \cdot b$)

Place a coordinate system on $ABCD$ so that $A(0, 0)$, $B(b, 0)$, $C(b + c, a)$ and $D(c, a)$.

Prove: Area of parallelogram $ABCD = a \cdot b$

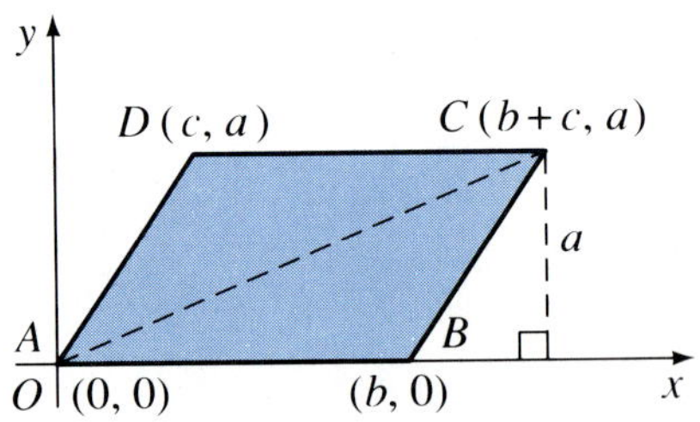

Proof

Statements	*Reasons*
1. $\overline{AC}$ is a diagonal of $ABCD$	1. Definition 7-2
2. $K_{ABC} = K_{CDA}$	2. Diagonal separates parallelogram into two congruent triangles (Theorem 7-1), and congruent triangles

Statements	*Reasons*
	determine regions having equal areas (Post. 7-3).
3. $K_{ABC} = \frac{1}{2}ab$	3. Theorem 7-12
4. $K_{ABCD} = K_{ABC} + K_{CDA}$	4. Post. 7-2
5. $K_{ABCD} = \frac{1}{2}ab + \frac{1}{2}ab = a \cdot b$	5. Axioms for real numbers

Proofs of the following corollaries are left to the reader.

Corollary 7-13(a) The area (K) of a rectangle is equal to the product of the length of its base (b) and the length of its altitude (a). ($K = a \cdot b$)

Corollary 7-13(b) The area (K) of a square with side of length s is s^2. ($K = s^2$)

Theorem 7-14 The area (K) of a trapezoid is equal to one-half the product of the length of its altitude (a) and the sum of the lengths of its bases (b and b'). ($K = \frac{1}{2}a(b + b')$)

Given: Trapezoid $ABCD$ with $\overline{AB} \parallel \overline{CD}$
Prove: $K_{ABCD} = \frac{1}{2}a(b + b')$

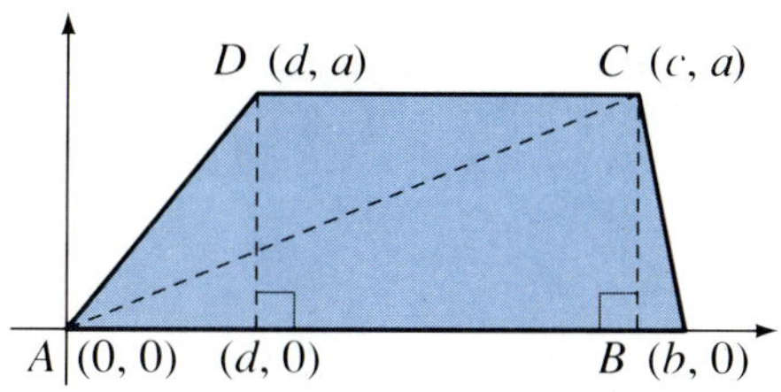

Proof Let $b = m\overline{AB}$ and $b' = m\overline{CD} = c - d$. Since $K_{ABC} = \frac{1}{2}ab$ and $K_{CDA} = \frac{1}{2}a(c - d) = \frac{1}{2}a \cdot b'$, $K_{ABCD} = \frac{1}{2}ab + \frac{1}{2}ab' = \frac{1}{2}a(b + b')$.

Theorem 7-15 The area of a rhombus is equal to one-half the product of the lengths of its diagonals.

Examples

1. Find the area of a parallelogram with consecutive sides of lengths 13 and 20 and an altitude of 12.

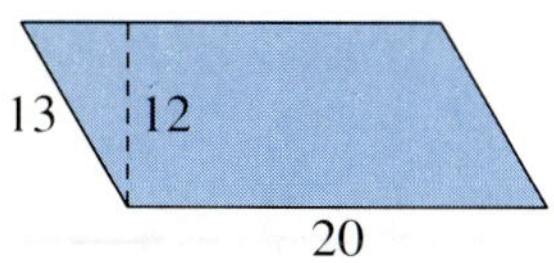

Solution By Theorem 7-13, we know the area of a parallelogram is $K = a \cdot b$; hence $K = 12(20) = 240$.

2. Find the area of the parallelogram with consecutive sides of lengths 12 and 20 and an included angle of 30°.

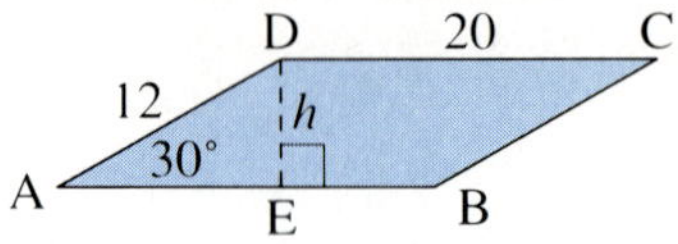

Solution In order to find the area we require the length of the base and the altitude to that base. We are given that $b = m\overline{AB} = 20$, and we need to find the value of h. Since $\triangle AED$ is a right triangle and $m\angle A = 30$, by Theorem 6-11, $h = \frac{1}{2}m\overline{AD} = 6$. $K = 6(20) = 120$.

3. Find the area of trapezoid $ABCD$ if
   (a) $h = m\overline{DE} = m\overline{CF} = 8$, $m\overline{AB} = 20$ and $m\overline{DC} = 15$
   (b) $m\overline{AD} = 12$, $m\overline{BC} = 12$, $m\overline{DC} = 20$ and $m\angle A = 60$.

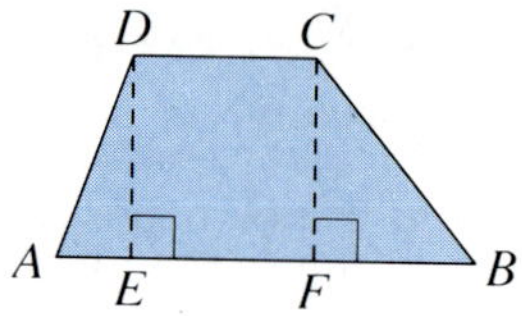

Solution
(a) By Theorem 7-14, $K = \frac{1}{2}a(b + b') = \frac{1}{2}(8)(20 + 15) = 4(35) = 140$.
(b) The formula $K = \frac{1}{2}a(b + b')$ requires that we find $m\overline{DE}$ and $m\overline{AB}$. Since $\triangle AED$ is a 30°, 60° right triangle with hypotenuse 12, by Theorem 6-11 we know that $m\overline{AE} = 6$ and $m\overline{DE} = 6\sqrt{3}$. Since $\overline{DE} \cong \overline{CF}$ (Corollary 5-19), $\triangle ADE \cong \triangle BCF$ and $\overline{AE} \cong \overline{FB}$; hence $m\overline{FB} = 6$. Also $\overline{CD} \cong \overline{EF}$, so $m\overline{EF} = 20$ and $m\overline{AB} = 32$. Now we can find the area:

$$K = \tfrac{1}{2}(6\sqrt{3})(32 + 20) = 3\sqrt{3}(52) = 156\sqrt{3}$$ ◁

Exercise Set 7.6

1. Find the area of parallelogram $ABCD$ if
   (a) $m\overline{AB} = 8$, $m\overline{DE} = 10$
   (b) $m\overline{AB} = 20$, $m\overline{DE} = 6.4$
   (c) $m\overline{AD} = 15$, $m\overline{AB} = 20$, $m\angle A = 45$
   (d) $m\overline{AB} = 12$, $m\overline{BC} = 20$, $m\angle B = 120$
   (e) $m\overline{AB} = m\overline{BC} = 8$ and $m\angle A = 45$
   ▶ **(f)** $m\overline{AD} = 12$, $m\overline{AB} = 16$, $m\angle A = 40$

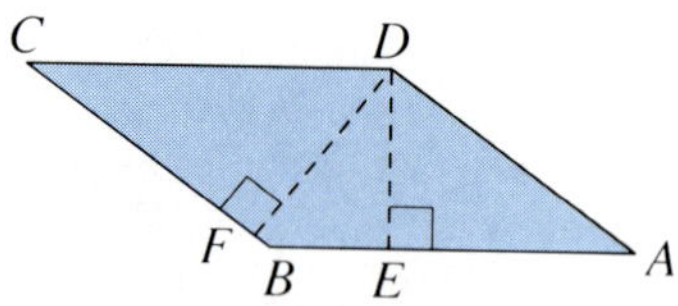

2. Given the parallelogram of Exercise 1 with $\overline{DE} \perp \overline{AB}$ and $\overline{DF} \perp \overline{BC}$, $m\overline{AB} = 18$, $m\overline{DE} = 10$, $m\overline{DF} = 12$. Find $m\overline{BC}$.
3. Find the area of a trapezoid with altitude of length 8 and median of length 20. (See Exercise 13, page 201.)

4. Find the area for the trapezoid $HIJK$ if
   (a) $m\overline{HI} = 12$, $m\overline{KJ} = 4$, $m\overline{KL} = 5$
   (b) $m\overline{JM} = 8$, $m\overline{KJ} = 6$, $m\overline{HI} = 14$
   (c) $m\overline{HK} = 16$, $m\overline{HI} = 24$, $m\angle H = 30$, $m\overline{KJ} = 10$
   (d) $m\overline{KH} = m\overline{JI} = 5$, $m\angle I = 45$, $m\overline{KJ} = 20$
   (e) $\angle H \cong \angle I$, $m\overline{JM} = 6$, $m\overline{JI} = m\overline{KJ} = 10$

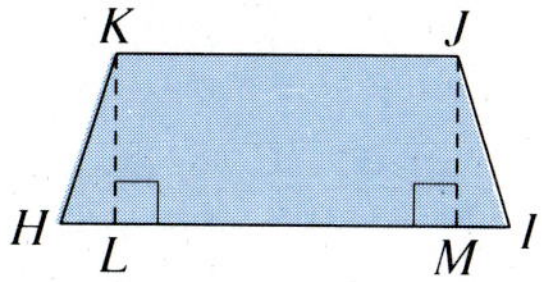

5. (a) Prove Corollary 7-13(a).
   (b) Prove Corollary 7-13(b).
6. Prove that the diagonals of a parallelogram separate it into four triangles of equal area.
7. How are the areas of a triangle and a parallelogram with bases of equal length and altitudes of equal length related?
8. Prove that if the diagonals of a quadrilateral are perpendicular, its area is equal to one-half the product of the lengths of the diagonals.
9. Given a square with sides of length $(a + b)$, as shown below. *Prove:*
   (a) $\triangle EAF \cong \triangle FBG \cong \triangle GCH \cong \triangle HDE$
   (b) Angles EFG, FGH, GHE, and HEF are right angles
   (c) $EFGH$ is a square
   (d) $(a + b)^2 = c^2 + 2ab$
   (e) From (d) $a^2 + b^2 = c^2$

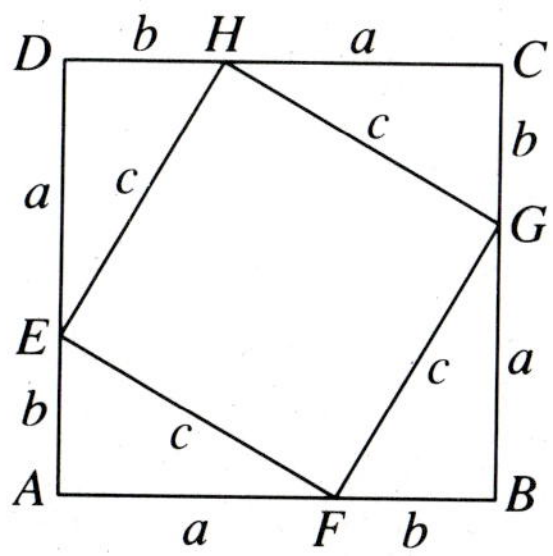

10. The nonparallel sides of an isosceles trapezoid are each 8 in. in length, the upper base is 10 in. and the lower base is 18 in. What is the area of the trapezoid?
11. The length of a rectangle is 6 units more than its width. If its area is 91 square units, find the lengths of its sides.
12. Prove Theorem 7-15.
13. Find the area of a rhombus that has diagonals 8 cm and 15 cm in length.

REVIEW EXERCISES

1. Without referring to the text, define (a) parallelogram, (b) trapezoid, (c) isosceles trapezoid.
2. How many sides has a polygon if the sum of its interior angles is 360°? 540°? 720°?

3. Prove that the segments joining the midpoints of a rhombus, taken in order, form a rectangle.
4. Prove that the segments joining the midpoints of the sides of an isosceles trapezoid, taken in order, form a rhombus.
5. Prove that if two parallel lines are cut by a transversal, the bisectors of the interior angles form a rectangle.
6. Prove that if the base angles of a trapezoid are congruent, the trapezoid is isosceles.
7. If the perimeters of two similar polygons are 240 ft and 580 ft and a side of the first measures 48 ft, find the length of the corresponding side of the second polygon.
8. Prove that the segments joining the midpoint of a diagonal of a quadrilateral to the remaining two vertices separate the quadrilateral into two regions with equal areas.
9. One diagonal of a rhombus is 3 in. longer than the other diagonal. If the area of the rhombus is 27 in.^{2}, find the lengths of the diagonals.
10. Find the area of a rhombus if the sum of the lengths of its diagonals is 12 ft and their ratio is $3:5$.
11. If the perimeter of a rectangle is 72 ft and the length is twice the width, find the area.
12. Find the ratio of the altitudes of two triangles of equal area if the base of one is three times that of the other.
13. Find the area of an isosceles right triangle if the hypotenuse is 20 ft in length.
14. Find the area of an equilateral triangle if its side measures 8 in.
15. Show that the area of an equilateral triangle with side of length s is $\frac{1}{4}s^2\sqrt{3}$.
16. The sides of a right triangle are in the ratio of $3:4:5$. The altitude to the hypotenuse measures 12 ft. Find the area of the triangle.
17. The area of a trapezoid is 700 ft^2. The lengths of the bases are 30 ft and 40 ft, respectively. Find the length of the altitude.
18. At 90¢ a square foot, how much will it cost to tile a wall 8 ft high and 17 ft long?
19. The dimensions of a tennis court are 50 by 90 ft. If the cost of resurfacing is $9.30 a square yard, what is the total cost to resurface the court?
20. A triangular pattern being made by a template maker was to have an area of 75 in.^{2} The area was found by dividing the product of the lengths of two of its sides by 2. What kind of triangle was the pattern? The sides used in finding the area had a ratio of $2:3$. Find the lengths of all three sides.

8

Circles

When asked what was the object of being born, the philosopher Anaxagoras (?500–428 B.C.) replied, "The investigation of the sun, moon and heaven."

Certainly of all geometric figures the circle is among the most common and useful. The circle literally makes our civilization "go," with its wheels, gears, jet fans, etc. Before reading further, try to describe a circle without comparing it to a physical object.

8-1 BASIC DEFINITIONS

We need first to define the figure we call a circle.

Definition 8-1 A **circle** is a set of points in the plane a given distance from a given point. The given point is the **center,** and the given distance is the length of a **radius** of the circle. Two circles are said to be **congruent** if they have congruent radii. (Radii is the plural of radius.)

Definition 8-2 The **interior of a circle** is the set of points in the plane of the circle whose distance from the center is less than the length of the radius. The **exterior of a circle** is the set of points in the plane of the circle whose distance from the center is greater than the length of a radius of the circle (Fig. 8-1).

When we speak of a circle, the set of points does *not* include the center nor does it include a radius. A radius of a circle will be some segment with length r that is congruent to any segment OX, where O is the center of the circle and X is a point on the circle. We will use the word radius to mean both the segment and the length of the segment.

In Fig. 8-2 we see some of the segments and lines that have special relationships to circles.

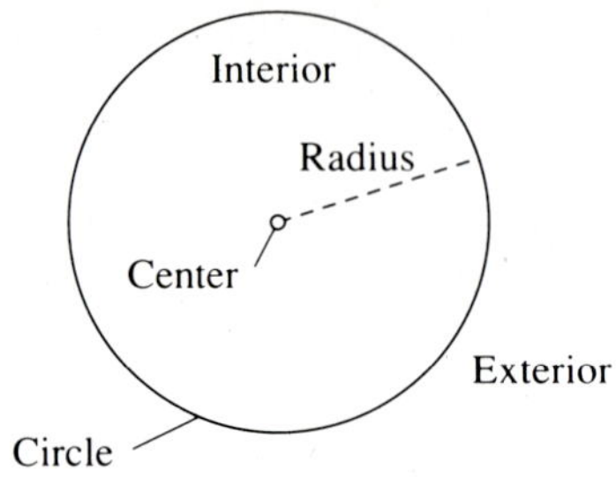

Figure 8-1

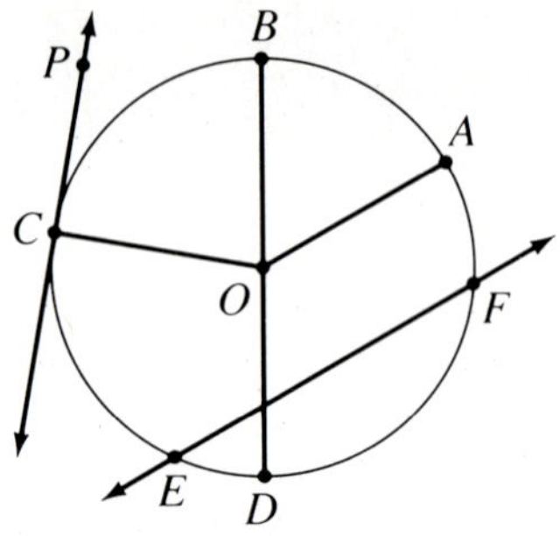

Figure 8-2

Definition 8-3 A **secant** ($\overleftrightarrow{EF}$) is a line that intersects the circle in two points. A **chord** ($\overline{EF}$) is a segment with its endpoints on the circle. A **diameter** ($\overline{BD}$) is a chord that contains the center (O) of the circle (i.e., B-O-D). A **radius** ($\overline{OA}$, $\overline{OB}$, $\overline{OC}$, $\overline{OD}$) is a segment with one endpoint at the center of the circle and the other endpoint on the circle. A **tangent** ($\overleftrightarrow{CP}$) is a line in the plane of the circle that intersects the circle in only one point, the **point of tangency.**

Definition 8-4 Two circles in the same plane are **tangent circles** if they are tangent to the same line at the same point. Circles are **internally tangent** if their centers are on the same side of the tangent line, and they are **externally tangent** if their centers are on opposite sides of the tangent line (Fig. 8-3).

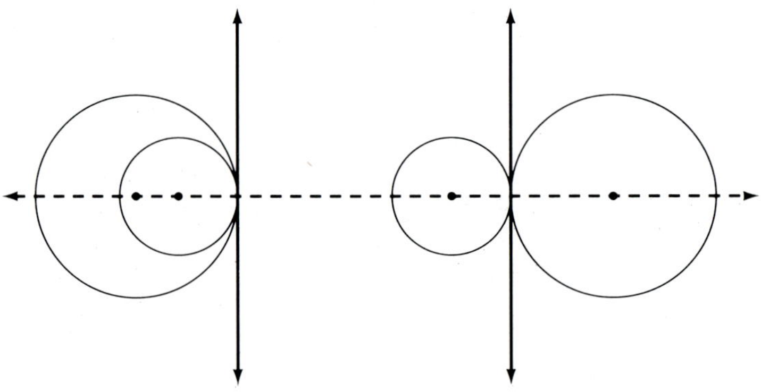

Figure 8-3

Definition 8-5 Two or more circles with the same center are said to be **concentric circles** (Fig. 8-4).

We shall refer to a circle as "a circle at O," or simply "O." When we write "circle A," we are referring to a circle with its center at some point A.

The following theorems can be proved easily with the definitions just stated, together with facts from previous chapters.

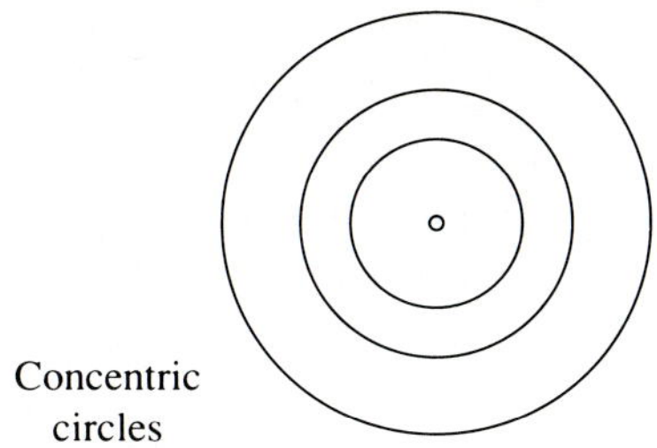

Figure 8-4

Theorem 8-1 A line through the center of a circle perpendicular to a chord bisects the chord.

Theorem 8-2 In the same or congruent circles, chords equidistant from the center of the circle are congruent.

Theorem 8-3 In the same or congruent circles, congruent chords are equidistant from the center of the circle.

Theorem 8-4 A line bisecting a chord that is not the diameter of a circle is perpendicular to the chord.

Theorem 8-5 In the plane of a given circle, the perpendicular bisector of a chord passes through the center of the circle.

Equation of a Circle*

The definition of circle tells us that it is a set of points everywhere equidistant from its center. Place a coordinate system on the center of a circle O, as in Fig. 8-5, and let $A(a, 0)$ be the endpoint of radius $\overline{OA}$. Then any other point $B(x, y)$ on the circle is such that $m\overline{OB} = m\overline{OA}$. By the Distance Formula we have $m\overline{OB} = \sqrt{(x-0)^2 + (y-0)^2} = \sqrt{x^2 + y^2}$, and $m\overline{OA} = a$, so $\{(x, y) | x^2 + y^2 = a^2\}$ is the set of all points on the circle that has radius a and center at $(0, 0)$. The equation $x^2 + y^2 = a^2$ is the equation of the circle.

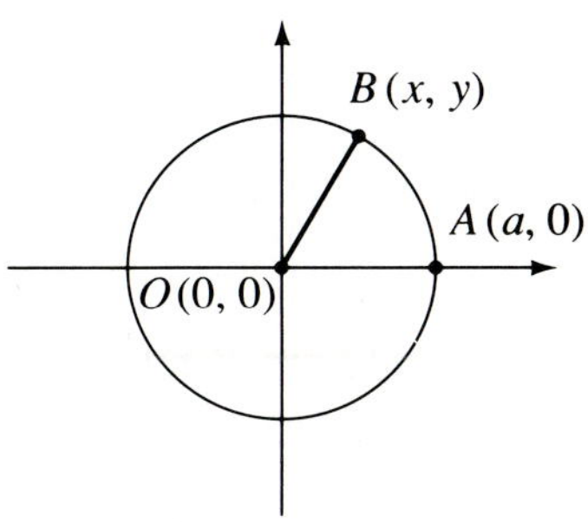

Figure 8-5

* This section may be omitted.

Example

(a) What is the equation of the circle with center (0, 0), if (7, 24) is a point on the circle? (b) What is the radius of the circle $x^2 + y^2 = 49$?

Solution (a) The distance from (0, 0) to (7, 24) is $\sqrt{7^2 + 24^2} = \sqrt{49 + 576} = \sqrt{625} = 25$. Thus $x^2 + y^2 = 25^2$, or $x^2 + y^2 = 625$, is the equation. (b) Since $a^2 = 49$, $a = 7$ is the radius of the circle. ◁

Exercise Set 8.1

1. Based on your previous knowledge and the information in Sec. 8-1, decide which of the following statements are true.
  (a) Circles with congruent radii will have the same centers.
  (b) A secant will contain two points of the circle.
  (c) A center of a circle may bisect only one chord of the circle.
  (d) A chord of a circle may contain one, two, or three points of the circle.
  (e) A radius will contain two points of the circle.
  (f) All radii of a circle or congruent circles are congruent.

2. Draw circle A with radius of length $2\frac{1}{2}$. If A, B, C, D, and E are distinct points, through point B on the circle draw
  (a) a radius $\overline{AB}$
  (b) a tangent $\overleftrightarrow{BC}$
  (c) a secant $\overleftrightarrow{BD}$
  (d) a chord $\overline{BE}$

3. Prove Theorem 8-1.

4. Prove Theorem 8-3.

5. Prove Theorem 8-5.

6. If $\overline{CD}$ is a diameter and $\overline{AB}$ is a chord of circle O and $\overline{CD} \perp \overline{AB}$ at E, prove
  (a) $\triangle ABO$ is isosceles
  (b) $\triangle AEO \cong \triangle BEO$
  (c) $\overline{AE} \cong \overline{BE}$
  (d) $\overline{AD} \cong \overline{BD}$

7. Find the length of the radius of a circle if a chord 6 in. long is 2 in. from the center of the circle.

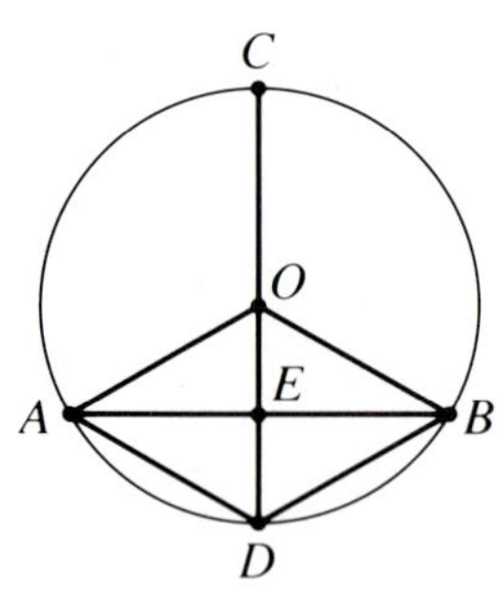

8. If diameter $\overline{BD} \perp \overline{AC}$ at E, prove $\overline{AD} \cong \overline{CD}$.

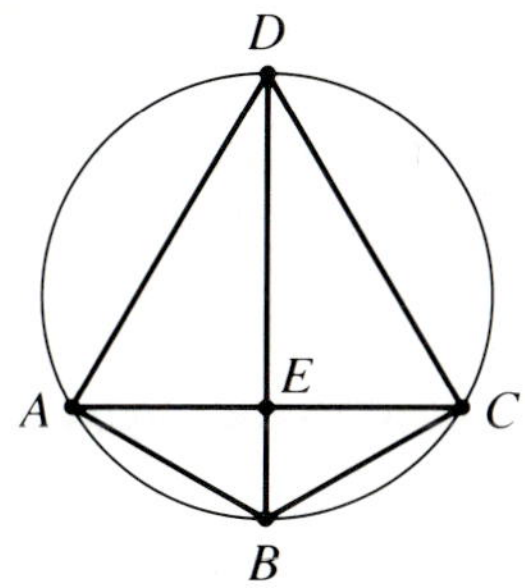

9. If a circle has a diameter of 12 in., how long is a chord that is 3 in. from the center?

***10.** Write the equation of a circle with center at the origin if point $A(8, 15)$ is on the circle.

***11.** Write an equation of the circle with center at (3, 4) if a point $P(4, 8)$ is on the circle.

8-2 ARCS AND CENTRAL ANGLES

We need now to consider some angles formed when segments and lines intersect in a circle.

Definition 8-6 A **central angle** of a circle is an angle with its vertex at the center of the circle.

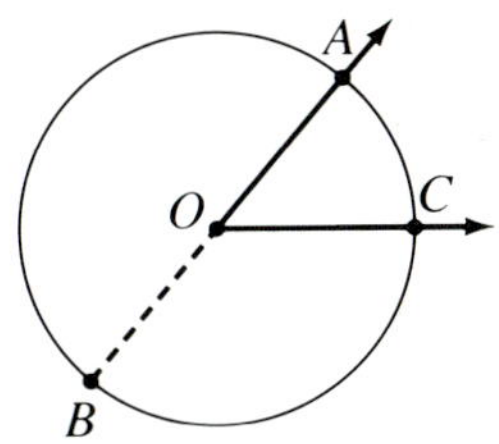

Figure 8-6

In Fig. 8-6 the diameter $\overline{AB}$ and the point C divide the circle into distinct sets of points, which we will define as follows.

Definition 8-7(a) If A and C are points on a circle O with A and C not endpoints of a diameter, an **arc** AC is the union of points A and C with the set of points of the circle between A and C.

1. The union of A and C with the set of points of the circle contained in the interior of $\angle AOC$ is a **minor arc** ($\overset{\frown}{AC}$) of the circle.
2. The union of A and C with the set of points of the circle exterior to $\angle AOC$ is a **major arc** ($\overset{\frown}{ABC}$) of the circle.

* The more difficult problems are indicated by an asterisk.

Definition 8-7(b) If A and C are endpoints of a diameter, the union of A, C and the set of points of the circle on one side of $\overline{AC}$ form a **semicircle.**

Since two points A and B on a circle could be the endpoints of either a major or a minor arc, we will indicate the minor arc by two letters as $\widehat{AB}$ and the major arc by three letters, such as $\widehat{AYB}$, in which Y is a point on the circle between A and B. We may find it useful to use three letters for a minor arc (such as $\widehat{AXB}$ in Fig. 8-7) but a major arc will not be designated using only two letters.

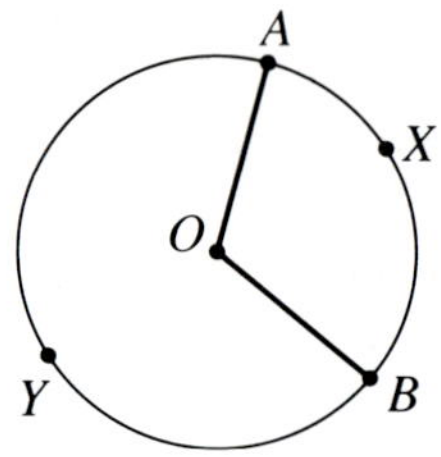

Figure 8-7

The measure of an arc relates to the measure of its central angle in the following way.

Definition 8-8 The **degree measure** of $\widehat{AXB}$ ($m°\widehat{AXB}$) is
(a) the measure of its corresponding central angle if $\widehat{AXB}$ is a minor arc ($m°\widehat{AXB} = m\angle AOB$);
(b) 180 if A and B are endpoints of a diameter;
(c) $360 - m°\widehat{AYB}$ if $\widehat{AXB}$ is a major arc and $\widehat{AYB}$ is the corresponding minor arc.

Examples

1. If $m\angle AOB = 70$ and $\overline{AB} \parallel \overline{CD}$, find $m°\widehat{AC}$, $m°\widehat{AB}$, and $m°\widehat{BD}$.

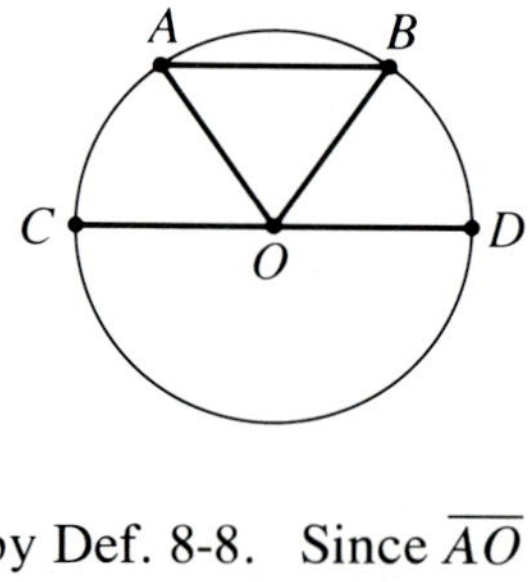

Solution $m°\widehat{AB}$ is 70 by Def. 8-8. Since $\overline{AO} \cong \overline{BO}$, $\angle A \cong \angle B$ and $m\angle A = m\angle B = 55$; but $\angle A \cong \angle AOC$ and $\angle B \cong \angle BOD$, since they are alternate-interior angles. Thus $m°\widehat{AC} = m°\widehat{BD} = 55$.

2. In the following figure, chord $\overline{AB}$ is congruent to radius $\overline{AC}$. What is the degree measure of $\widehat{AB}$?

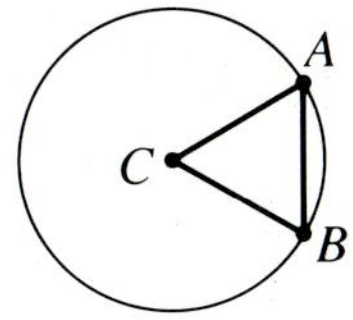

Solution Since $\overline{AC}$ and $\overline{BC}$ are radii, the triangle is equilateral, equiangular, and $m\angle ACB = 60$. Thus $m°\overset{\frown}{AB} = 60$.

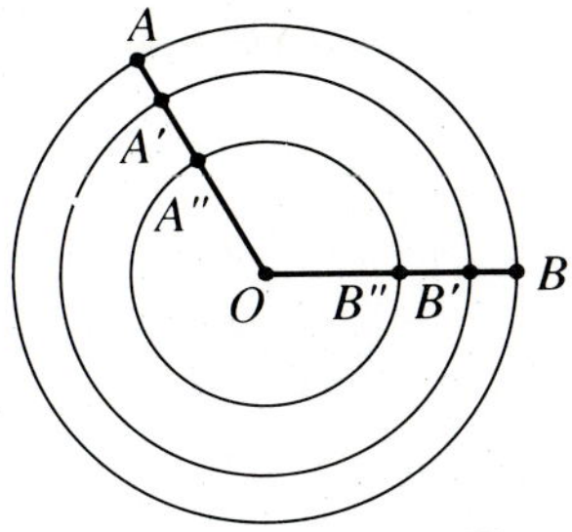

In the figure we see that $m°\overset{\frown}{AB} = m°\overset{\frown}{A'B'} = m°\overset{\frown}{A''B''}$, since all have the same central angle AOB, but the distances from A to B, from A' to B', and from A'' to B'' on the circles are different. We will discuss the measure of this distance in a later section. ◁

Theorem 8-6 If A, B, and C are points on a circle such that $\overset{\frown}{ACB}$, then $m°\overset{\frown}{AC} + m°\overset{\frown}{CB} = m°\overset{\frown}{AB}$. (The proof of this theorem follows from the addition of angular measure, Post. 3-3, and the definition of degree measure of arcs.)

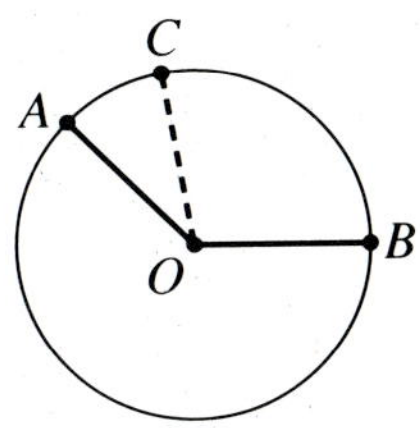

Definition 8-9(a) In the same or congruent circles, two **arcs are congruent** if they have the same degree measure.

Definition 8-9(b) The **midpoint of an arc** separates the arc into two congruent arcs.

If D is the midpoint of $\overset{\frown}{AB}$, then $\overset{\frown}{AD} \cong \overset{\frown}{DB}$ or $m°\overset{\frown}{AD} = m°\overset{\frown}{DB}$ (Fig. 8-8).

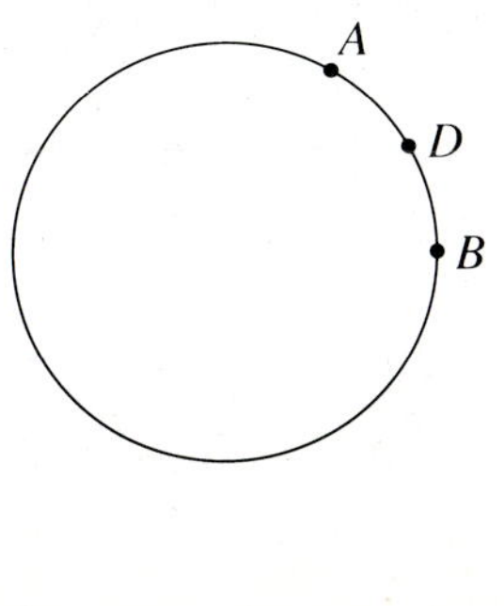

Figure 8-8

Theorem 8-7 In the same or congruent circles, if two arcs are congruent, their corresponding chords are congruent.

Given: Congruent circles P and Q with $\widehat{AB} \cong \widehat{CD}$
Prove: $\overline{AB} \cong \overline{CD}$

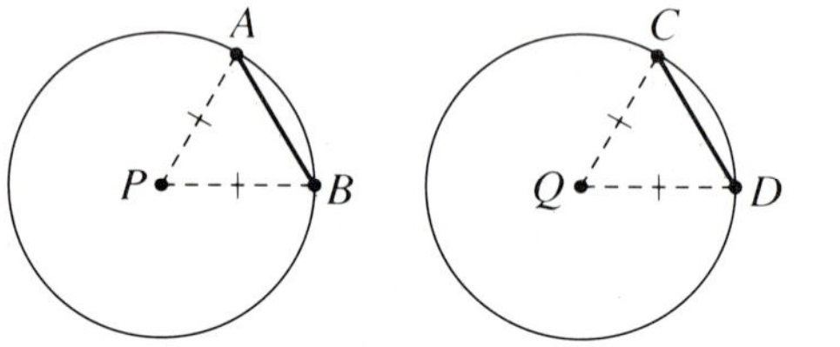

Proof The radii of circles P and Q are congruent (Def. 8-4). Since $\widehat{AB} \cong \widehat{CD}$, $m°\widehat{AB} = m°\widehat{CD}$ and $m\angle APB = m\angle CQD$ by Defs. 8-9(a) and 8-8. Thus $\triangle APB \cong \triangle CQD$ by the SAS Postulate, so $\overline{AB} \cong \overline{CD}$.

The converse of this theorem can be proved in a similar fashion and is left to the reader.

Theorem 8-8 In the same or congruent circles, if two chords are congruent, their corresponding minor arcs are congruent.

Examples

1. If, in a circle P, the endpoints of the diameters $\overline{AB}$ and $\overline{CD}$ are joined consecutively by segments, what figure is formed?

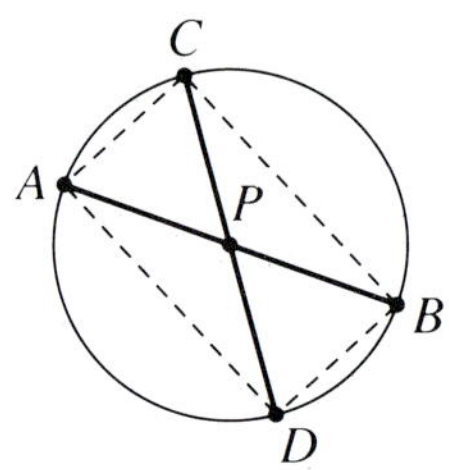

Solution $\angle APD \cong \angle CPB$ and $\angle APC \cong \angle BPD$ since they are vertical angles, their respective arcs and chords are therefore congruent, and the figure is a parallelogram. (Is it more than just a parallelogram?)

▶ **2.** Given a circle with radius of length 2. What is the length of the chord that is intercepted by a central angle of 70°?

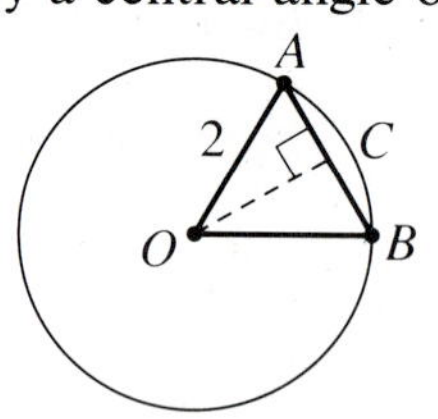

Solution Let $m\angle AOB = 70$ and $m\overline{AO} = 2$. Draw $\overline{OC} \perp \overline{AB}$ at C. Then $\triangle AOC \cong \triangle BOC$ and $\overline{AC} \cong \overline{BC}$, $\angle AOC \cong \angle BOC$, so $m\angle AOC = 35$.

Sin 35° = $m\overline{AC}/2$, or $m\overline{AC} = 2 \sin 35° \approx 2(.5736) = 1.1472$; $m\overline{AB} = 2m\overline{AC}$, so $m\overline{AB} \approx 2(1.1472) = 2.2944$. ◁

Exercise Set 8.2

1. If $\overline{AC}$ is a diameter, find
   (a) $m°\widehat{EDC}$ = ______ **(b)** $m°\widehat{ECB}$ = ______ **(c)** $m°\widehat{BAD}$ = ______

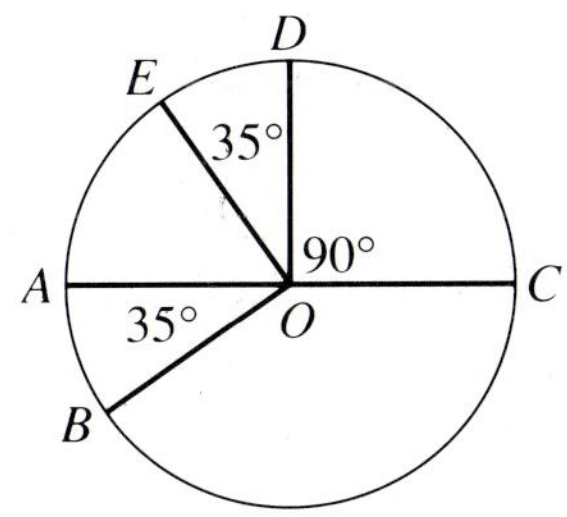

2. If a circle is separated into six arcs with the same degree measure, what is the measure of each of the congruent central angles?
3. Given $\widehat{AD} \cong \widehat{BC}$ and $\widehat{AB} \cong \widehat{DC}$ in the following figure. Prove $ABCD$ is a parallelogram.

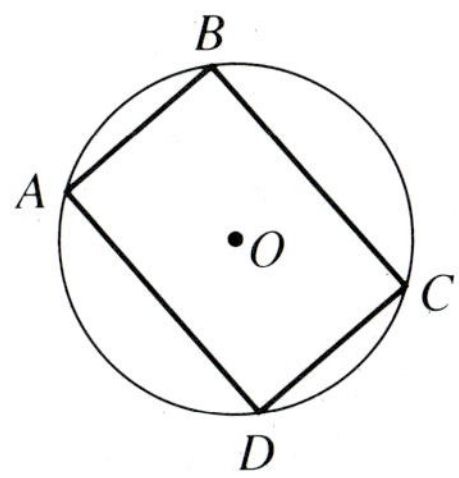

4. If the degree measure of an arc is doubled, is the measure of its central angle doubled? Is the length of its chord doubled? Explain.
5. Given B the midpoint of $\widehat{CD}$, $\overline{AB}$ a diameter. Prove $\overline{AC} \cong \overline{AD}$.

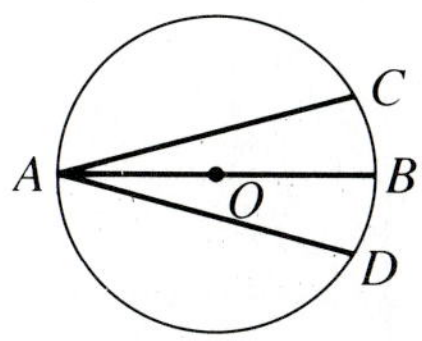

6. Suppose $\overline{UT} \parallel \overline{RS}$ and $\overline{RS}$ a diameter of circle O.
   (a) Prove $m°\widehat{RU} = m°\widehat{TS}$.
   (b) If the above is given and $m\angle ROT = 115$, find $m\angle TOS$, $m\angle ROU$ and $m\angle TOU$.

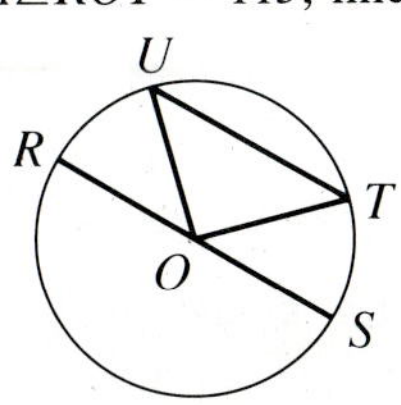

7. If $ABCD$ is a rhombus and $\overline{AC}$ and $\overline{BD}$ are diameters, prove that $ABCD$ is a square.

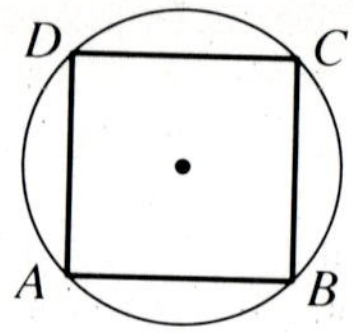

8. In the following figure, $m°\widehat{AB} = 3x + 10$, $m\angle BOC = 2x - 5$, $m\angle COD = 40$, and AOD is a diameter. Find: (a) $m\angle AOB$, (b) $m°\widehat{AC}$, (c) $m°\widehat{BD}$.

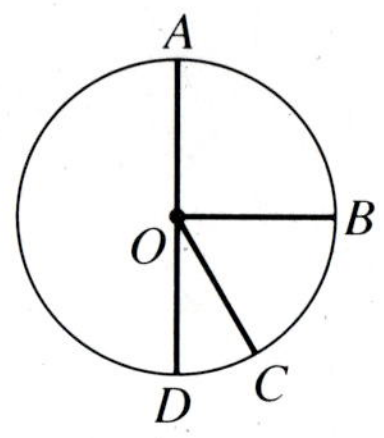

9. How long is the chord of a circle with radius of length r if the central angle of the chord is as given?
(a) 120° and $r = 9$
(b) 60° and $r = 2\sqrt{3}$
(c) 90° and $r = 2$
▶**(d)** 70° and $r = 10$

10. **(a)** Prove Theorem 8-6.
(b) Prove Theorem 8-8.

11. Prove that a diameter perpendicular to a chord bisects the chord and its arcs.

In Exercises 12–15, use the following figure and determine the measures by means of a protractor.

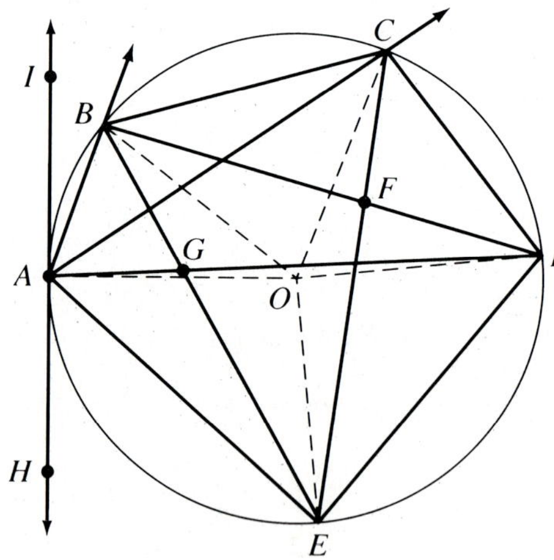

12. Find the degree measure of each of the following:
(a) $\widehat{AB}$ **(b)** $\widehat{BC}$ **(c)** $\widehat{CD}$ **(d)** $\widehat{DE}$ **(e)** $\widehat{AE}$

13. Find each of the following:

(a) $m\angle AEB$, $m\angle BCA$, $m\angle ADB$
(b) $m\angle BAC$, $m\angle BEC$, $m\angle BDC$
(c) $m\angle CBD$, $m\angle CAD$, $m\angle CED$
(d) $m\angle DCE$, $m\angle DBE$, $m\angle DAE$
(e) $m\angle ABE$, $m\angle ACE$, $m\angle ADE$
(f) Compare the preceding answers to those in the corresponding parts of Exercise 12.

14. (a) $m°\widehat{CD} + m°\widehat{BE} =$ ___?___; $m\angle CFD =$ ___?___
(b) $m°\widehat{AB} + m°\widehat{DE} =$ ___?___; $m\angle AGB =$ ___?___.

15. (a) $m°\widehat{AB} =$ ___?___; $m\angle IAB =$ ___?___
(b) $m°\widehat{AE} =$ ___?___; $m\angle HAE =$ ___?___
(c) $m°\widehat{ABD} =$ ___?___; $m\angle IAD =$ ___?___
(d) $m°\widehat{AED} =$ ___?___; $m\angle HAD =$ ___?___.

8-3 OTHER ANGLES RELATED TO THE CIRCLE

Arcs of circles are sometimes cut off by angles other than the central angle. One such angle is shown in Fig. 8-9.

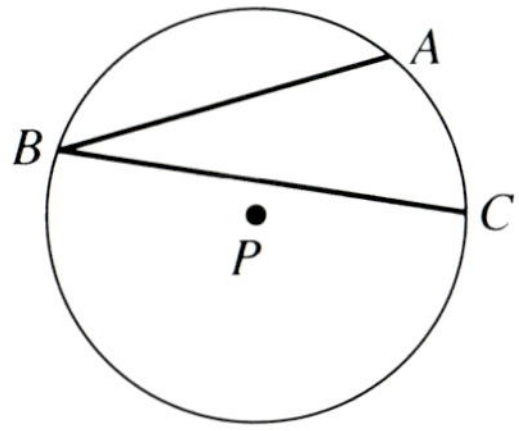

Figure 8-9

Definition 8-10 An angle ABC is **inscribed** in an arc ABC if the vertex B lies on the circle and endpoints A and C of the arc lie on the sides of the angle. The angle ABC is called an **inscribed angle.**

In Fig. 8-10(a) the angle ABC is inscribed in major arc ABC, whereas in Fig. 8-10(b) angle ABC is inscribed in minor arc ABC. In Fig. 8-10(c) angle ABC is inscribed in a semicircle. In all the figures in Fig. 8-10, the angles are said to *intercept* the arcs AXC.

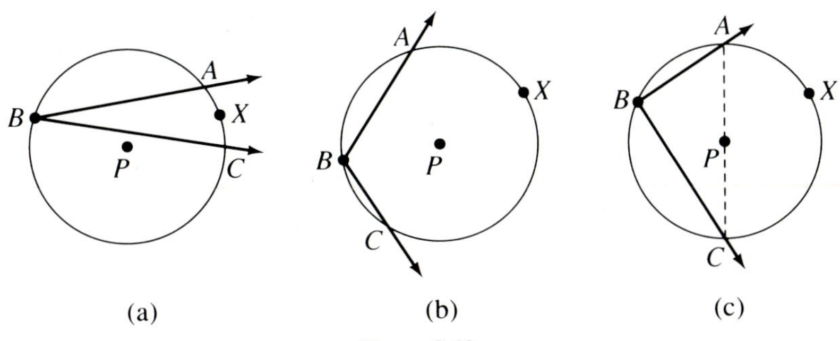

Figure 8-10

Definition 8-11 An angle **intercepts** an arc if, except for its endpoints, the arc lies in the interior of the angle.

We will use the arc intercepted by angles because under some conditions there is a simple relationship between the measure of the angle and the measure of the arc. The illustrations in Fig. 8-11 show some of the various ways an angle might intercept an arc of a circle.

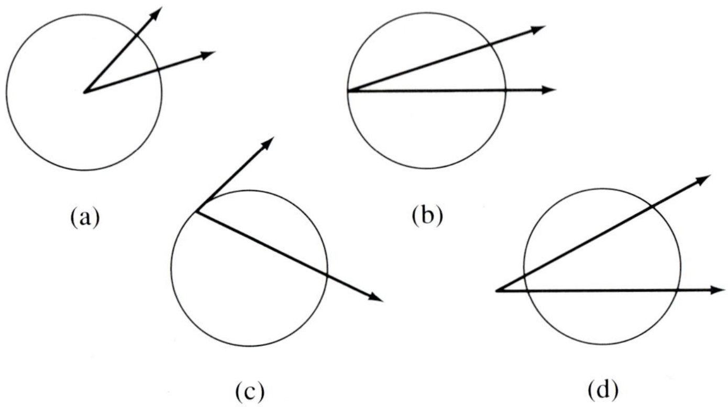

Figure 8-11

Theorem 8-9 The measure of an angle inscribed in a circle is one-half the degree measure of its intercepted arc.

Given: $\angle ABC$ inscribed in a circle P

Prove: $m\angle ABC = \frac{1}{2}m^\circ\widehat{AC}$

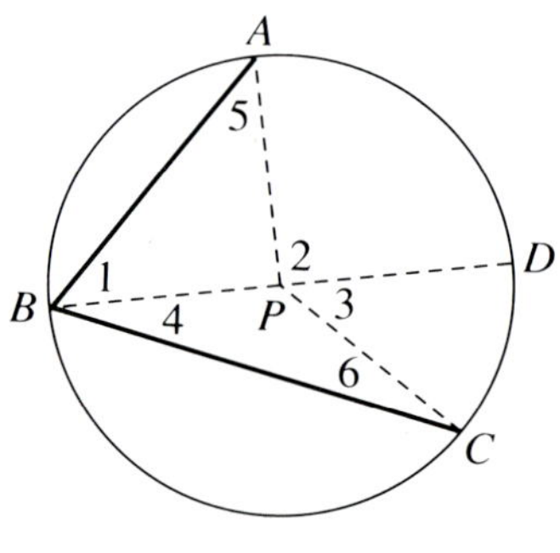

Proof

Statements	*Reasons*
1. From B draw diameter $\overline{BD}$	1. Definition 8-3
2. $m\angle 1 + m\angle 5 = m\angle 2$, and $m\angle 4 + m\angle 6 = m\angle 3$	2. The measure of an exterior angle of a triangle is equal to the sum of the measures of the two nonadjacent interior angles (Corollary 5-21(d))

Statements	*Reasons*
3. $\overline{PA} \cong \overline{PB} \cong \overline{PC}$	3. Definition 8-1
4. $\angle 1 \cong \angle 5$ and $\angle 4 \cong \angle 6$	4. Angles opposite congruent sides of a triangle are congruent (Theorem 4-5)
5. $m\angle 1 = \frac{1}{2}m\angle 2$ and $m\angle 4 = \frac{1}{2}m\angle 3$	5. Statements 2 and 4 and Multiplication Law of Equals
6. $m\angle ABC = \frac{1}{2}m\angle APC$	6. If P is in the interior of $\angle ABC$, then $m\angle ABC = m\angle ABP + m\angle PBC$. (Post. 3-3)
7. $m\angle ABC = \frac{1}{2}m°\widehat{AC}$	7. Definition 8-8

(If one side of the inscribed angle is a diameter, the proof is simplified in that only $\angle 1$, $\angle 5$, and $\angle 2$ need be considered. If A and C are on the same side of $\overline{BD}$, the proof is similar to the one above.)

The following important corollaries can be derived from this theorem.

Corollary 8-9(a) An angle inscribed in a semicircle is a right angle (Fig. 8-12).

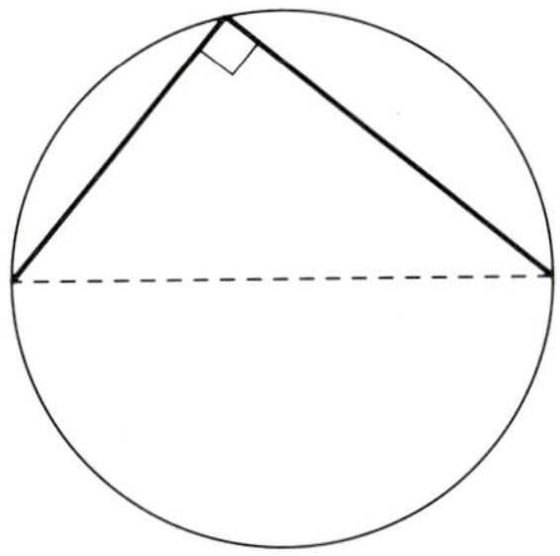

Figure 8-12

Corollary 8-9(b) Angles inscribed in the same arc are congruent (Fig. 8-13).

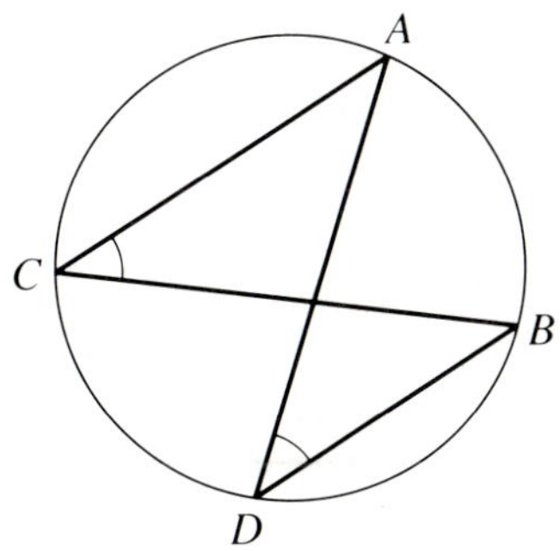

Figure 8-13

Theorem 8-10 The measure of an angle formed by two secants intersecting in the interior of a circle is one-half the sum of the degree measures of the arcs intercepted by the angle and its vertical angle.

Given: $\overleftrightarrow{AB}$ and $\overleftrightarrow{CD}$ intersect at E, a point in the interior of a circle
Prove: $m\angle AED = \frac{1}{2}(m°\widehat{AD} + m°\widehat{CB})$

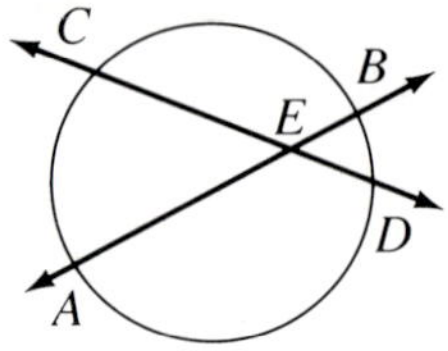

Exercise Set 8.3

1. A quadrilateral $ABCD$ is inscribed in circle Q, with $m\angle B = 80$, $m°\widehat{AB} = 60$, and $m°\widehat{DC} = 40$. Find:
   (a) $m°\widehat{AD}$ **(b)** $m°\widehat{BC}$
   (c) $m\angle A$ **(d)** $m\angle C$
   (e) $m\angle D$

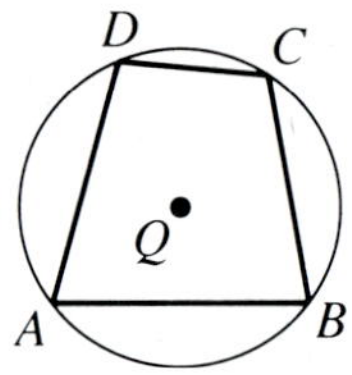

2. **(a)** Prove Corollary 8-9(a).
   (b) Prove Corollary 8-9(b).
3. Prove that the opposite angles of an inscribed quadrilateral are supplementary.
4. If chords $\overline{AC}$ and $\overline{BD}$ intersect at E, prove that $\triangle AED \sim \triangle BEC$.

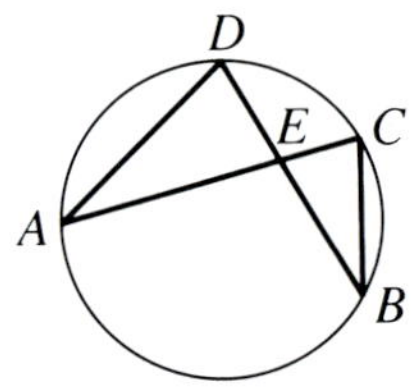

5. Prove Theorem 8-10. (*Hint:* Draw chord $\overline{AD}$ in the figure accompanying the theorem.)
6. In the following figure $m\angle DFB = 50$, $m°\widehat{AE} = 70$, and $m°\widehat{ED} = 120$. Find $m°\widehat{DB}$, $m°\widehat{AB}$, $m\angle A$, $m\angle EDA$, $m\angle C$, and $m\angle EFD$.

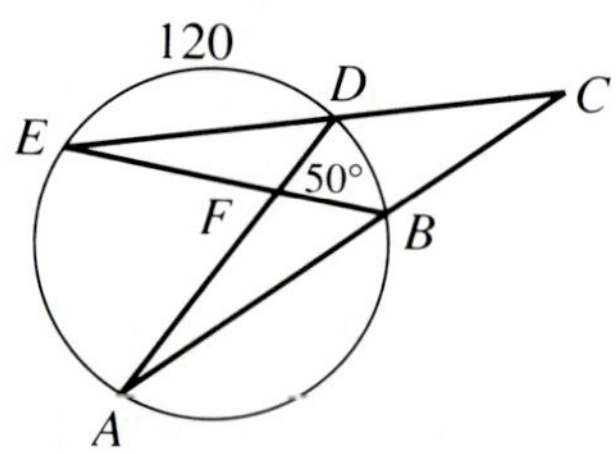

7. If in the following figure $\overline{AC}$ is a diameter, $m°\widehat{DC} = 40$ and $m\angle ACB = 70$, find $m°\widehat{AD}$, $m°\widehat{AB}$, $m°\widehat{BC}$, $m\angle ADC$, $m\angle ACD$, $m\angle BAC$, and $m\angle ABC$.

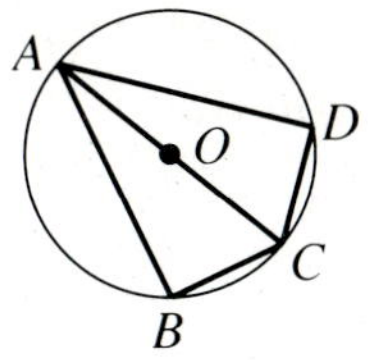

8. $\overline{AB}$ is a diameter and chord $\overline{CE} \perp \overline{AB}$ at D. If $m\overline{AD} = 4$ and $m\overline{DB} = 9$, find $m\overline{CE}$.

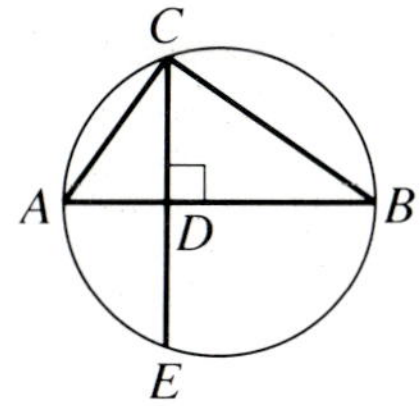

9. *Given:* Chords $\overline{AC}$ and $\overline{BD}$ intersect at E
Prove: $a(b) = c(d)$

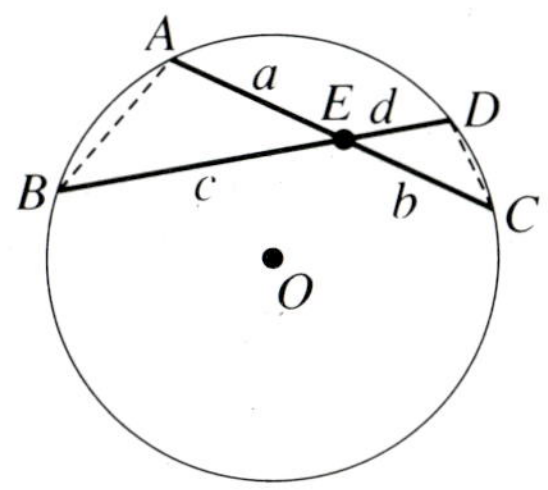

10. Refer to the figure for Exercise 9.
(a) If $m\overline{AE} = 24$, $m\overline{EC} = 10$ and $m\overline{BE} = 15$, find $m\overline{ED}$.
(b) If $m\overline{AC} = 22$, $m\overline{BE} = 8$ and $m\overline{ED} = 12$, find $m\overline{EC}$.

11. *Given:* $\overrightarrow{AB}$ and $\overrightarrow{ED}$ are secants of circle O that intersect at C
Prove: $m\overline{AC} \cdot m\overline{BC} = m\overline{EC} \cdot m\overline{DC}$ (Supply reasons for the following statements.)

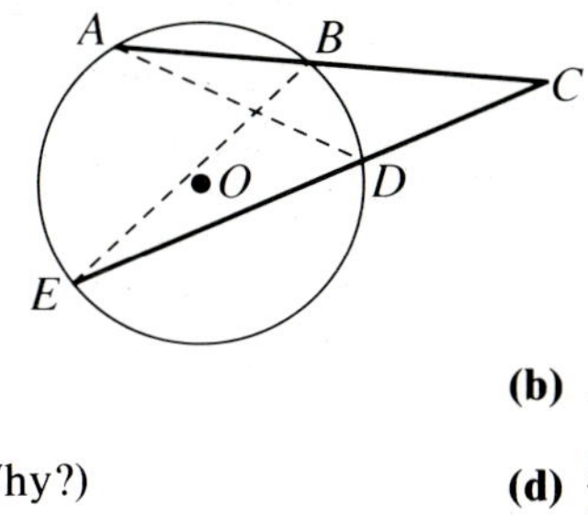

(a) $\angle C \cong \angle C$ (Why?)
(b) $\angle BAD \cong \angle DEB$ (Why?)
(c) $\triangle ADC \sim \triangle EBC$ (Why?)
(d) $\dfrac{m\overline{AC}}{m\overline{EC}} = \dfrac{m\overline{DC}}{m\overline{BC}}$ (Why?)
(e) $m\overline{AC} \cdot m\overline{BC} = m\overline{EC} \cdot m\overline{DC}$ (Why?)

12. Refer to the figure for Exercise 11. If $m\overline{AC} = 45$, $m\overline{BC} = 14$, and $m\overline{CD} = 18$, find $m\overline{DE}$.

Now let us consider what happens when lines or segments intersect from the exterior of the circle.

Theorem 8-11 If a line is tangent to a circle, it is perpendicular to the radius drawn to the point of contact.

Given: Circle P with $\overleftrightarrow{AB}$ tangent at B
Prove: $\overleftrightarrow{AB} \perp \overline{PB}$

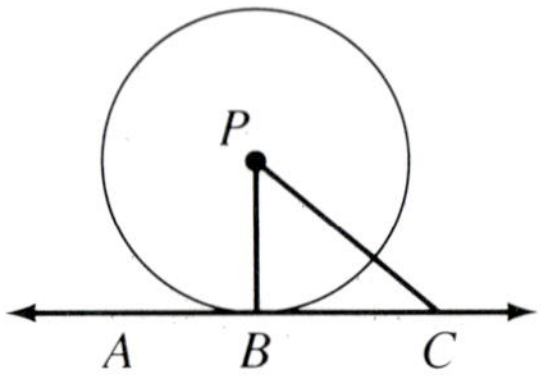

Proof

Statements	*Reasons*
1. $\overleftrightarrow{AB}$ is tangent to P at B	1. Given
2. From a point C on $\overleftrightarrow{AB}$, $B \neq C$, draw $\overline{PC}$	2. Post. 2-1(a)
3. C is in the exterior of circle P.	3. Definition of tangent (Def. 8-3)
4. $\text{m}\overline{PC} > r$, or $\text{m}\overline{PC} > \text{m}\overline{PB}$	4. Definition of exterior of circle (Def. 8-2)
5. $\overleftrightarrow{AB} \perp \overline{PB}$	5. The shortest segment joining a point to a line is the perpendicular segment (Theorem 5-5)

Corollary 8-11 The two tangent segments from a point of the exterior of a circle are congruent (Fig. 8-14).

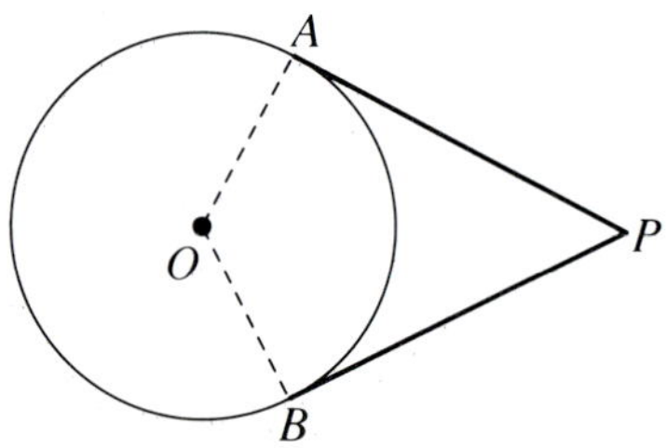

Figure 8-14

Theorem 8-12 The measure of an angle formed by a secant ray and a tangent ray with its vertex on the circle is equal to one-half the degree measure of its intercepted arc.

Given: Circle P with $\angle ABC$ formed by secant $\overrightarrow{BA}$ and tangent $\overrightarrow{BC}$
Prove: $\text{m}\angle ABC = \frac{1}{2}\text{m}°\widehat{AB}$

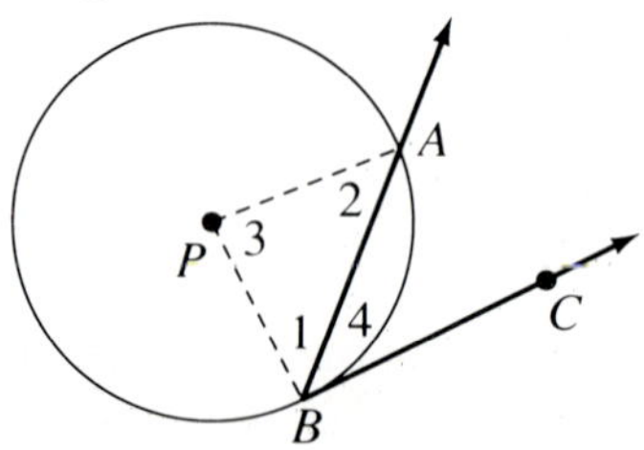

Proof Triangle APB is isosceles; hence $\angle 1 \cong \angle 2$ and $2\text{m}\angle 1 + \text{m}\angle 3 = 180$. Thus $\text{m}\angle 1 + \frac{1}{2}\text{m}\angle 3 = 90$. From Theorem 8-11, $\overline{PB} \perp \overline{BC}$ and $\text{m}\angle 1 + \text{m}\angle 4 = 90$. Applying the Transitive Axiom, $\text{m}\angle 1 + \text{m}\angle 4 = \text{m}\angle 1 + \frac{1}{2}\text{m}\angle 3$ and then $\text{m}\angle 4 = \frac{1}{2}\text{m}\angle 3$. Since $\text{m}\angle 3 = \text{m}^\circ\overset{\frown}{AB}$, $\text{m}\angle 4 = \frac{1}{2}\text{m}^\circ\overset{\frown}{AB}$.

Example

In the following figure $\overline{EB}$ is a diameter, $\overline{DB}$ and $\overline{CB}$ are chords, $\overline{PC}$ a radius and $\overrightarrow{BA}$ a tangent to circle P at B. If $\text{m}\angle BPC = 70$ and $\text{m}^\circ\overset{\frown}{ED} = 20$ find

(a) $\text{m}\angle ABC$

(b) $\text{m}\angle ABD$

(c) $\text{m}\angle CBE$

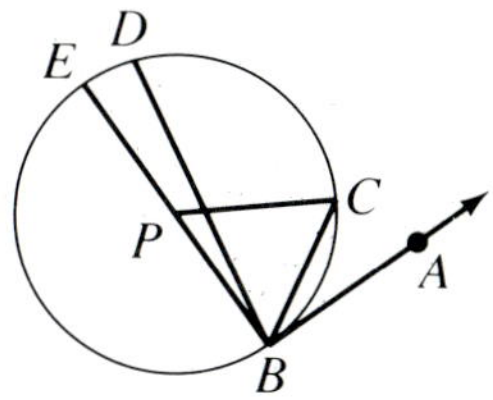

Solution

(a) $\text{m}^\circ\overset{\frown}{BC} = 70$, $\text{m}\angle ABC = 35$ (Theorem 8-12)

(b) $\text{m}\angle PCB = \text{m}\angle PBC = \frac{1}{2}(180 - 70) = 55$; $\text{m}\angle EBD = \frac{1}{2}\text{m}^\circ\overset{\frown}{ED} = 10$, so $\text{m}\angle ABD = 90 - 10 = 80$

(c) $\text{m}^\circ\overset{\frown}{CD} = 90$, so $\text{m}^\circ\overset{\frown}{CD} + \text{m}^\circ\overset{\frown}{DE} = \text{m}^\circ\overset{\frown}{CE} = 110$ (or $\overset{\frown}{BCE}$ is a semicircle so that $\text{m}^\circ\overset{\frown}{CE} = 180 - \text{m}^\circ\overset{\frown}{BC} = 110$, $\text{m}\angle CBE = 55$ ◁

Theorem 8-13 The measure of the angle formed by two secants intersecting in the exterior of a circle is one-half the difference between the degree measures of the larger and the smaller of the two intercepted arcs.

Given: Secants AB and CD intersect at point P in the exterior of the circle

Prove: $\text{m}\angle APC = \frac{1}{2}(\text{m}^\circ\overset{\frown}{AC} - \text{m}^\circ BD)$

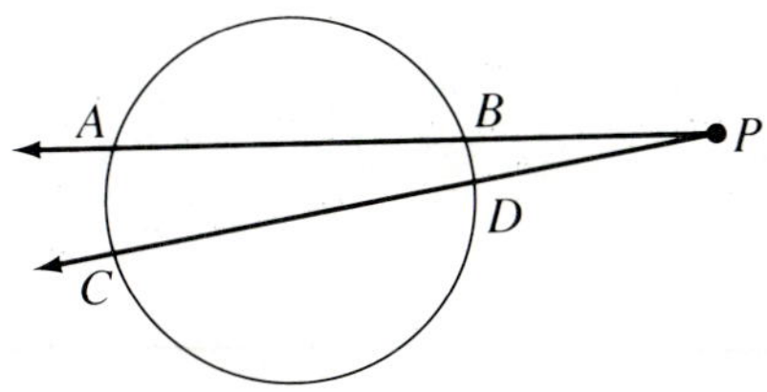

Exercise Set 8.4

1. Prove Corollary 8-11. (*Hint:* Draw a segment from the external point to the center of the circle.)

2. In the following figure $\overleftrightarrow{AB}$ is tangent to circle O at P, and Q, R, and S are points on the circle. Complete the following:
(a) $\angle RPS \cong$ ______ (b) $\angle QPR \cong$ ______
(c) $\angle BPS$ is supplementary to ______ (d) $\angle QRS$ is supplementary to ______
(e) $\angle APQ \cong$ ______ $\cong$ ______ (f) $\angle PQS$ is supplementary to ______.

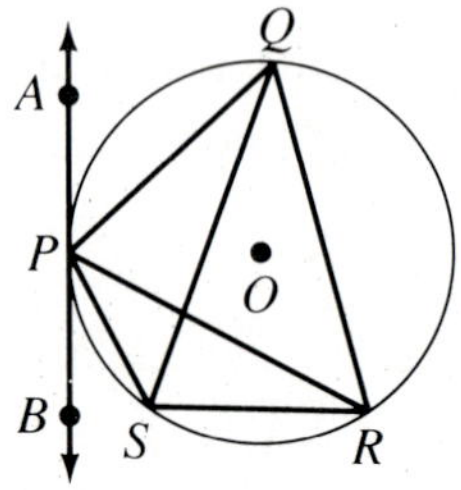

3. In the following figure $\overleftrightarrow{AB}$ and $\overleftrightarrow{CD}$ are tangent to circle O. $\overline{BC}$ is a diameter 8 units in length, and $m°\widehat{EC} = 120$. Find the length of (a) $\overline{BE}$ and (b) $\overline{AC}$.

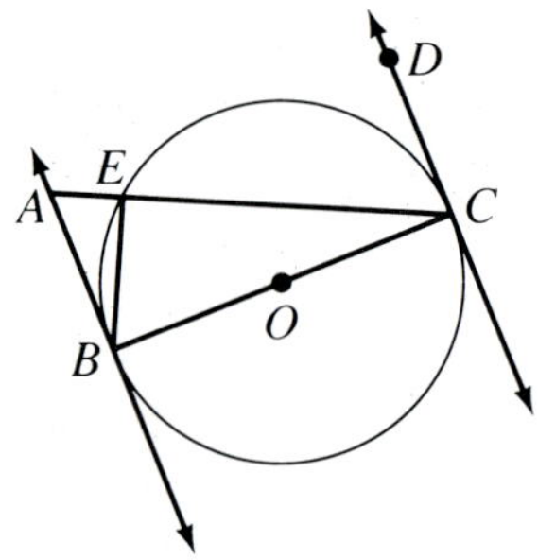

4. In the following figure tangent $\overleftrightarrow{AB}$ and secant $\overleftrightarrow{BD}$ intersect at B, $m°\widehat{AD} = 100$, and $m°\widehat{AC} = 60$. Find $m\angle B$.

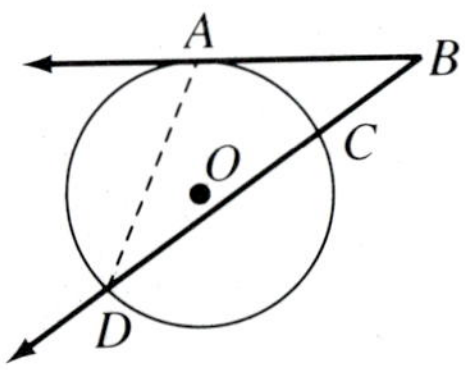

5. Using the figure for Exercise 4, prove that $m\angle B = \frac{1}{2}(m°\widehat{AD} - m°\widehat{AC})$.
6. In the following figure $m°\widehat{EB} = 106$, $m°\widehat{ED} = 80$, and $\overline{BD}$ bisects $\angle ABC$. Find $m°\widehat{DF}$, $m°\widehat{BF}$, $m\angle A$, $m\angle ABD$, $m\angle DBC$, $m\angle EGB$, and $m\angle EGD$.

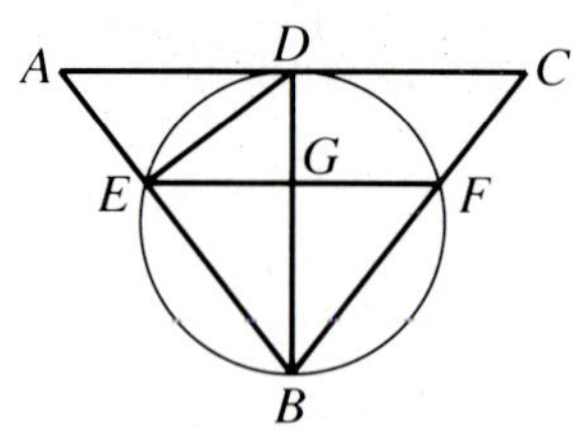

7. Given $\overrightarrow{BA}$ tangent to circle O at A, and $\overrightarrow{BD}$ a secant intersecting the circle at C and D, prove $(m\overline{AB})^2 = m\overline{BD} \cdot m\overline{BC}$.

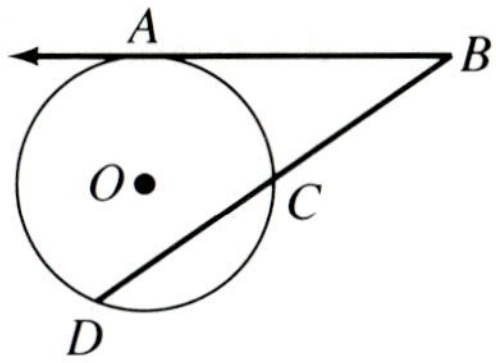

8. Refer to the figure of Exercise 7. If $m\overline{BC} = 8$ and $m\overline{BD} = 12$, find $m\overline{AB}$.
9. If circles O_1 and O_2 are tangent at the point C, prove that $\overline{AB} \cong \overline{AD}$.

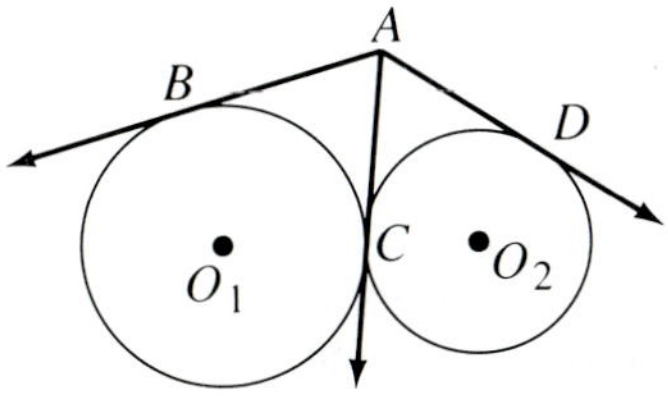

10. How far from the center of a circle with radius 8 must a point be so that the tangents from the point to the circle form a right angle at the point?
11. Prove Theorem 8-13. (*Hint:* Draw chord $\overline{BC}$ in the figure accompanying the theorem.)
12. Refer to the answer for Exercise 6 on page 230. Is the measure of angle C consistent with what you would expect from Theorem 8-13?
13. In the following figure secants $\overrightarrow{MN}$ and $\overrightarrow{RQ}$ intersect at P, which is in the exterior of the circle. If $m\angle P = 35$ and $m°\widehat{NQ} = 40$, find $m°\widehat{MR}$.

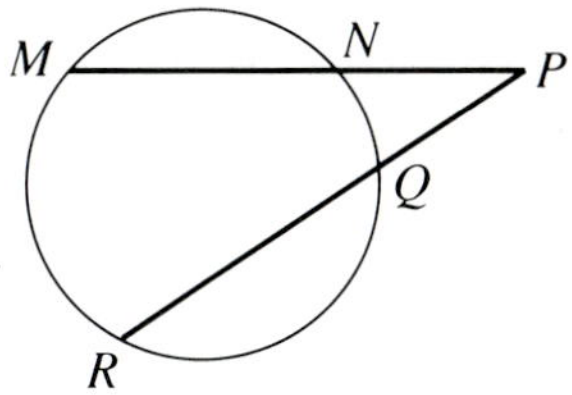

8-4 CIRCLES AND REGULAR POLYGONS

In Chapter 7 convex polygons were defined and some of their properties were discovered. A number of additional properties relate to special polygons which can be inscribed in a circle.

Definition 8-12 A polygon is **inscribed in a circle** if all of its vertices lie on the circle. The circle is then said to be **circumscribed** around the polygon and is called the **circumcircle** of the polygon.

Definition 8-13 A convex polygon is said to be **regular** if all its sides are congruent and all its angles are congruent. A **central angle** of a regular polygon is the angle determined by the center of the circumscribed circle and the endpoints of a side of the polygon. The segment joining the center of the circumscribed circle to a side of the regular polygon is the **apothem** of the polygon (Fig. 8-15).

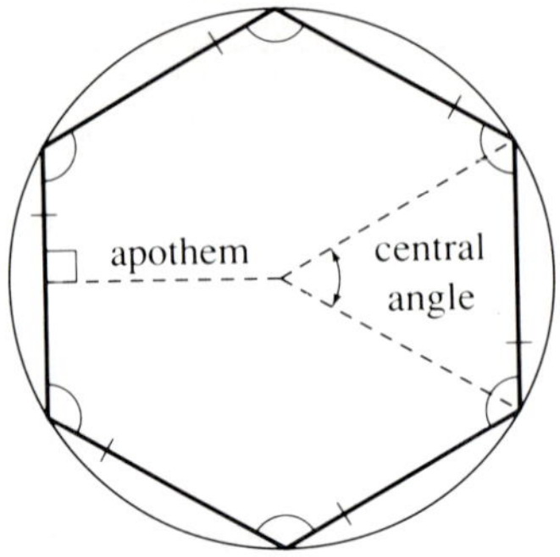

Regular Polygon

Figure 8-15

If we place a coordinate system on a circle with the origin at its center, as in Fig. 8-16, and then bisect each of the angles formed by the axes, we find that the circle is separated into 8 congruent arcs. If we draw the chords of these arcs, we have congruent sides for the polygon. If we draw radii from the center to each of the endpoints of the arcs, we have formed isosceles triangles. Since the central angles are congruent the angles of the polygon are congruent and the polygon is a regular octagon.

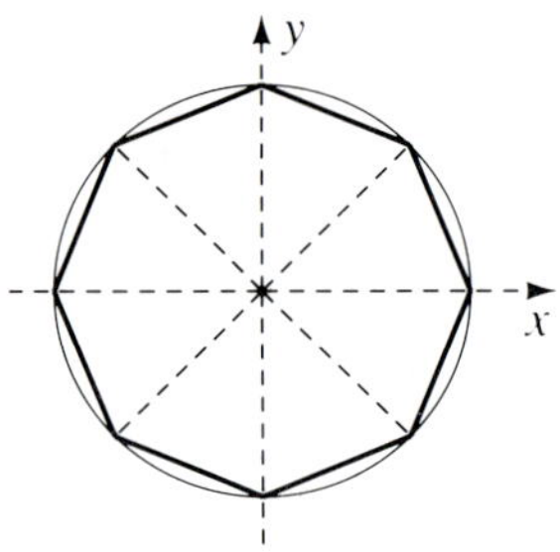

Figure 8-16

In Fig. 8-16 and the previous discussion, we determined that the measure of each central angle is 360/8, and 8 is the number of congruent sides of the inscribed regular polygon.

Likewise we can find the measure of the central angle of any regular polygon, as indicated in the following theorem.

Theorem 8-14 The measure of a central angle of a regular polygon of n sides is $360/n$.
Given: Circle P with a regular n-gon inscribed
Prove: $m\angle APB = 360/n$

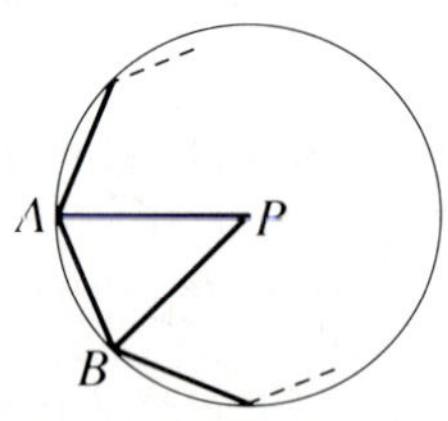

Proof

Statements	*Reasons*
1. $m\angle APB = m°\widehat{AB}$	1. Definition of degree measure of an arc (Def. 8-8(a))
2. $n \cdot m°\widehat{AB} = 360$	2. Theorem 8-6, Def. 8-8(b)
3. $m°\widehat{AB} = \frac{360}{n}$	3. Multiplication Law of Equals
4. $\therefore m\angle APB = \frac{360}{n}$	4. Axiom E-4, Statements 1 and 3

Example

If a regular hexagon is inscribed in a circle of radius r, find the measure of (a) its central angle, (b) each side, (c) angle ABC.

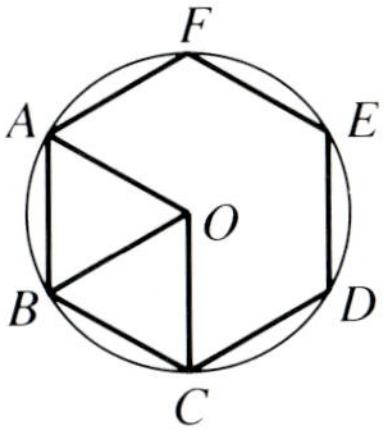

Solution (a) A regular hexagon has six congruent sides and angles; $\therefore m\angle AOB = 360/6 = 60$.

(b) Since $\triangle AOB$ is isosceles and $m\angle AOB = 60$, $m\angle OAB = m\angle OBA = \frac{1}{2}(180 - 60) = 60$ and the triangle is equilateral. Thus $m\overline{AB} = r$.

(c) $m\angle ABC = m\angle ABO + m\angle OBC = 120$. ◁

The area of a regular polygon may be found by summing the areas of the triangles formed by the radii from the center of the circumscribed circle and the sides of the polygon. The area of each triangle is one-half of the product of the lengths of a side and the apothem, since the apothem is perpendicular to the side. Thus if a is the length of the apothem, b is the length of a side, and n is the number of sides, the area of the polygon is $n \cdot \frac{1}{2}ab = \frac{1}{2}a \cdot nb$. Since the perimeter of the polygon is nb, we can state this as a theorem.

Theorem 8-15 The area (K) of a regular polygon is equal to one-half the product of the length of its apothem (a) and its perimeter (p). ($K = \frac{1}{2}ap$)

Examples

▶ **1.** Find the area of a regular pentagon inscribed in a circle with radius 10.

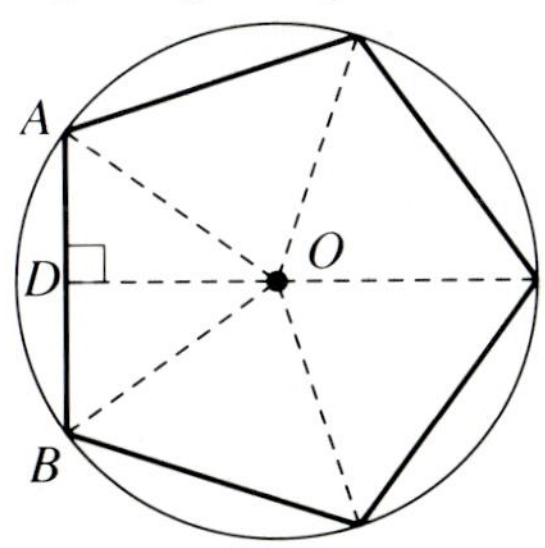

Solution $m\angle AOB = 360/5 = 72$. Draw $\overline{OD} \perp \overline{AB}$, and $m\angle AOD = m\angle BOD = \frac{1}{2} \cdot 72 = 36$. Then $\cos 36° = a/10$, so $a = 10 \cos 36°$, and $\sin 36° = m\overline{AD}/10$, so $m\overline{AD} = 10 \sin 36°$. Since $p = 5 \cdot 2m\overline{AD}$, $K = \frac{1}{2}(10 \cos 36°)(10 \cdot 10 \sin 36°) \approx 500(0.8090)(0.5878) \approx 239$.

2. Find the area of a regular hexagon inscribed in a circle of radius r.

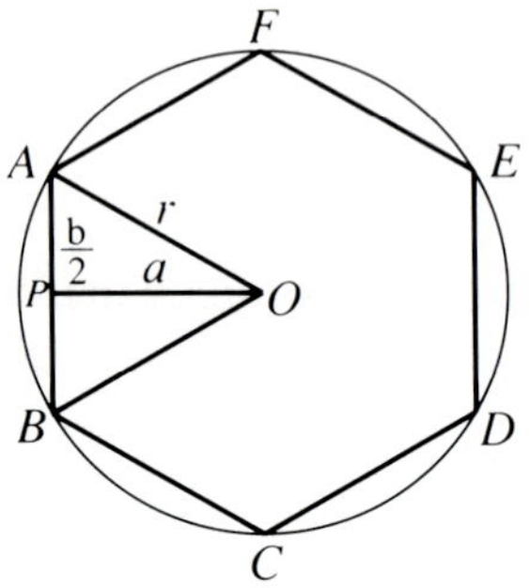

Solution In $\triangle APO$, $m\angle OAP = 60$ and $m\angle AOP = 30$. (Why?) Hence $a = \frac{1}{2}\sqrt{3} \cdot r$ and $b/2 = r/2$ or $b = r$, so $K = \frac{1}{2}(\frac{1}{2}\sqrt{3} \cdot r)(6r) = \frac{3}{2}\sqrt{3}\, r^2$. ◁

Exercise Set 8.5

1. Use a figure that is similar to the following to show that the sum of the measures of the angles of a convex polygon with n sides is $(n - 2)180$.

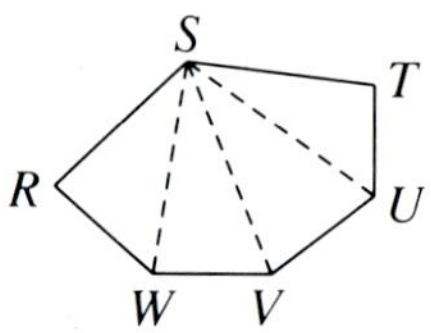

2. Verify the conclusion of Exercise 1 using the following figure.

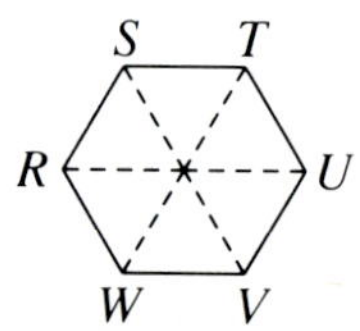

3. Prove that the measure of an angle of a regular polygon of n sides is $(n - 2) \cdot 180/n$.
4. A polygon is *circumscribed about a circle* if each side is tangent to the circle. Prove that the area of a polygon circumscribed about a circle is equal to one-half its perimeter times the radius of the circle.
5. Prove that the area of a square inscribed in a circle with radius r is $2r^2$.
6. Prove that the area bounded by a square circumscribed about a circle with radius r is $4r^2$.

7. If the radius of a circle is 8:
   (a) What is the area of the inscribed square?
   (b) What is the area of the circumscribed square?
   (c) What is the length of the apothem of the circumscribed square?

8. Show that the measure of the sum of the exterior angles of a convex polygon is 360. (*Hint:* Refer to the measure of the interior angles.)

9. If the length of the apothem of a regular hexagon is $2\sqrt{3}$, find **(a)** the radius of the circumscribed circle, **(b)** the perimeter of the hexagon, and **(c)** the area of the hexagon.

10. Repeat Exercise 9 if the length of the apothem is 6.

11. Determine the measure of each angle of a regular polygon of
   (a) 5 sides **(b)** 6 sides
   (c) 9 sides **(d)** 10 sides
   (e) 15 sides **(f)** 18 sides
   (g) 20 sides

12. How many sides does a regular polygon have if the measure of one of its angles is
   (a) 135 **(b)** 144
   (c) 150 **(d)** 120
   (e) $128\frac{4}{7}$

13. How many sides does a regular polygon have if the measure of each of its exterior angles is
   (a) 24 **(b)** 36
   (c) 45 **(d)** 72

14. Determine the area of a regular hexagon with side of length 8.

▶ 15. Determine the area of a regular dodecagon if each side has length 10.

▶ 16. Find the length of a side and the area of a regular decagon inscribed in a circle of radius 20.

▶ 17. Show that the area of a regular n-gon inscribed in a circle of radius r is $nr^2\left(\sin\frac{180}{n}\right)\left(\cos\frac{180}{n}\right)$.

▶ 18. Refer to Exercise 15, page 211, and prove that the area of a regular n-gon inscribed in a circle of radius r is $\frac{1}{2}nr^2\sin\frac{360}{n}$.

19. The cam for the distributor of a six-cylinder engine (a regular hexagon) is 0.25 in. between the high points such as A and B. What is the distance OC from the center to the edge of the cam?

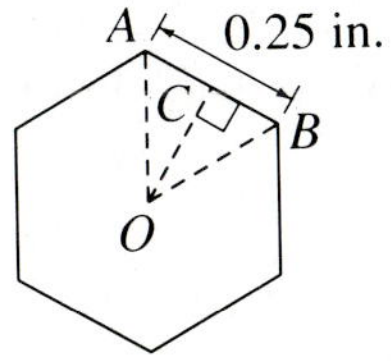

▶ The cam for the distributor of an eight-cylinder engine (a regular octagon) measures 0.375 in. from the center O to a vertex A. The shaft opening has a 0.250-in. radius. What is the area of one face of the cam?

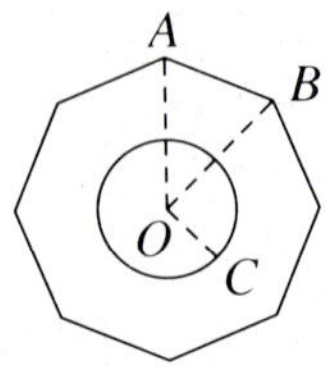

8-5 CIRCUMFERENCE AND AREAS OF CIRCLES

Let us now consider a regular n-gon inscribed in a circle of circumference C. If the regular polygon inscribed in this circle has a very large number of sides, we can use its perimeter P to approximate the value of C. The more we increase the number of sides of the polygon, the closer our approximation will be to the value of C. We describe this situation by saying that the perimeter P of the inscribed regular polygon will approach the circumference C as the number of sides of the polygon is allowed to increase without limit. We write this as

$$P \to C$$

and say that "P approaches C as a limit." If we let this idea be the definition for circumference of a circle, we then may state a postulate in terms of this concept (Fig. 8-17).

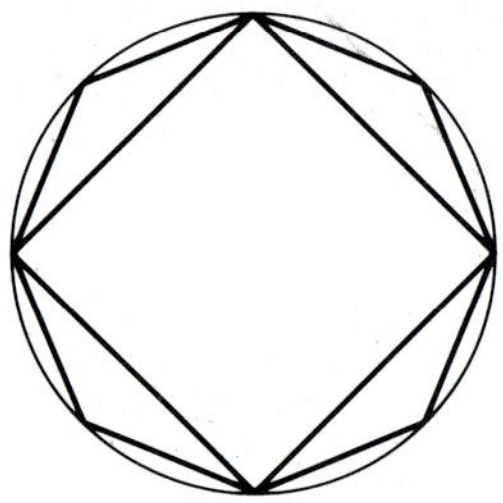

Figure 8-17

Postulate 8-1 The circumference of a circle is the limit of the perimeters of the inscribed regular polygons.

Recorded history tells us that man has been aware of and has used a constant to relate the circumference and diameter of a circle for over 3500 years. The Papyrus of Ahmes (1550 B.C.) made reference to this ratio as being 3.16. In prior courses you probably used the rational approximation of 22/7 or perhaps 3.1416 for this number, which we symbolize by π. To show that this ratio is the same for all circles, we need the following theorem.

Theorem 8-16 The ratio $\frac{C}{d}$ of the circumference to the diameter is constant for all circles.

ARCHIMEDES (287?–212 B.C.) has been called The Great Geometer. In his book, *Sand Reckoner,* he speculates on the amount of sand required to fill the universe. It is through this work that the theories of the contemporary astronomer Aristarchus were recorded. Aristarchus believed that the earth revolves about the sun, a theory that fell into disfavor when another astronomer named Ptolemy claimed that the earth was the center of the universe and that the planets, the sun, and the stars revolved around the earth. This theory was maintained for many centuries until another astronomer, Copernicus, held sway and established the foundations of modern astronomy. Archimedes, though famous for his geometry and applied mathematics, is also known for his inventions of machines of war. These machines were so successful that besieging armies were held off from Syracuse for three years. When the city was finally captured, the triumphant soldiers of Claudius Marcellus were under orders not to kill Archimedes because of his scientific genius. Unfortunately, a soldier found Archimedes drawing figures in the sand and ordered him to move, and when the preoccupied scientist failed to respond he was killed. (*Historical Pictures Service, Chicago*)

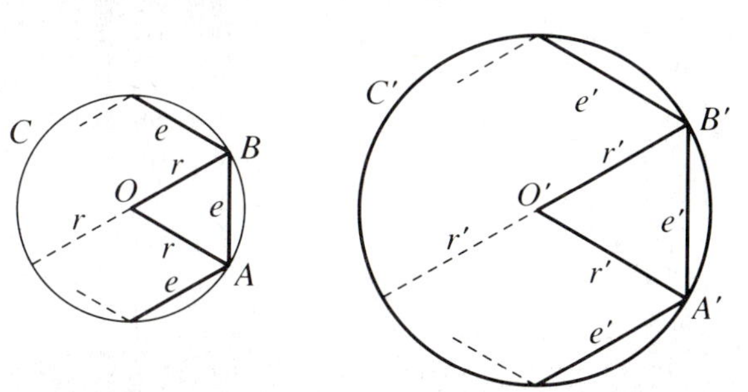

In circles O and O' inscribe regular n-gons (n the same in each circle). The central triangles will be similar, since the central angles are congruent and adjacent sides proportional. (Why?) Therefore $e/r = e'/r'$ or $e/e' = r/r'$, and $(n \cdot e)/(n \cdot e') = r/r'$. Since $P = n \cdot e$ and $P' = n \cdot e'$, $P/P' = r/r'$ and $P/P' = d/d'$. As $P \to C$ and $P' \to C'$, by Post. 8-1 we have

$$\frac{C}{C'} = \frac{2r}{2r'}, \quad \text{or} \quad \frac{C}{2r} = \frac{C'}{2r'}, \quad \text{so} \quad \frac{C}{d} = \frac{C'}{d'}$$

The symbol π is used to denote this constant ratio, C/d, so if $C/d = \pi$, $C = \pi d$ or $C = 2\pi r$ are formulas for determining circumference when the diameter or radius is given. The number π is irrational, but we can approximate its value to as many decimal places as necessary, i.e., 3, 3.14, 3.1416, 3.14159265358979. . . . The approximation 22/7 is reasonably "good," since $22/7 \approx 3.1428$ and is actually closer to the "true" value of π than 3.14. (See Exercise 1 in the following set of exercises.)

Example

What is the circumference of a circle that contains an inscribed square of side 5?

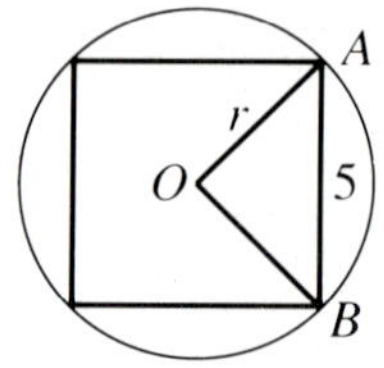

Solution Since $C = 2\pi r$, we need to find r. The triangle AOB is isosceles since $\overline{OA} \cong \overline{OB}$ and $m\angle AOB = 360/4 = 90$ by Theorem 8-14. Therefore, $m\overline{OA} = 5\sqrt{2}/2$ and $C = 2\pi \cdot 5\sqrt{2}/2 = 5\sqrt{2}\pi$. Note that we did not replace π by an approximate value. This can always be done later if needed. Always leave your answer in terms of π unless requested to compute it to a particular number of decimal places or significant digits. ◁

Exercise Set 8.6

1. Which is the closer approximation to the value of π, 3.14 or 22/7? Prove your answer.

In Exercises 2–6, leave your answers in terms of π when appropriate.

2. Find the circumference of a circle of radius 5 in.
3. Find the circumference of a circle with diameter of length 28 ft.
4. Find the circumference of the path that the moon describes each month, assuming the orbit to be circular and the distance from the center of the earth to the moon to be 240,000 mi.
5. If the circumference of a circle is 24π ft, what is the radius?
6. If the circumference of a circle of radius 1000 mi. is increased by 10 ft, what is the radius of the new circle?
7. The radius of a circle is 25 in. How much is the circumference changed by increasing the radius by 1 in.? By 5 in.?

*8. If the equations of two circles are $C_1 = \{x, y) | x^2 + y^2 = 36\}$ and $C_2 = \{(x, y) | x^2 + y^2 = 64\}$, what is the ratio of their circumferences?

9. What is the circumference of a circle circumscribed about a hexagon with side of length 8?

▶ 10. What is the circumference of a circle circumscribed about a pentagon with side of length 12 (to the nearest tenth).

11. A rectangular swimming pool 12 ft by 16 ft is to be enclosed by a circular fence, which must be at least 3 ft from any point of the pool. What are the least radius and circumference possible for the fence?
12. If a bicycle wheel of radius 14 in. revolves 1500 times, how many feet has it traveled? (Use $\pi = 22/7$.)

▶ 13. A pentagonal garden 6 ft wide is to be built around a pool of radius 12 ft.

* The more difficult problems are indicated by an asterisk.

(a) What is the area of the garden?
(b) If a sprinkler is to be placed in the center of the pool, what is the minimum radius required to ensure water coverage of the garden?

Length of Arc

We have seen that an arc is a part of a circle, hence there is some real number that can be determined as a fractional part of the circumference which we will call the *length* of the arc.

Recall, the degree measure of an arc is that number associated with its central angle. We now need to develop the relationship between the degree measure and arc length. Consider an arc of degree measure 45. This arc is one-eighth the circumference of the circle, hence its length must be $\frac{1}{8}(2\pi r)$ or $\pi r/4$. We state this relation in the following postulate.

Postulate 8-2 The ratio of the length of an arc of a circle to its circumference is equal to the ratio of the measure of the central angle to 360.

In Fig. 8-18, if s is the length of $\widehat{AB}$, then $s/C = m\angle AOB/360$ or $s = C \cdot m\angle AOB/360 = \pi r \cdot m\angle AOB/180$.

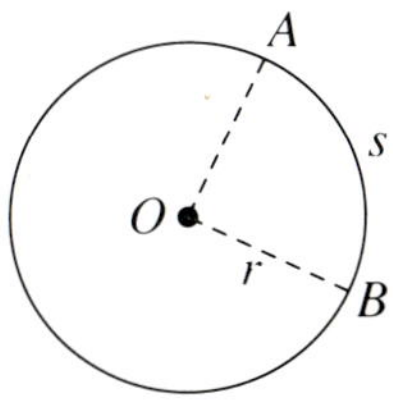

Figure 8-18

Examples

1. In a circle of radius 8, what is the length of an arc if its central angle is (a) 60°, (b) 75°?

Solution (a) $s = \dfrac{\pi \cdot 8 \cdot 60}{180} = \dfrac{8\pi}{3}$ (b) $s = \dfrac{\pi \cdot 8 \cdot 75}{180} = \dfrac{10\pi}{3}$

2. If the degree measure of an arc is 45 and its length is 50 in., find its radius.

Solution $\dfrac{\pi \cdot r \cdot 45}{180} = 50$ in., $r = \dfrac{200 \text{ in.}}{\pi}$ ◁

Exercise Set 8.7

1. In a circle of radius 10 in., what is the length of an arc whose degree measure is
   (a) 30 **(b)** 60 **(c)** 75 **(d)** 135 **(e)** $22\frac{1}{2}$
2. If the length of an arc of degree measure 60 is 6 in., what is the radius of the arc? What is the length of the chord of the arc?
3. Find the length of an arc of an inscribed regular quadrilateral with a diagonal of 20.
4. Find the measure of the central angle of a circle having an arc of length 8π if the circle has a radius of 12.

5. If the minute hand of a clock is 9 in. long, how far will the tip travel in 25 min?

▶ 6. Find the lengths of the arcs of a circle with radius 4 in. that are determined by a chord of length 6 in.

▶ 7. A regular pentagon with sides 8 in. long is inscribed in a circle. What is the length of the arc cut off by each side?

▶ 8. A belt connects two pulleys with diameters 10 in. and 8 in., as shown. What is the length of the belt if the centers are 12 in. apart?

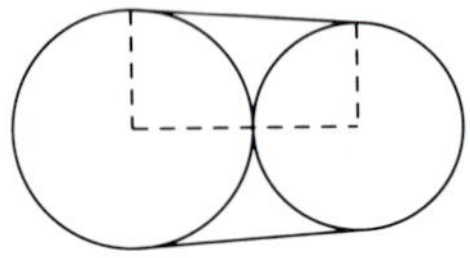

9. A dial indicator must show 15 equally spaced intervals. If the length of the indicator is 5.2 in., what is the distance traveled by the tip in moving from 1 to 2 (to the nearest tenth of an inch)?

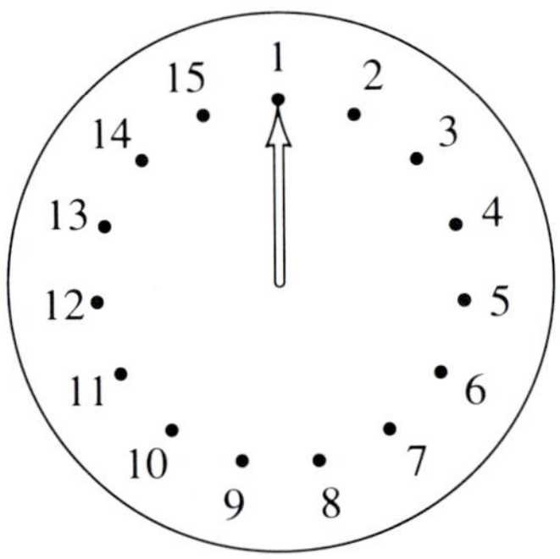

To complete this discussion on circles, we need to consider areas. As with polygons, we will define the circular region, for it is a region that has area.

Definition 8-14 A **circular region** is the union of the circle and its interior.

Recall that the area of an inscribed polygonal region is one-half the product of the measure of its apothem and perimeter. Referring to Fig. 8-19, we see that

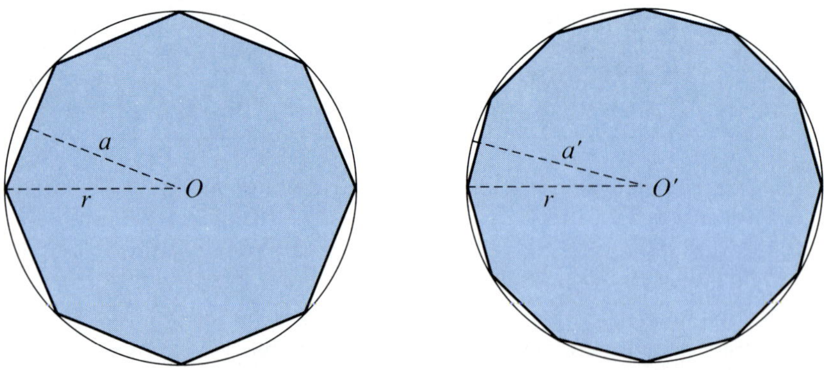

Figure 8-19

as we increase the number of sides of the polygon the length of the apothem approaches the radius as a limit as the perimeter approaches the circumference. Consider a square and a regular octagon inscribed in a circle of radius r. The square has apothem $\frac{1}{2}r\sqrt{2}$ (approximately $0.707r$) and the octagon has apothem of length $\frac{1}{2}r(\sqrt{2 + \sqrt{2}})$ (approximately $0.924r$). Here we can observe that, in this case, as we increased the number of sides of the inscribed polygon, the apothem more nearly approximated the radius. Thus the area of the polygon is approaching the area of the circle.

Postulate 8-3 The area of a circular region is the limit of the areas of the inscribed polygons.

From our discussion leading to the statement of Post. 8-3, we see that if K_n is the area of the polygon with apothem of length a and perimeter p, and if K is the area of the circumscribed circular region with radius r and circumference C, then $K_n \rightarrow K$. Since $K_n = \frac{1}{2}ap$, as $a \rightarrow r$ and $p \rightarrow C$, $\frac{1}{2}ap \rightarrow \frac{1}{2}rC$. But $C = 2\pi r$, so $\frac{1}{2}rC = \frac{1}{2}r \cdot 2\pi r = \pi r^2$, and $K_n \rightarrow K = \pi r^2$.

Following the practice adopted in the previous chapter, we will simply refer to the area of the circle rather than area of the circular region.

Theorem 8-17 The area (K) of a circle of radius r is πr^2.

Examples

1. Find the area of a circle with

(a) radius 6

(b) circumference 16π

Solution (a) $K = \pi r^2 = \pi \cdot 6^2 = 36\pi$ square units

(b) $C = 2\pi r$; $\therefore 16\pi = 2\pi r$. Multiplying by $1/2\pi$, we obtain $r = 8$. Then $K = \pi \cdot 8^2 = 64\pi$ square units.

2. Find the radius of a circle with area 36π square units.

Solution $K = \pi r^2$; $\therefore 36\pi = \pi r^2$ and $r^2 = 36$, or $r = 6$. ◁

Definition 8-15 The region of a **sector of a circle** is the union of two radii, their intercepted arc, and the set of points that are both in the interior of the circle and in the interior of the angle formed by the radii.

Postulate 8-4 The ratio of the area of a sector of a circle to the area of the circle is equal to the ratio of the measure of the central angle of the sector to 360.

In Fig. 8-20, if K is the area of sector AOB, $K/\pi r^2 = m\angle AOB/360$, or $K = \pi r^2 \cdot m\angle AOB/360$.

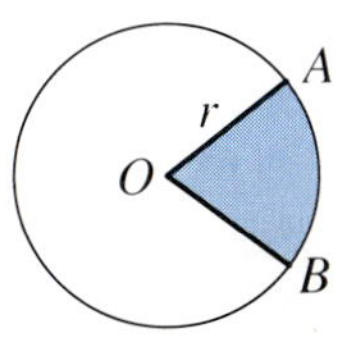

Figure 8-20

Example

Find the area of a sector of a circle with radius 5 whose arc has degree measure of 72.

Solution $K = \dfrac{\pi r^2 \cdot m\angle AOB}{360} = \dfrac{\pi(5)^2(72)}{360} = 5\pi$ ◁

Exercise Set 8.8

Give answers in terms of π when appropriate.

1. Find the circumference of a circle with radius
   (a) 4 **(b)** 5
   (c) 8 **(d)** 10
2. Find the area of a circle with diameter
   (a) 4 **(b)** 5
   (c) 8 **(d)** 10
3. Find the area of the shaded portion in the following figure if
   (a) $r_1 = 3$, $r_2 = 5$ **(b)** $r_1 = 4$, $r_2 = 6$
   (c) $r_1 = 2.25$, $r_2 = 3.75$ **(d)** $r_1 = 2/\pi$, $r_2 = 4/\pi$
   (e) $r_1 = \frac{14}{33}$, $r_2 = \frac{28}{55}$

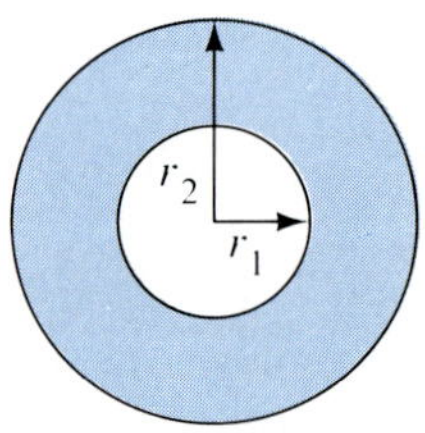

4. Find the area of a circle with circumference
   (a) 16π **(b)** 24π
   (c) 48π **(d)** 50π
5. If the ratio of the areas of two circles is 4 to 1, what is the ratio of their circumferences?
6. What is the area of a circle circumscribed about a square with sides of length (a) 6, (b) 8?
7. If the radius of a circle is 8, find the area of a sector with an arc which has degree measure
   (a) 30 **(b)** 45
   (c) 72 **(d)** 120
8. What is the area swept by the 9-in.-long minute hand of a clock in 25 min. (to the nearest tenth)?
9. Find the measure of the central angle of a sector of a circle of radius 3 if the area of the sector is
   (a) $\dfrac{5\pi}{4}$ **(b)** 4π
10. If the length of the arc of a sector of a circle with radius 10 is 8, find the area of the sector.

11. If the area of a sector of a circle with radius 15 is 45, find the length of the arc of the sector.

12. A *segment* of a circle is the region bounded by an arc and its chord. The area of the segment may be found by subtracting the area of triangle AOB from the area of sector AOB, as indicated in the following figure. Find the area of the segment if
   (a) $m°\widehat{AB} = 60$ and $r = 16$
   ▶ **(b)** $m\angle AOB = 150$ and $r = 12$
   ▶ **(c)** $m°\widehat{AB} = 70$ and $r = 6$

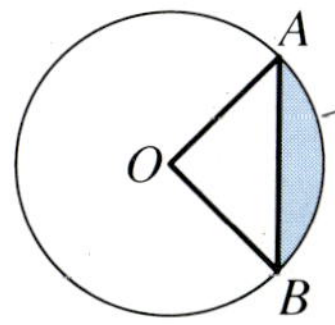

13. The following figure represents a cross section of a blower assembly. Using the dimensions given, find the area of the cross section to the nearest square inch.

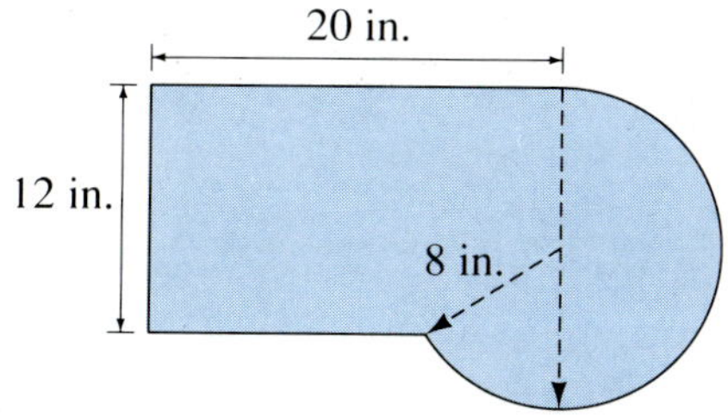

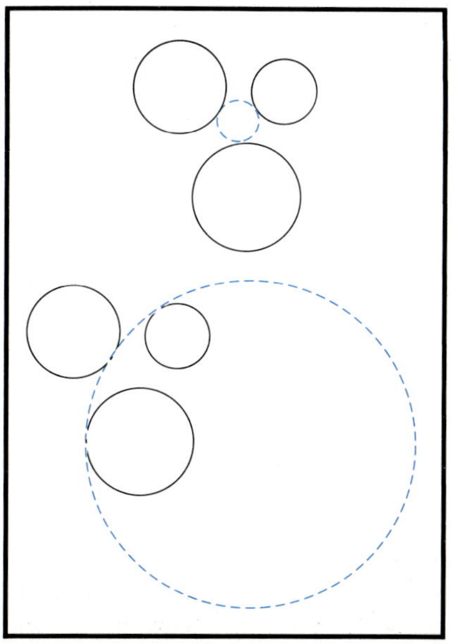

Figure 8-21

Pictured are two solutions of the problem of Apollonius, a Greek mathematician who contributed as much to three-dimensional geometry as Euclid did to plane geometry. The problem is: Given three fixed, nonintersecting circles, find a circle that touches each of them at one point. There are eight solutions. Can you find the others? (The heavy lines represent the three fixed circles and the broken lines, the solution circle.)

8-6 CONCURRENCE OF LINES; CONSTRUCTIONS

You have been asked to construct the angle bisectors of a triangle and the perpendicular bisectors of the sides of a triangle in exercises in previous chapters. In each case the three lines met at the same point.

Definition 8-16 **Concurrent lines** are two or more lines that intersect at a common point.

Theorem 8-18 The bisectors of the angles of a triangle are concurrent at a point equidistant from the sides of the triangle.

Proof In the figure that follows, we see that if bisectors $\overrightarrow{AF}$ and $\overrightarrow{BG}$ intersect at point P, then P is equidistant from $\overline{AC}$ and $\overline{AB}$ by Theorem 5-6; P is also equidistant from $\overline{AB}$ and $\overline{BC}$ by the same theorem. (Recall that the distance from a point to a line is the perpendicular distance.) Thus $\overline{PR} \cong \overline{PS}$ and $\overline{PS} \cong \overline{PF}$, so it follows that $\overline{PR} \cong \overline{PF}$, which means that P is on the bisector of angle ACB by Corollary

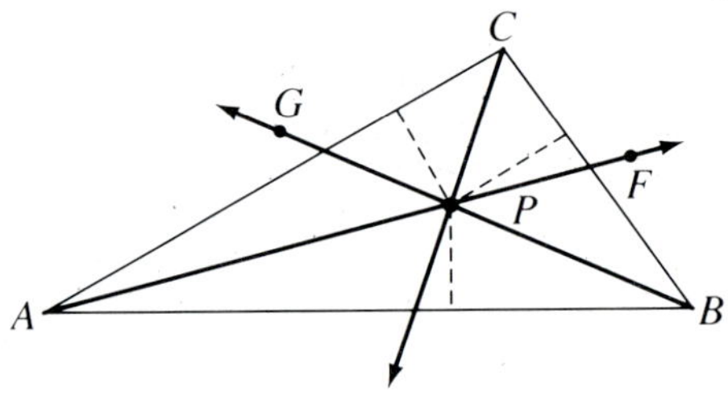

6-8(c). By means of this theorem we are able to construct a circle with center at P and tangent to the three sides of the triangle. The circle is said to be **inscribed** in the triangle, and its center is the **incenter** of the triangle. The construction requires just two angle bisectors, however, since the two are sufficient to determine the incenter.

Theorem 8-19 The perpendicular bisectors of the sides of a triangle are concurrent at a point equidistant from the vertices of the triangle.

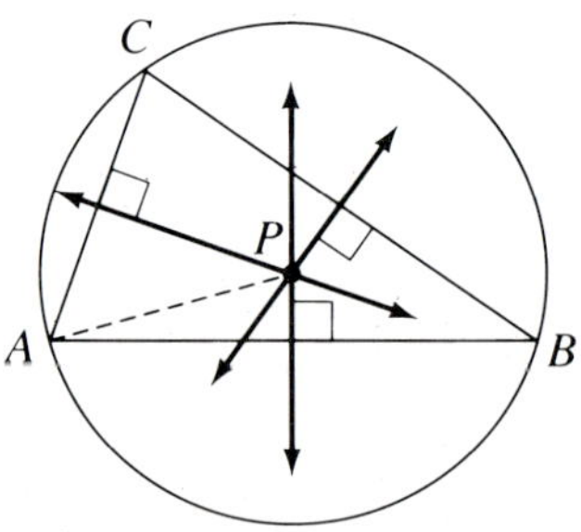

The proof is left to the student. The point found in Theorem 8-19 is the **circumcenter** of the triangle, and with this as center, a circle can be drawn that passes through the three vertices of the triangle. This circle is called the **circumcircle** of the triangle.

Theorem 8-20 The medians of a triangle are concurrent at a point that is two-thirds of the distance from any vertex to the midpoint of the opposite side.

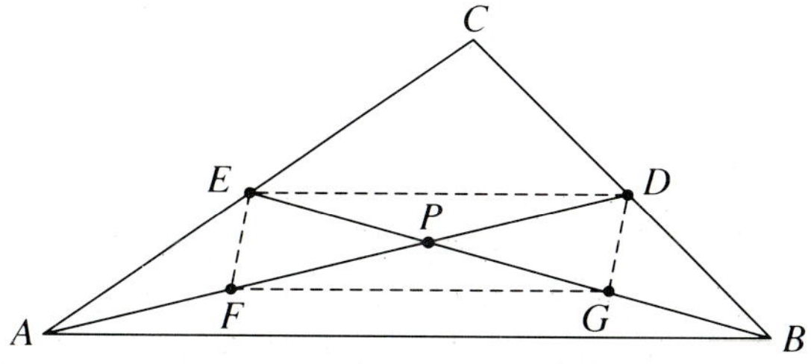

Proof Let $\overline{AD}$ and $\overline{BE}$ be medians that intersect at point P. Find F and G, the midpoints of $\overline{AP}$ and $\overline{BP}$, respectively. Then by Theorem 5-6, $\overline{ED}$ and $\overline{FG}$ are both parallel to $\overline{AB}$, so they are parallel to each other, and $m\overline{ED} = m\overline{FG} = \frac{1}{2}m\overline{AB}$, so $\overline{ED} \cong \overline{FG}$. Thus $DEFG$ is a parallelogram and its diagonals bisect each other, so $\overline{AF} \cong \overline{FP} \cong \overline{PD}$, which means that $m\overline{AP} = \frac{2}{3}m\overline{AD}$. Similarly $\overline{BG} \cong \overline{GP} \cong \overline{PE}$, so $m\overline{BP} = \frac{2}{3}m\overline{BE}$. Using the median from C and median $\overline{AD}$ would produce a like result.

The point of intersection of the three medians is called the **centroid** of the circle. The centroid of a triangle cut out of cardboard or other material of uniform density is the center of balance of the triangle.

Theorem 8-21 The lines that contain the three altitudes of a triangle are concurrent. (The proof is left as an exercise.)

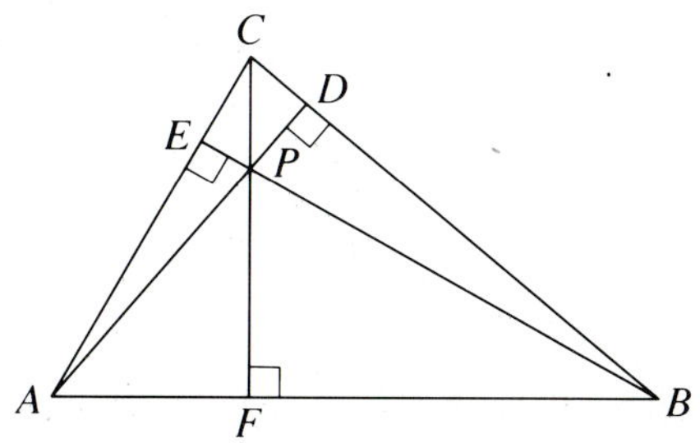

The point of intersection of the three altitudes is called the **orthocenter** of the triangle. It is interesting to note that the three points of concurrency we have just discussed, the circumcenter, centroid, and orthocenter of a given triangle, are collinear; the line through these three points is called the *Euler Line,* after the mathematician who proved that relationship. It is also remarkable that in any triangle the midpoints of the sides, the feet of the altitudes, and the midpoints of the segments connecting the orthocenter to the vertices all lie on a circle with its center located at the point that bisects the segment having the circumcenter and orthocenter as its endpoint. This circle is known as the *nine-point circle.*

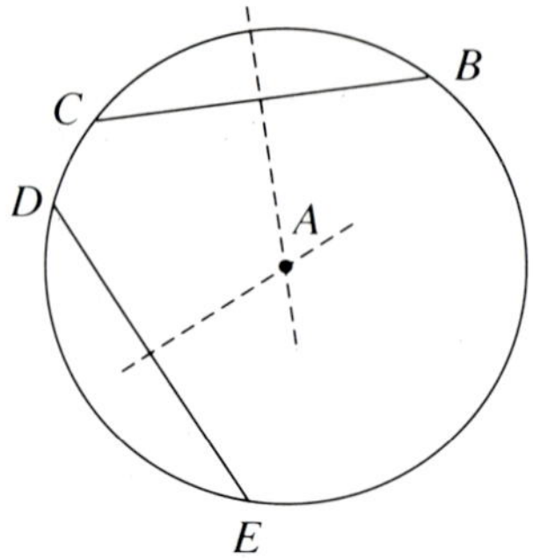

Figure 8-22

Theorem 8-5 can be used to locate the center of a circle if that point is unknown. In Fig. 8-22, if chords *BC* and *DE* are drawn and their perpendicular bisectors are constructed, then their point of intersection, *A*, must be the center of the circle.

Exercise Set 8.9

1. Construct a triangle with sides of length 3 in., 3.5 in., and 4 in.; then construct the perpendicular bisectors of the sides, determine the circumcenter, and construct the circumcircle.
2. Construct a triangle with sides of length 5.5 in., 4 in., and 3.5 in.; then construct the angle bisectors of the triangle, determine the incenter, and construct the inscribed circle.
3. Construct a triangle with sides of length 2.5 in., 4 in., and 5 in.; then construct the medians and locate the centroid. Measure the distance from a vertex to the centroid and the length of that median, and verify that Theorem 8-18 is correct. (*Hint:* Use of a millimeter scale might be easier than using sixteenths of an inch.)
4. Draw a circle by tracing the bottom of a tin can or a small cooking utensil. Locate the center of the circle by using Theorem 8-5, as described before. Check to see how close your construction came by drawing the circle with the center point you constructed and radius the distance from the center point to a point on the circle. Indicate which circle is which, unless you were accurate enough to be able to locate *the* center.
5. Prove Theorem 8-19.
6. Construct a triangle similar to that in the following figure and then construct the Euler Line through circumcenter *L*, centroid *G*, and orthocenter *N*. Point *O*, the bisector of $\overline{LN}$, is the center of the nine-point circle. Be precise!

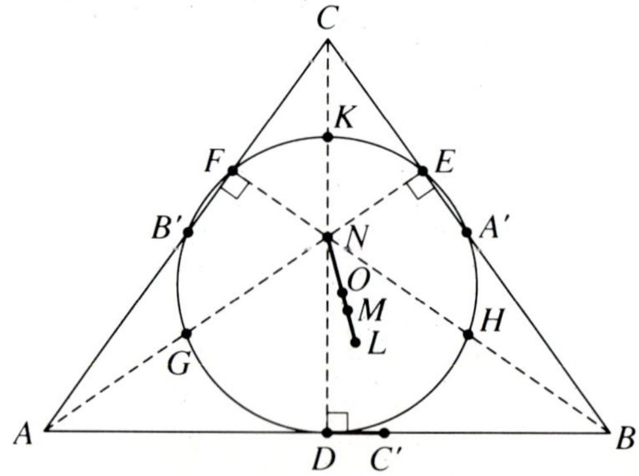

7. A circle passes through points with coordinates $(0, -1)$, $(4, 7)$, and $(9, 2)$. Locate these points, find the center of the circle, and then draw the circle.
8. Prove Theorem 4-4 analytically.
9. Prove Theorem 8-21. (*Hint:* Draw lines through C parallel to $\overline{AB}$, through A parallel to $\overline{BC}$, and through B parallel to $\overline{AC}$, forming triangle $A'B'C'$, as shown in the following figure.

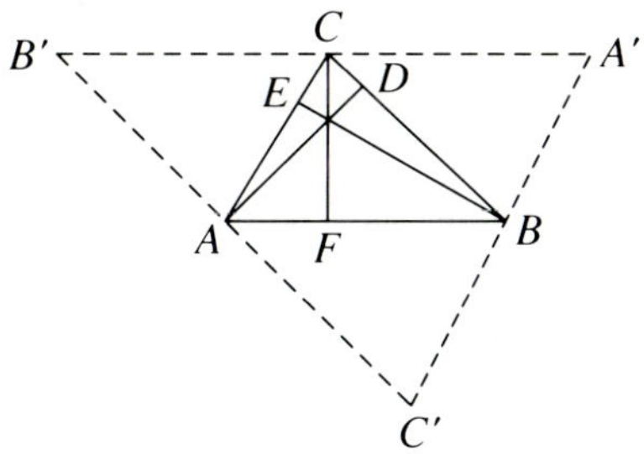

REVIEW EXERCISES

1. If O is the center of the circle in the following figure, name the following:
   (a) $\overline{AO}$ ________ **(b)** $\overline{CB}$ ________
   (c) $\overleftrightarrow{DE}$ ________ **(d)** $\overleftrightarrow{FG}$ ________

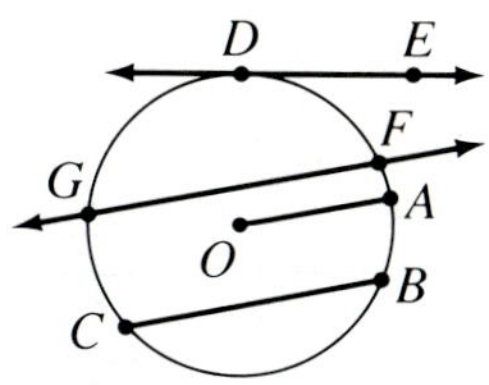

2. Two or more circles are said to be concentric if ________.
3. If $\overline{AB}$ and $\overline{CD}$ are chords of a circle and the distance from the center to $\overline{AB}$ is less than the distance from $\overline{CD}$ to the center, then $m\overline{AB}$ ________ $m\overline{CD}$. ($<, =, >$) Prove your result.
4. An angle inscribed in a semicircle is a(an) ________ angle.
5. In the following circle, $\overline{CE}$ is a diameter, O is the center, $m°\widehat{AB} = 50$, $m°\widehat{CB} = 40$, and $m°\widehat{DE} = 60$. Find
   (a) $m\angle AOB$ **(b)** $m\angle BOC$
   (c) $m\angle CDB$ **(d)** $m°\widehat{DC}$
   (e) $m°\widehat{AE}$ **(f)** $m\angle DCE$

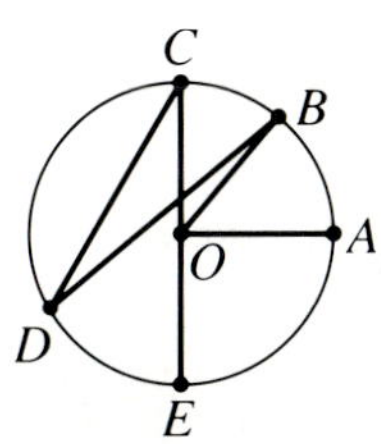

6. In the following figure $\overrightarrow{CA}$ is a secant, $\overrightarrow{CD}$ a tangent, $m\angle ABE = 40$, $m^\circ\overset{\frown}{ED} = 80$, and $\overline{AC} \parallel \overline{ED}$. Find
(a) $m\angle DEB$ **(b)** $m\angle C$ **(c)** $m^\circ\overset{\frown}{AB}$

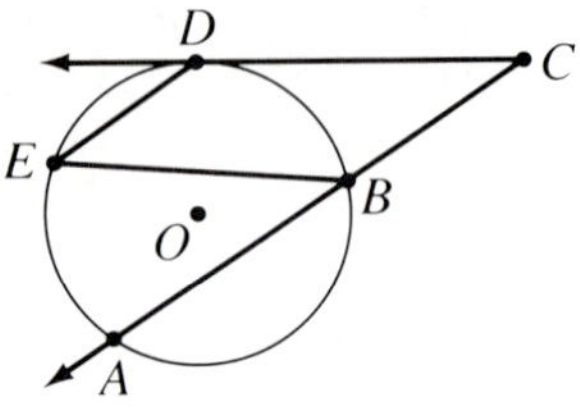

7. Chords $\overline{AB}$ and $\overline{CD}$ intersect at point E so that $m\overline{CE} = 3$ and $m\overline{BE} = 4$. If $m\overline{AC}$ is 15, find $m\overline{BD}$.

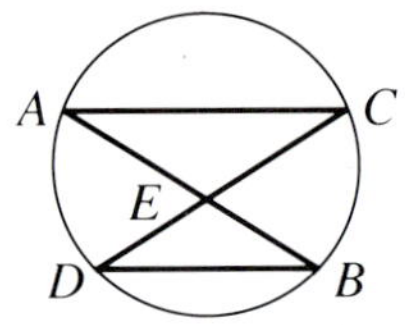

8. Chord $\overline{BC}$ is parallel to tangent $\overleftrightarrow{AD}$ and $\overline{BC}$ bisects radius $\overline{AO}$. If $\overline{BC}$ is 12 in. long, how long is $\overline{OA}$?

9. If $\overset{\frown}{ABC}$ is a semicircle and $\overline{BD}$ is perpendicular to $\overline{AC}$ at D, prove $m\overline{BD}$ is a geometric mean between $m\overline{AD}$ and $m\overline{DC}$.

10. If the difference between the areas of the two concentric circles in the following figure is 225π, find the length of the chord $\overline{RS}$ that is tangent to the smaller circle.

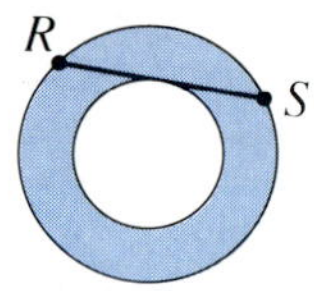

11. If two angles of an inscribed quadrilateral have measures 48 and 83, what is the measure of the other two angles of the quadrilateral?

12. Prove that if $ABCD$ is an inscribed quadrilateral with $m\overline{CD} < m\overline{AB}$ and $\overline{AD} \cong \overline{BC}$, then $ABCD$ is an isosceles trapezoid.

13. If $\overrightarrow{AB}$ and $\overrightarrow{CB}$ are tangents to circle O and $m^\circ\overset{\frown}{AC} = 75$, find $m\angle BAC$, $m\angle BCA$, $m\angle B$ and $m\angle OAC$.

14. Prove that if $\overrightarrow{AB}$ is a tangent and $\overrightarrow{AC}$ a secant with D the midpoint of $\overset{\frown}{AC}$, then $\overrightarrow{AD}$ is the bisector of $\angle BAC$.

15. The centers of two circles of radii 3 in. and 7 in. are 12 in. apart. What is the length of their common external tangent, $\overline{AB}$?

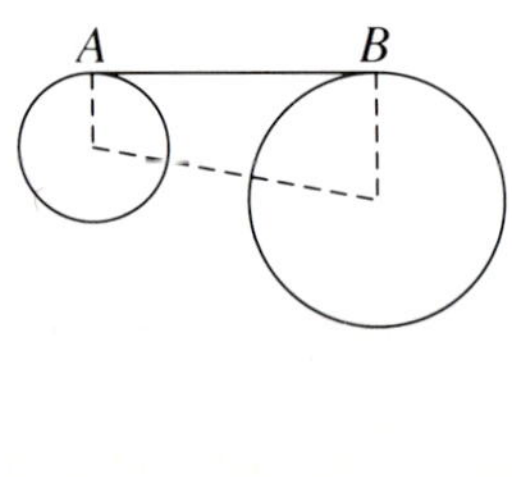

16. Prove that if two circles and a line intersect at the same point, as in the following figure, then the line bisects the common external tangent.

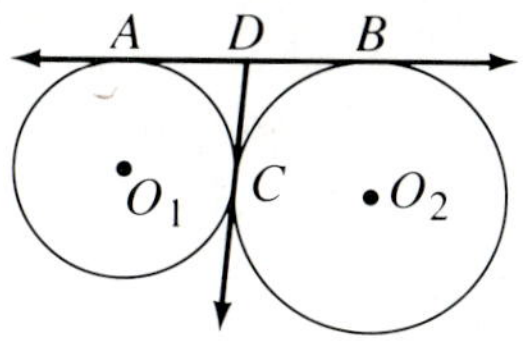

17. The circumference of a circle of radius 4 is exactly twice the circumference of a second circle. Find the length of the diameter of the second circle.

18. If the radius of a circle were increased by 2 ft, the area of the circle would be exactly 81π ft². Find the radius.

▶ 19. **(a)** Compute the area of a dodecagon inscribed in a circle with radius 10.
(b) Compute the area of the circle inscribed in the dodecagon.

20. A wall clock has a minute hand of length 15 in. **(a)** What is the distance traveled by the tip of the minute hand in 8 min? **(b)** What is the area swept by the minute hand during this time?

21. What is the maximum size for a square peg that can be inserted into a round hole of diameter 2 in.?

Cumulative Review—Chapters 2–8

Supply the word or phrase that will make each of the following correct.

1. If a segment is parallel to one side of a triangle and intersects the other two sides, then it ________.
2. In an isosceles right triangle with leg length a, the hypotenuse has length ________.
3. If the vertex angle of an isosceles triangle is 80, the measure of each base angle is ________.
4. If the diagonals of a parallelogram are congruent, the parallelogram is a ________.
5. A pentagon has ________ diagonals.
6. Another name for regular quadrilateral is ________.
7. If the area of a parallelogram is 50 in.², the area of a triangle with the same base and altitude is ________.
8. If the ratio of the radii of two circles is 1 to 4, the ratio of their areas is ________.
9. A ________ of a triangle is a segment joining a vertex to the midpoint of the opposite side.
10. If parallel lines are cut by a transversal, the interior angles on the same side of the transversal are ________.
11. If there is a point P on $\overline{AB}$ such that $\overline{AP} \cong \overline{PB}$, then P is said to be the ________ of $\overline{AB}$.
12. Points that lie on the same line are said to be ________.
13. Three noncollinear points determine a ________.
14. A set is called a convex set if, for any two points A and B in the set, ________ lies entirely in the set.
15. Two opposite rays form a convex set if their ________ is nonempty.

16. The distance from $P(a, b)$ to $Q(c, d)$ is ________.
17. The slope of the line passing through the points (2, 4) and (4, 6) is ________.
18. The coordinates of the midpoint of the segment joining the points (−5, 4) and (3, 2) are ________.
19. The measure of the central angle of a regular polygon of 9 sides is ________.
20. The ratio of an arc of a circle to the circumference is equal to the ratio of the area of the ________ to the area of the circle.

Problems

1. If the measure of one angle of a triangle is 65 and the difference in the measures of the other two angles is 15, what is the measure of each of these angles?
2. Given the following figure, prove $m\overline{BC}/m\overline{AB} = m\overline{CD}/m\overline{BE}$.

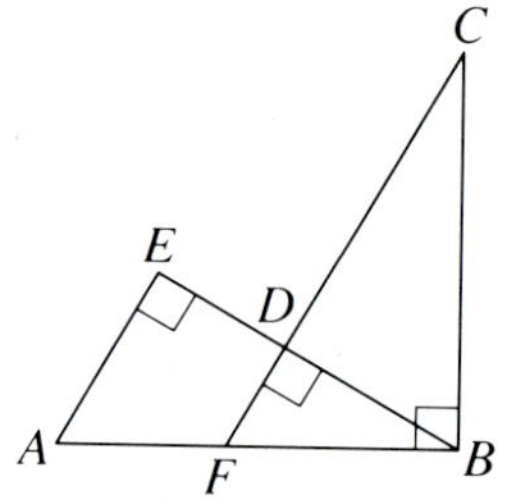

3. In the following triangle, $\angle A \cong \angle DEB$.
   (a) If $m\angle A = 50$ and $m\angle C = 60$, find $m\angle BDE$.
   (b) If $m\overline{AB} = 12$, $m\overline{AC} = 9$, and $m\overline{DE} = 8$, find $m\overline{BE}$.

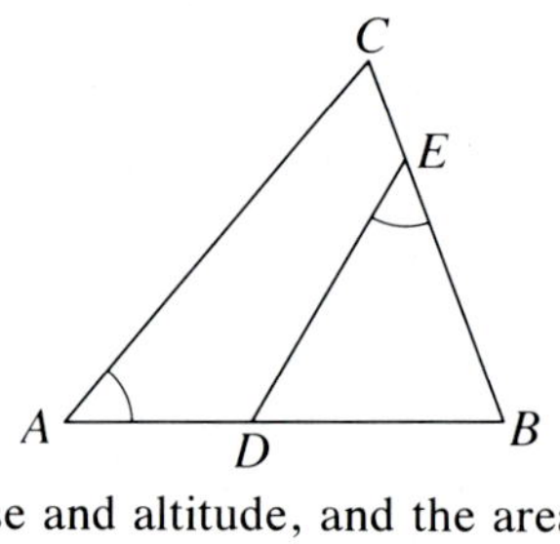

4. Find the lengths of the base and altitude, and the area of an isosceles triangle with a base angle of 40° opposite a side of length 12.
5. Prove that the bisectors of the base angles of an isosceles triangle cut off two pairs of congruent segments on the opposite sides.
6. In the following figure, $\overline{AB} \cong \overline{EF}$, $\overline{AB} \parallel \overline{EF}$, and $\overline{AD} \cong \overline{EC}$. Prove $\overline{BC} \parallel \overline{FD}$.

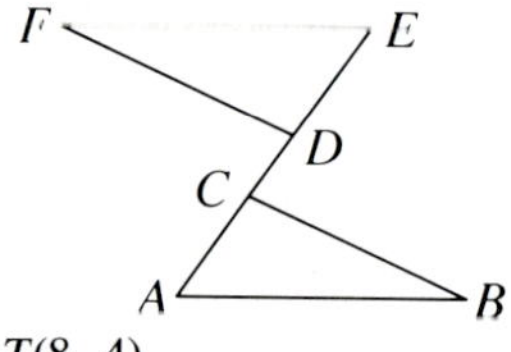

7. Given $R(-3, 6)$, $S(0, 0)$ and $T(8, 4)$.
   (a) Find the slopes of $\overline{RS}$ and $\overline{ST}$.
   (b) Give the coordinates of the center of the circle which passes through R, S, and T.

8. Identify the hypothesis and conclusion of each of the following statements.
   (a) Three collinear points determine a plane.
   (b) If it rains the roads will be slippery.
   (c) If two angles are supplementary, their sum is 180.
   (d) Complements of the same angle are congruent.
   (e) If an angle has measure 40, it is said to be an acute angle.
9. Locate points $G(5, 5)$, $T(-4, 8)$ and $O(0, 0)$; then draw $\overline{GT}$, $\overline{TO}$, and $\overline{OG}$.
   (a) Find $m\angle T$, $m\angle G$, and $m\angle O$ with your protractor.
   (b) Locate the midpoints A, K, and D of $\overline{GT}$, $\overline{TO}$, and $\overline{OG}$ respectively.
   (c) Find $m\angle AKD$, $m\angle KAD$ and $m\angle ADK$ with your protractor.
10. Repeat Problem 9 using $G(5, 1)$, $T(9, 10)$ and $O(0, 0)$.
11. If $\overline{AE}$ and $\overline{CD}$ are altitudes of triangle ABC, $m\overline{AB} = 12$, $m\overline{CD} = 10$, and $m\overline{AE} = 8$, find $m\overline{BC}$.
12. The length of a rectangle is 6 more than its width and its area is 91 square units. Find the measure of each side.
13. If $ABCD$ is a parallelogram, $m\overline{AB} = 25$, $m\overline{CD} = 2x + 3$ and $m\angle A = 5x$, find (a) x, (b) $m\angle D$, (c) $m\angle C$.

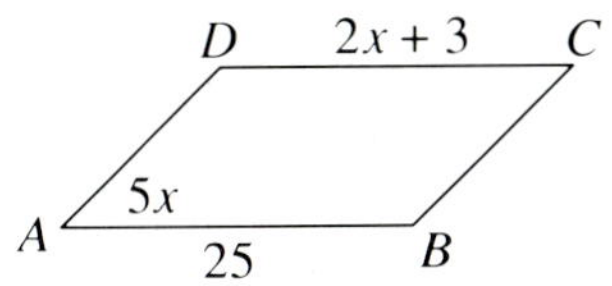

14. Two sides of a triangle measure 10 and 12 in., respectively, and they include an angle of 30°. What is the area of the triangle?
15. If $RSTU$ is a parallelogram, give the coordinates of point U.

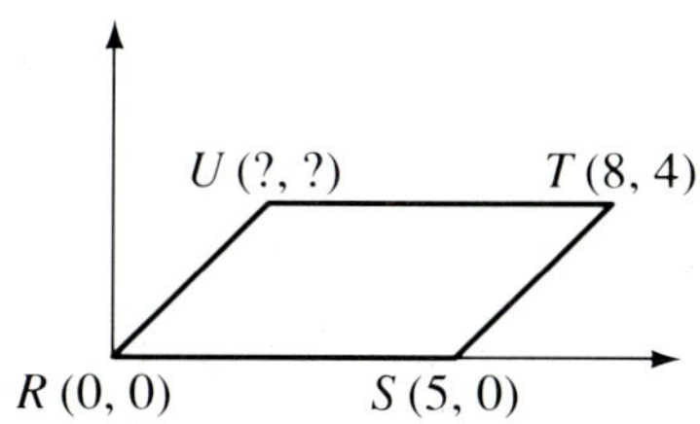

16. The area of a trapezoid is 150 ft^2. If the length of the altitude is 12 ft and one base measures 10 ft, find the length of the other base.
17. The two congruent sides of an isosceles trapezoid each measure 8 in. If the upper base and lower base measure 10 and 18 in. respectively, what is the area of the trapezoid?
18. If a square is to have the same area as a triangle with base 27 in. and altitude 24 in., what is the length of the side of the square?
19. ▶ A diagonal of a rectangle forms a 20° angle with one side. If the diagonal is 8 units in length, what is the exact area of the rectangle?
20. The diagonals of a rhombus are 7 ft and 8 ft in length. What is the area of the rhombus?
21. $\overline{AB}$ is a diameter of a circle O and $\overline{AB} \perp \overline{CD}$ at E. If $m\overline{AE} = 8$ and $m\overline{EB} = 12.5$, find $m\overline{CD}$.

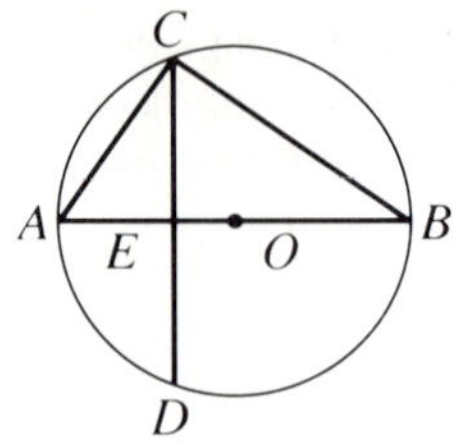

22. $\overrightarrow{PA}$ is tangent to circle O at point A, and secant $\overrightarrow{PB}$ intersects the circle at B and C.
    (a) If $m°\widehat{AB} = 70$, $m\angle ACB =$ ________ and $m\angle PAB =$ ________.
    (b) Triangles CAP and ABP are ________.
    (c) If $m\overline{AP} = 12$ and $m\overline{PC} = 16$, find $m\overline{PB}$.
23. A lamb is tied by a 40-ft rope to an outside corner of a rectangular garden, which is enclosed by a fence 30 ft by 40 ft. What is the grazing area?
24. A circle is circumscribed about an equilateral triangle, each side of which measures 12 in.
    (a) Find the area of the circle.
    (b) What is the total area of the circular segments cut off by the triangle? (To the nearest tenth.)
25. In the following figure, $\overrightarrow{PA}$ is tangent to circle O at A, $\overrightarrow{PE}$ and $\overrightarrow{PD}$ are secants. If $m°\widehat{AE} = m°\widehat{CD} = 100$, $m\angle APE = m\angle EPD = 20$, and $m\angle BED = 65$, find **(a)** $m°\widehat{AB}$, **(b)** $m°\widehat{BC}$, **(c)** $m°\widehat{ED}$.

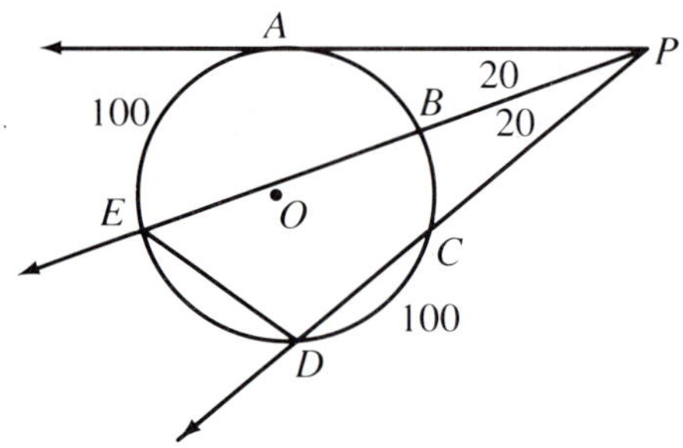

26. What is the minimum diameter required for a single pipeline to replace two pipes with 12-in. and 16-in. diameters?
27. The radius of each of the four circles in the following figure is 5 in. Find the total shaded area. (To the nearest tenth.)

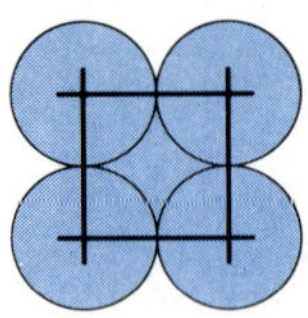

28. A rectangle with sides 8 in. and 15 in. is inscribed in a circle. Find the area of the circular region.
29. A pendulum 15 ft in length swings through an arc of 5°. What is thc lcngth of the arc (to the nearest inch)?
30. A regular hexagon has a side of length 18 in. Find the difference between the areas of the circumscribed and the inscribed circles.

9

Space Geometry

Mighty is Geometry; joined with Art, resistless.

EURIPEDES (480–406 B.C.)

9-1 INTRODUCTION

In previous chapters we have for the most part confined our attention to sets of points within one plane. A figure which contains only points which lie in one plane is said to be two dimensional, i.e., it may have length or width (or both), but no *depth* or thickness.

The set of all points is called *space,* and the study of figures in space is known as *solid geometry* or *space geometry*. In this chapter we will introduce a number of terms, definitions, postulates and theorems in an ordered way as was done earlier for plane geometry, but few theorems will be proved since the intent is to consider only the fundamentals in this discussion.

Although we usually represent a plane by means of a parallelogram intended to lie in the plane, as in Fig. 9-1, a plane is understood to be infinite in extent just

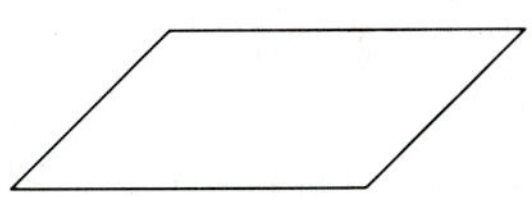

Figure 9-1

as a line is infinite in extent. A figure in space may be a union of points, lines, and surfaces. The "box" in Fig. 9-2 consists of points, segments such as $\overline{AB}$ and $\overline{CD}$, and *faces* such as rectangle $ABCD$, which are flat surfaces or parts of planes.

9-2 LINES AND PLANES IN SPACE

Recall the following postulate from Chapter 2.

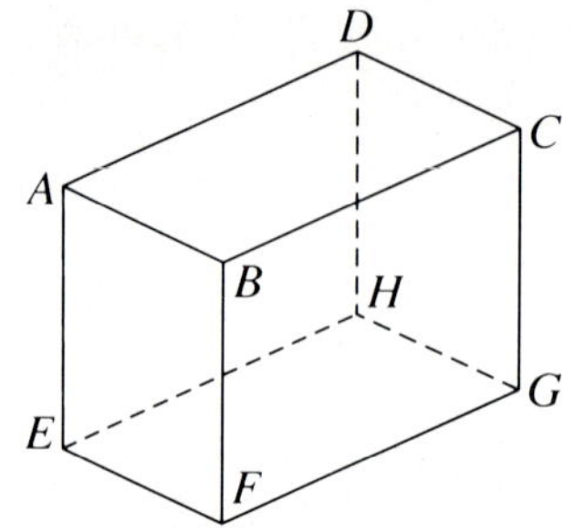

Figure 9-2

Postulate 2-8 A plane is uniquely determined by three noncollinear points.

If we can "get off of" a single plane, we will have what is called a *three-dimensional Euclidean space*.

Postulate 9-1 For every plane $\mathscr{E}$ there is at least one point X not on $\mathscr{E}$. (From this it follows that the number of planes in this geometry is infinite.)

Postulate 9-2 If two distinct points of a line lie in a plane, then all the points of the line lie in that plane.

We can now prove some useful theorems that involve determination of planes.

Theorem 9-1 A plane is uniquely determined by a line and a point not on the line.

As in Fig. 9-3, we have the line ℓ and the point P. According to Post. 2-1(b) there are at least two points on the line, so now we have the three points required by Post. 2-8.

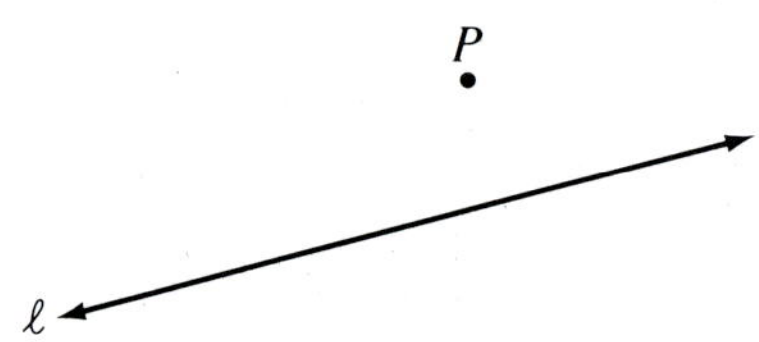

Figure 9-3

Theorem 9-2 A plane is uniquely determined by two intersecting lines.

In Fig. 9-4 $\overleftrightarrow{AB}$ and $\overleftrightarrow{CD}$ intersect at E. Points A, C, and E determine a plane $\mathscr{E}_1$. Is there only one? How do we know that A, C, and B, for example, do not form another plane $\mathscr{E}_2$? From Post. 9-2 we know that if A and E are in $\mathscr{E}_1$, B must

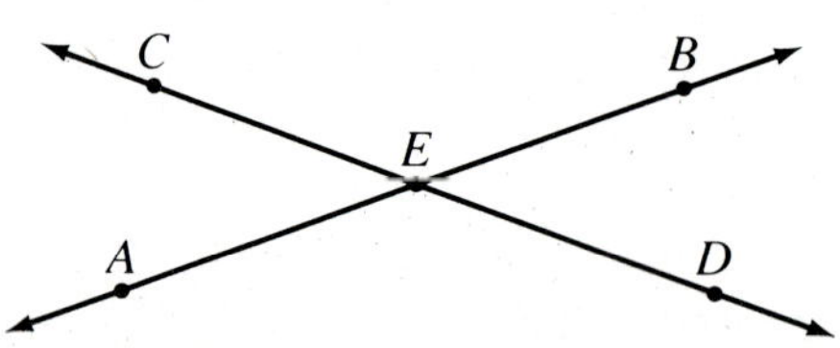

Figure 9-4

also be in $\mathscr{E}_1$. Since A, C, and B are all elements of $\mathscr{E}_1$, this is the only plane possible.

Theorem 9-3 The intersection of two distinct planes is a straight line.

Refer to Fig. 9-5. The *intersection* of two surfaces, or of a surface and a line, is the set of points common to both of them.

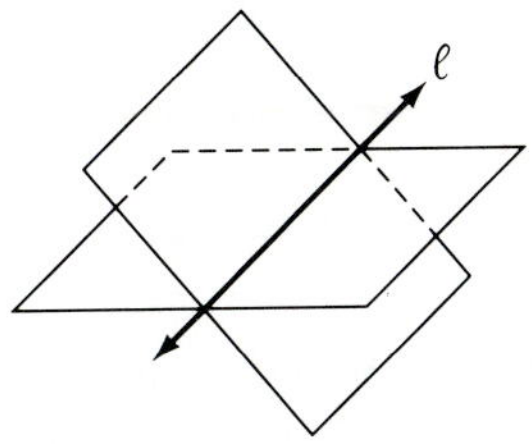

Figure 9-5

Definition 9-1 Two planes are said to be **parallel** if their intersection is the empty set. Two lines are said to be **skew lines** if they do not lie in the same plane.

In Fig. 9-6 we see that in a single plane two lines are either parallel (ℓ_1 and ℓ_2) or intersecting (ℓ_2 and ℓ_3), whereas in Fig. 9-7 we see that two lines in parallel planes may be either parallel or skew lines.

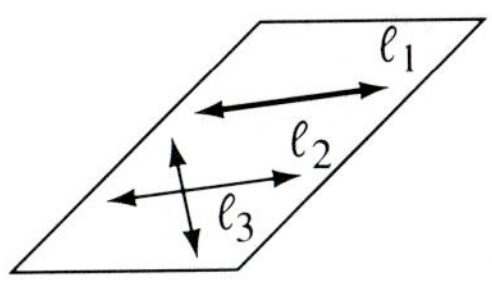

Figure 9-6

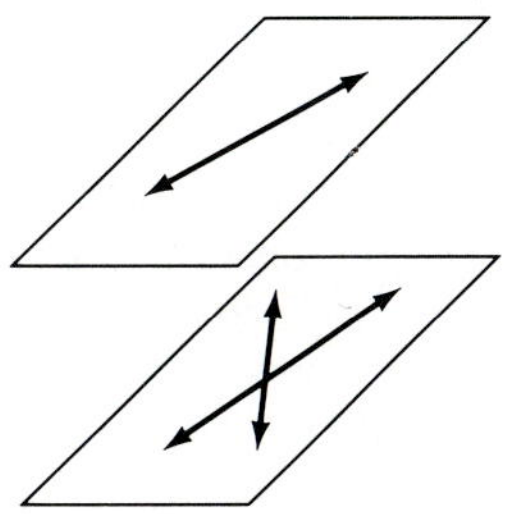

Figure 9-7

Definition 9-2 A line ℓ and a plane $\mathscr{E}$ are said to be **perpendicular** if ℓ intersects $\mathscr{E}$ at some point R and is perpendicular to every line in $\mathscr{E}$ passing through R.

In Fig. 9-8 ℓ is perpendicular to $\overleftrightarrow{RS}$ and $\overleftrightarrow{RT}$ at point R. Could there be other lines perpendicular to $\overleftrightarrow{RT}$ at R if we do not require that they lie in the same plane?

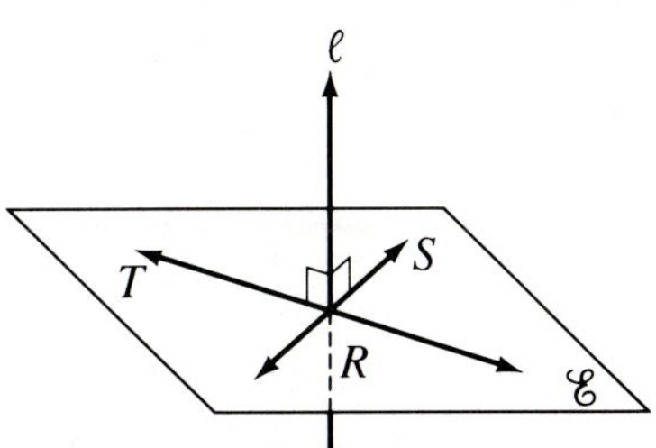

Figure 9-8

Example

(a) If two points of a straightedge touch a plane, how many other points of the straightedge touch the plane?

(b) Can a straight line be perpendicular to a line in a plane without being perpendicular to the plane? If so, illustrate with a figure.

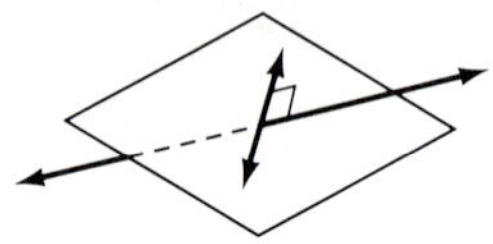

(c) Is it possible for the intersection of two planes to be a line segment? Explain.

Solution (a) All of the other points should touch the plane according to Post. 9-2, which states that if two points lie in a plane, the line containing them lies in the plane.

(b) Yes, as shown in the following figure.

Also consider an A-frame cabin, as in the following figure, in which the edges of the roof line are not perpendicular to the floor but are perpendicular to the line of intersection of the floor and the plane of one side of the roof.

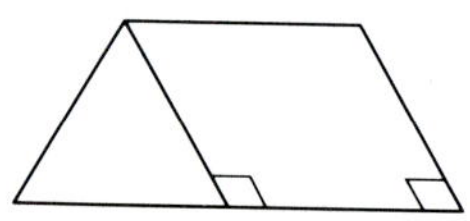

(c) No. The intersection of two planes is a line. Part of the intersection might be considered a segment. ◁

Exercise Set 9.1

Draw the figures for each problem.

1. How many planes are determined by
   (a) one point?
   (b) two points?
   (c) three points?
2. How many planes contain a given line AB and a point C not on $\overleftrightarrow{AB}$?
3. If two points A and B lie in a plane $\mathscr{E}$, what can be said about $\overleftrightarrow{AB}$?
4. Can two lines in space be nonparallel and yet not intersect?
5. If a plane Q and a line AB have points R and S in common, what can you conclude about the plane and the line?
6. A line ℓ_1 is contained in a plane $\mathscr{E}_1$, and a line ℓ_2 is perpendicular to ℓ_1 and to $\mathscr{E}_1$. Show that a plane $\mathscr{E}_2$ distinct from $\mathscr{E}_1$ is determined.
7. How many different planes may be determined by **(a)** three distinct lines AD, BD, and CD? **(b)** four distinct lines EF, EG, EH, and EI?
8. How many planes can pass through two parallel lines?

9. In the following figure, A is not in $\mathscr{E}$, B and C are in $\mathscr{E}$. $\overleftrightarrow{AB} \perp \mathscr{E}$, but $\overleftrightarrow{AC}$ is not perpendicular to $\mathscr{E}$.
   (a) Compare the lengths of $\overline{AB}$ and $\overline{AC}$.
   (b) Compare the measures of $\angle ABC$ and $\angle ACB$.

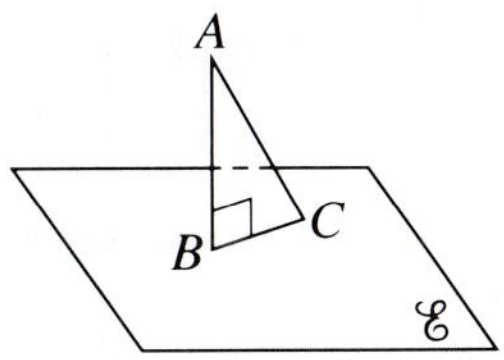

10. If planes $\mathscr{E}$ and $\mathscr{Q}$ intersect in $\overleftrightarrow{AB}$ and points R and S lie in both planes $\mathscr{E}$ and $\mathscr{Q}$, where must R and S lie?

Some additional line and plane relationships are given in the following theorems.

Theorem 9-4 If a line is perpendicular to two intersecting lines, then it is perpendicular to the plane containing the two intersecting lines (Fig. 9-9).

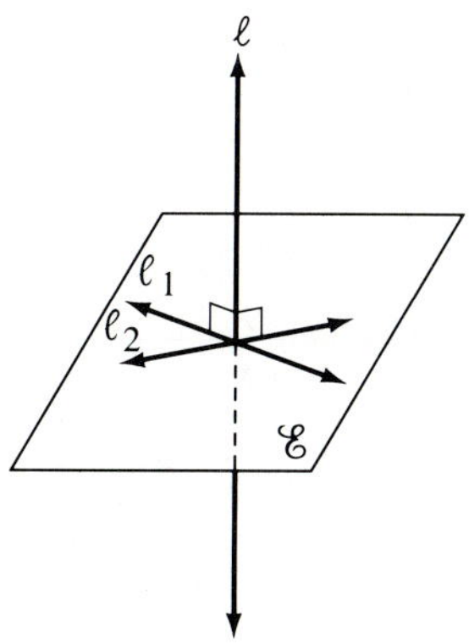

Figure 9-9

Theorem 9-5 There is exactly one plane perpendicular to a line at a given point on the line (Fig. 9-10).

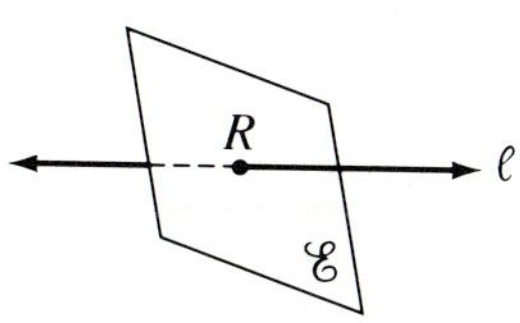

Figure 9-10

Theorem 9-6 Two lines perpendicular to the same plane are parallel (Fig. 9-11).

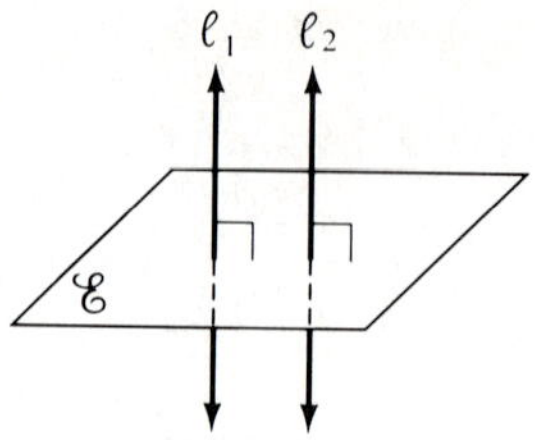

Figure 9-11

Theorem 9-7 Two planes perpendicular to the same line are parallel (Fig. 9-12).

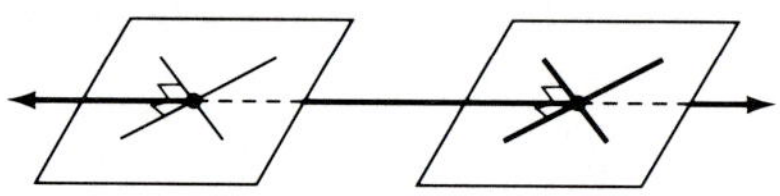

Figure 9-12

Theorem 9-8 The intersections of two parallel planes cut by a third plane are parallel lines (Fig. 9-13).

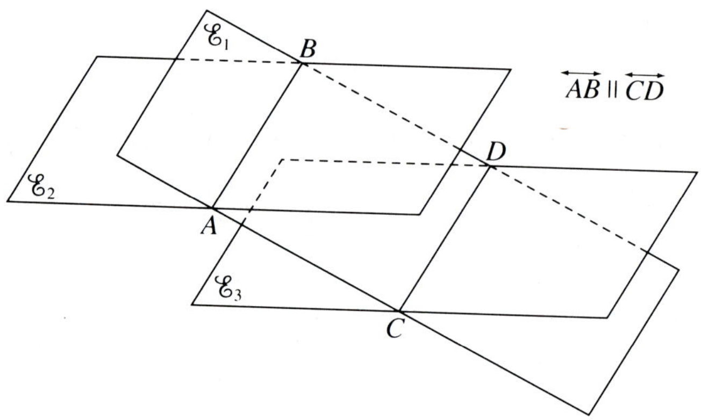

Figure 9-13

Example

(a) At a point on a line, how many lines can be drawn perpendicular to the line?

(b) Is it possible for a line to be parallel to two planes if the planes are not parallel?

(c) Describe the shortest distance from a point to a plane.

Solution (a) An infinite number, some of which are shown in the following figure. They would all be in the same plane.

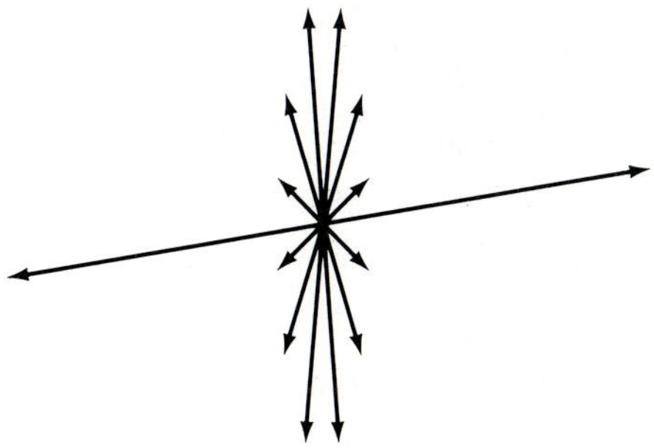

(b) Yes, any line such as $\ell_2 \parallel \ell_1$ and not in $\mathscr{E}_1$ or $\mathscr{E}_2$, as shown in the following figure.

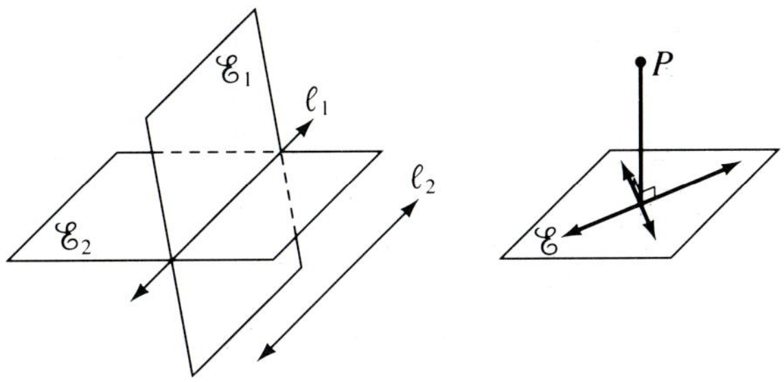

(a) (b)

(c) It will be the perpendicular from the point to the plane (refer to part (b) of the preceding figure).

Exercise Set 9.2

1. Is it possible for three lines to intersect at a point so that each of the lines is perpendicular to the others? If so, illustrate.
2. If a line ℓ is perpendicular to a plane $\mathscr{Q}$, how many planes may be drawn perpendicular to $\mathscr{Q}$ and containing ℓ? Illustrate your answer.
3. If two lines are parallel to a plane, need they be parallel to each other? Illustrate.
4. How many planes can be drawn parallel to a plane $\mathscr{R}$ through a point A not in $\mathscr{R}$?
5. Is it possible to draw a line perpendicular to two lines that are not parallel? Illustrate.
6. In the following figure, A is not in plane $\mathscr{E}$ and $\overline{AO} \perp \mathscr{E}$ at O. Circle O lies in plane $\mathscr{E}$. Prove that $\angle ABO \cong \angle ACO$.

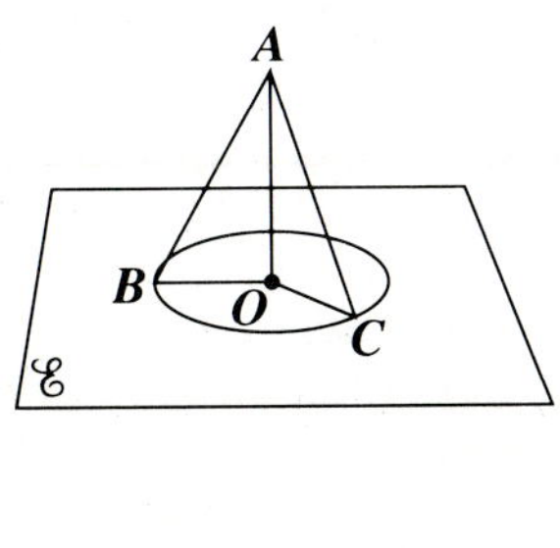

HYPATIA (370–415 A.D.) was the only woman among the Greek mathematicians whose work has been recorded in the history books. The daughter of Theon, a mathematician and astronomer of the Alexandrian school, she too became a popular lecturer at the school. Little of her written work has been preserved, but she wrote a treatise, *On the Conics of Apollonius,* about another mathematician of the school who preceded her by five hundred years. The age of Greek mathematics ended not long after Hypatia's death, and the thousand-year decline in European civilization called the Dark Ages came to pass. (*The Bettmann Archive*)

9-3 MEASUREMENT IN SPACE

Definition 9-3 If two planes intersect, the figure formed by the union of the two noncoplanar half planes (faces) and the line of intersection (edge) is called a **dihedral angle.**

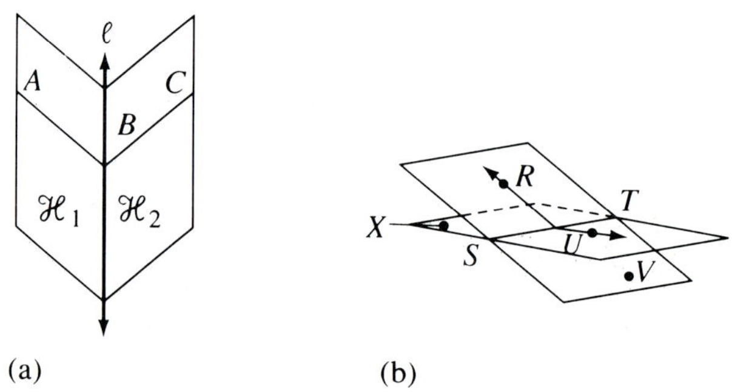

Figure 9-14

In Fig. 9-14(a), $\mathcal{H}_1 \cup \ell \cup \mathcal{H}_2$ form a dihedral angle. The angles are named by an edge and a point in each face, and in Fig. 9-14(b) we have $\angle R\text{-}ST\text{-}U$, $\angle X\text{-}ST\text{-}R$, $\angle V\text{-}ST\text{-}X$, and $\angle V\text{-}ST\text{-}U$.

Definition 9-4 The **measure of a dihedral angle** is the measure of the plane angle formed by two rays (one in each face) with a common end point on the edge and each perpendicular to the edge.

In Fig. 9-14(a), if $\overrightarrow{BA}$ and $\overrightarrow{BC}$ are perpendicular to ℓ at B, then m$\angle ABC$ is the measure of dihedral angle A-ℓ-C.

Theorem 9-9 The plane angles of a dihedral angle are congruent.

Example

(a) Name the dihedral angles indicated in the following figure.
(b) What is the interior of a dihedral angle?
(c) If two planes are perpendicular to a third plane, what can be said about the dihedral angles that are formed?

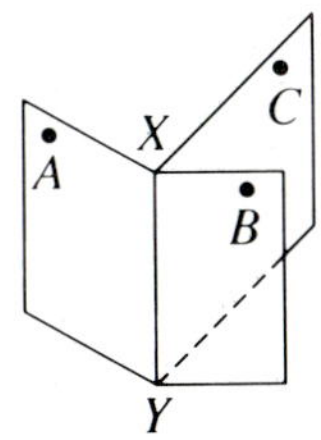

Solution (a) $\angle A$-XY-B, $\angle B$-XY-C, and $\angle C$-XY-A are dihedral angles.
(b) If the dihedral angle has faces $\mathscr{E}_1'$ and $\mathscr{E}_2'$, then the interior is the set of points in space on the same side of $\mathscr{E}_1$ as $\mathscr{E}_2$ and on the same side of $\mathscr{E}_2'$ as $\mathscr{E}_1'$. See the following figure.

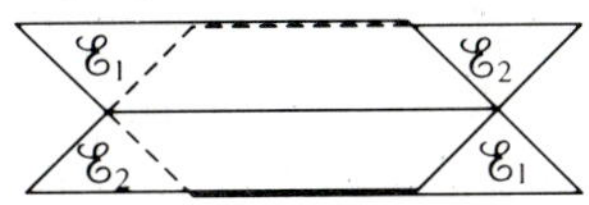

(c) They are right dihedral angles. ◁

Exercise Set 9.3

1. Name the dihedral angles in the following figure.

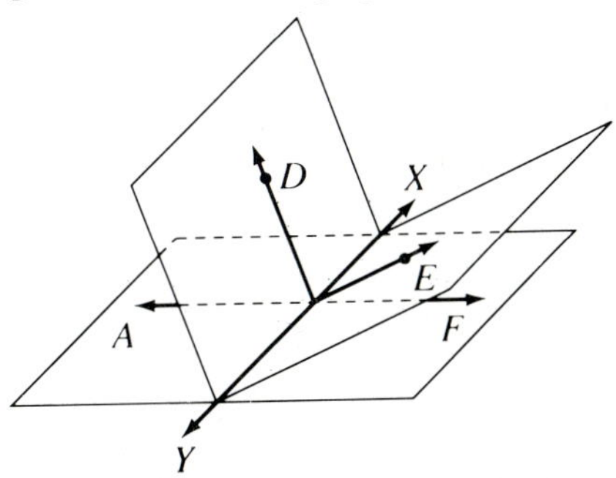

2. Name each dihedral angle of the tetrahedron in the following figure.

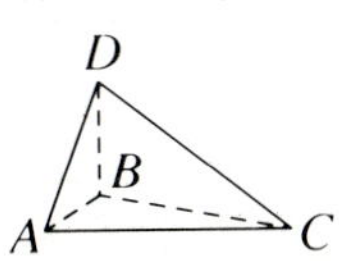

3. Draw an angle with three faces (a trihedral angle) and an angle with four faces (tetrahedral angle).
4. In each of the following, fill in the missing word or words.
   (a) A plane perpendicular to the faces of a dihedral angle is perpendicular to the ________ of the dihedral angle.
   (b) If a plane is perpendicular to one of two intersecting lines, it is ________ to the other.
   (c) If a line is drawn in one face of a right dihedral angle and is perpendicular to the edge, it is ________ to the other face.
   (d) Two dihedral angles are congruent if the measures of their ________ angles are equal.
5. Write a definition for each of the following:
   (a) right dihedral angle
   (b) acute dihedral angle
   (c) obtuse dihedral angle
6. Define the set of points common to both faces of a dihedral angle.
7. Prove that if $\overleftrightarrow{AO}$ and $\overleftrightarrow{BO}$ are lines in a plane $\mathscr{E}$ and each is a perpendicular bisector of a segment RS, $\overleftrightarrow{RS}$ intersecting $\mathscr{E}$ at O, then $\angle ABR \cong \angle ABS$.

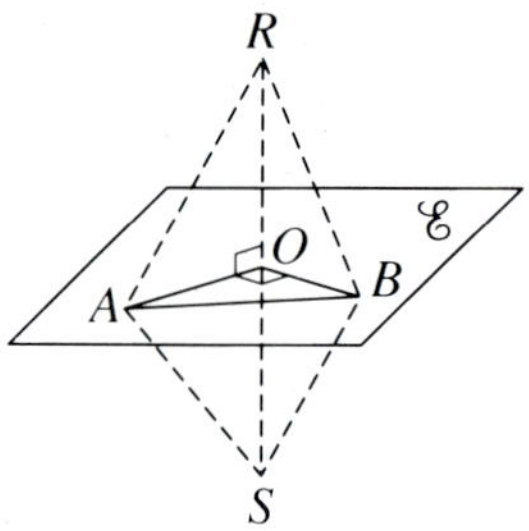

▶ 8. If a line segment 6 in. long is in one face of a dihedral angle of 36°, what is the length of its projection on the other face of the angle, correct to the nearest tenth?

▶ 9. If two points P and T are 6 and 9 in., respectively, from a given plane and $\overline{PT}$ is 4 in. in length, find the angle of inclination of the segment to the plane to the nearest degree.

10. Prove Theorem 9-9.

9-4 POLYHEDRONS

Figure 9-15 shows examples of familiar and useful geometric figures in three dimensions.

Definition 9-5 A **polyhedron** is a figure determined by the union of a finite number of polygonal regions so that (a) any two polygons have no points in common, exactly one vertex in common, or exactly one segment (edge) in common, and (b) every segment (edge) is on exactly two polygons.

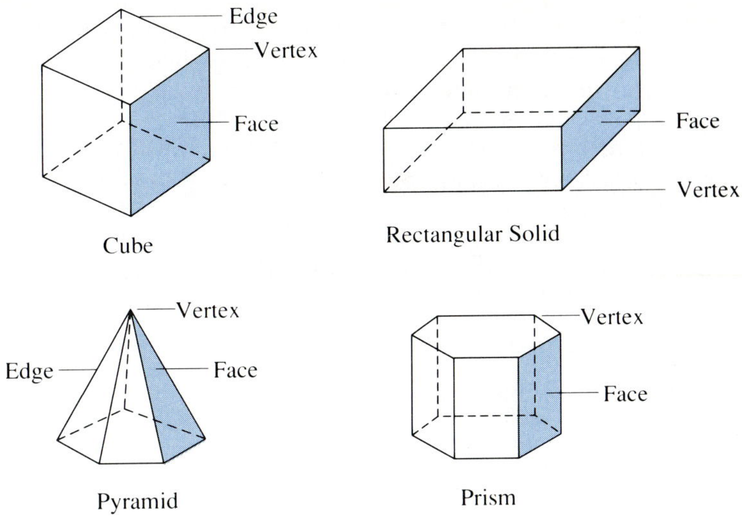

Figure 9-15

Definition 9-6 A **polyhedral angle** is the figure formed by the union of a point and the rays joining that point to the sides of a polygon in a plane not containing the point. $\angle P - ABCD$ is the polyhedral angle with P as its vertex in Fig. 9-16.

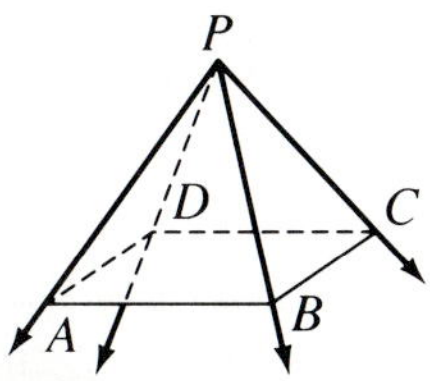

Figure 9-16

PLATO (427?–347 B.C.) The Greek philosopher drew a connection between the geometry of Pythagoras and the biology of Empedocles, a fifth century, B.C. philosopher and statesman, by means of a mathematical construction of the elements. Beginning with an isosceles right triangle, Plato constructed four of the regular solids: the cube, the tetrahedron, the octahedron, and the icosahedron. These four solids were then assumed to be the shapes of the corpuscles of the earth, fire, air, and water. These ideas affected the study of physiology and medicine for many centuries, since they were considered the basic building blocks of all organic and inorganic compounds. (*The Vatican Museums*)

Definition 9-7 A polyhedron separates space into two distinct regions, an interior region and an exterior region. The union of the polyhedron and its interior region is a **solid.** The polygons which form the solid are called its **faces.** The segments determined by the intersection of the faces are called the **edges.** The point of intersection of three or more faces is a **vertex of a polyhedral angle,** and also is called a **vertex of the polyhedron.**

Definition 9-8 A polyhedron is a **regular polyhedron** if each face is a regular polygon and all polyhedral angles are congruent.

It is interesting that there are only five regular polyhedra, the ones shown in Fig. 9-17.

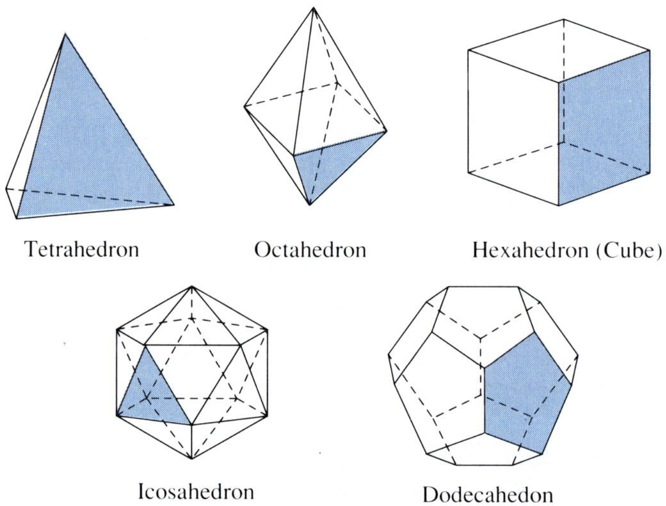

Figure 9-17

Exercise Set 9.4

1. Can a polyhedron have **(a)** three faces? **(b)** four edges? **(c)** five vertices?
2. An *octahedron* has eight faces, usually all triangular regions. **(a)** How many edges would such an octahedron have? **(b)** How many vertices would it have?
3. A *hexahedron* is usually formed by six quadrilateral regions. How many edges does it have? How many vertices? If each of the faces is a regular quadrilateral region, what is the figure commonly called?

9-5 AREA AND VOLUME

A polyhedron with two congruent parallel faces is called a *prism*. The parallel faces are the *bases* of the prism and the remaining faces are all formed by parallelograms. These are called *lateral faces;* their intersections are *lateral edges*. The *altitude* of a prism is the perpendicular distance between the planes of the parallel faces (Fig. 9-18). In a *right prism* the faces are perpendicular to the

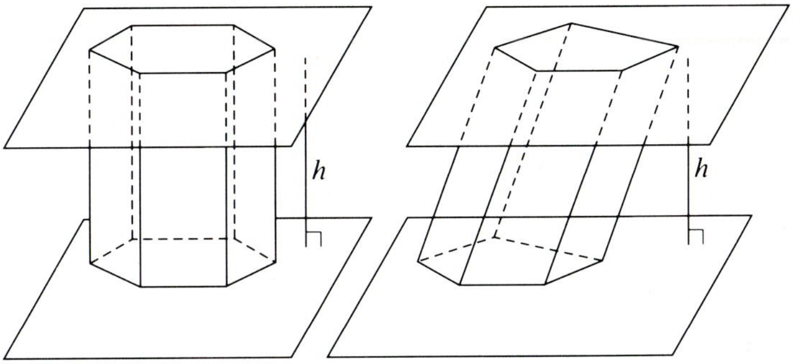

Figure 9-18

bases, the lateral faces are rectangular regions, and the lateral edges have the same length as the altitude. A right prism with bases that are rectangular regions, which is the familiar shape of the box, is called a *rectangular parallelepiped* (Fig. 9-19).

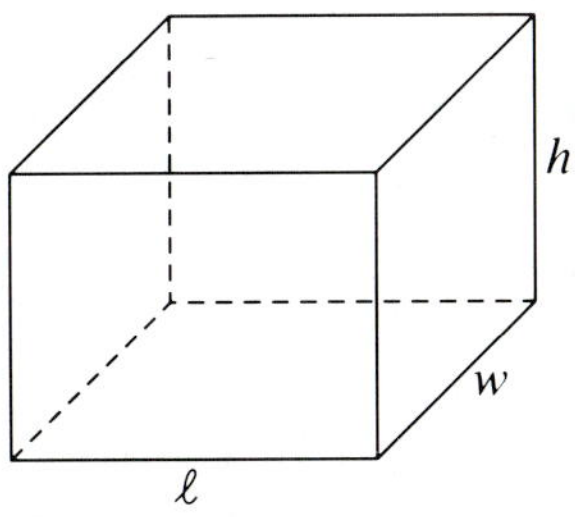

Figure 9-19

A *cube* is a rectangular parallelepiped with all bases and faces square regions.

Definition 9-9 The **surface area of a polyhedron** is the sum of the areas of its faces.

To find the surface area of the box we need only to find the sum of the areas of the rectangular faces, so

$$\begin{aligned} A_T &= 2(\ell \times w) + 2(\ell \times h) + 2(w \times h) \\ &= 2(\ell w + \ell h + wh) \end{aligned}$$

If we are intersected in only the lateral area of the box, then

$$A_L = 2(\ell h + wh) = h(2\ell + 2w)$$

which is the product of the altitude and the perimeter of the base. Thus $A_L = h \cdot p$ and $A_T = hp + 2A_B$, if A_B is the area of the base.

Exercise Set 9.5

1. A *parallelepiped* is a prism with bases that are parallelograms.
   (a) Draw a parallelepiped that is not rectangular.
   (b) How many faces does a parallelepiped have? How many edges? How many vertices?
2. Find the lateral area of a box with an altitude of 16 in. and a base with perimeter of 28 in.
3. Find the total surface area of a cube with an edge of 5 in.
4. Find the total surface area of the right prism in the following figure if the base is a right triangle with legs 9 and 12 units in length.

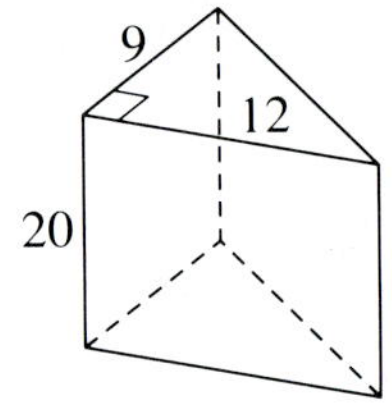

5. In the following figure, the base of the right prism is a rhombus with dimensions as shown.
   (a) Find its lateral area.
   (b) Find the total area.

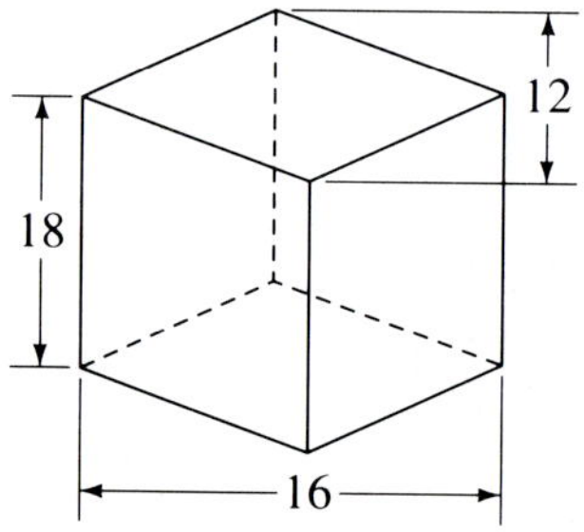

6. Write a formula for the total surface area of a cube if the length of an edge is e.
7. The base of a right prism 25 in. high is a region formed by a parallelogram with perpendicular diagonals 24 in. and 10 in. long. Find its total surface area.
8. ▶ A prism has bases that are equilateral triangles with sides 4 units long. Its lateral edges are 8 units in length and these make an angle of 45° with the plane of the base. Find the lateral area.
9. Find the dimensions of a box if its sides have lengths which are consecutive integers and the total surface area is 52 square units.
10. In the rectangular solid of the following figure, d is the length of diagonal $\overline{AD}$, b is the length of side $\overline{AB}$, w is the length of $\overline{CD}$, and h is the length of altitude $\overline{BC}$. Show that $d^2 = b^2 + w^2 + h^2$.

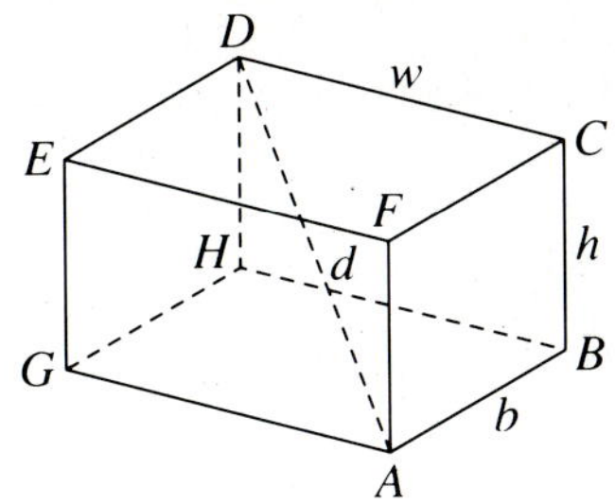

11. Find the length of the diagonal of a cube if its edge is 2 units in length.

12. If the diagonal of a cube is a, find its surface area.

Volume

The unit of measure for a *solid* is usually stated in *cubic units*. If measures of edges and altitudes of the solid are in feet, the *volume* will be stated in cubic feet; if given in inches, the volume will be in cubic inches.

Consider a simple box (Fig. 9-20) with edges of 2, 3, and 2 units. Its faces can be separated into square regions and its interior into cubes. We see 3 rows of 2 cubes each form a layer of 6 cubes, and 2 layers produce a total of 12 cubes. Thus a box with dimensions 2, 3, and 2 units has a volume of 12 cubic units.

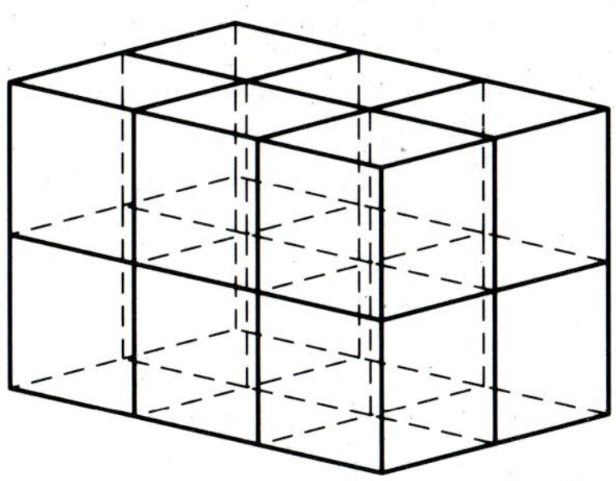

Figure 9-20

Postulate 9-3 The volume (V) of a rectangular parallelepiped (box) is equal to the product of its length (ℓ), width (w) and height (h). ($V = \ell wh$)

Example

Find the volume of the solid in the following figure if m$\overline{AD}$ = 5, m$\overline{BD}$ = 3, and m$\overline{BC}$ = 4.

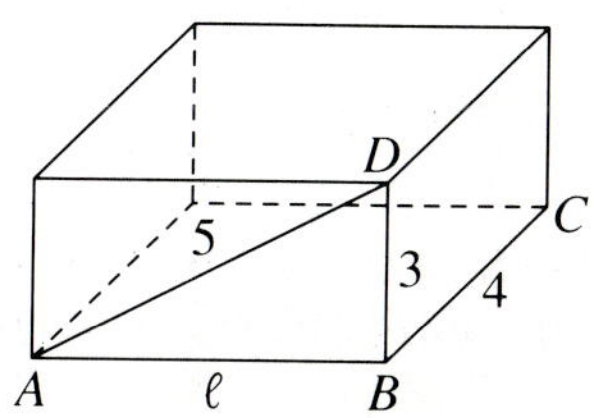

Solution We need to know the length of $\overline{AB}$, so if m$\overline{AB} = \ell$, $\ell^2 = 5^2 - 3^2 = 16$, and $\ell = 4$. Thus $V = 4 \cdot 4 \cdot 3 = 48$ cubic units.

Postulate 9-4 The volume of a prism is equal to the area of a base (K) times the length of the altitude to the base (h). ($V = K \cdot h$)

In Fig. 9-21 we see that if the end section of the parallelepiped is "removed" and placed at the opposite end, a box is formed that has the same volume as the parallelepiped.

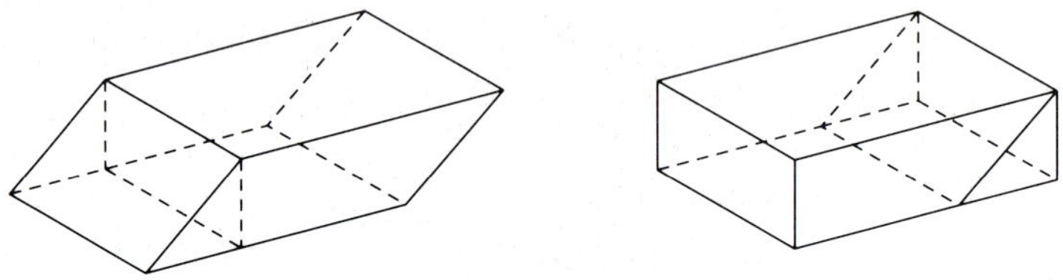

Figure 9-21

Examples

▶ **1.** The base $ABCD$ of the parallelepiped in the following figure has diagonals each 18 ft in length, which intersect to form an angle of 30°. Its height is 8 ft. The lateral edge $\overline{AH}$ forms an angle of 60° with the base $\overline{AD}$. Plane $ADEH$ is perpendicular to the plane of the base. Find the volume and surface area.

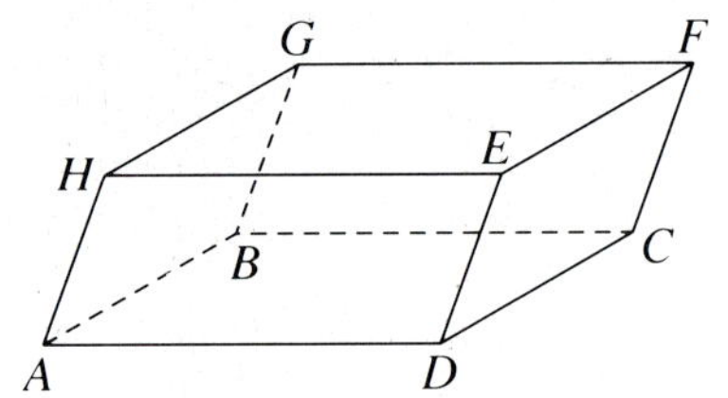

Solution Since the diagonals are equal in length, the base must be a rectangle (as in the following figure). We can determine the volume, then, if we find its area. Since m$\overline{AC} = 18$ and m$\angle CAD = 15$, m$\overline{CD} = w = 18 \sin 15° \approx 18(0.2588) \approx 4.66$ and m$\overline{AD} = \ell = 18 \cos 15° \approx 18(0.9659) \approx 17.39$. Thus $K_B = \ell w \approx (17.39)(4.66) \approx 81.04$ and the volume is $K_B \cdot h \approx (81.04)(8) \approx 648.30$.

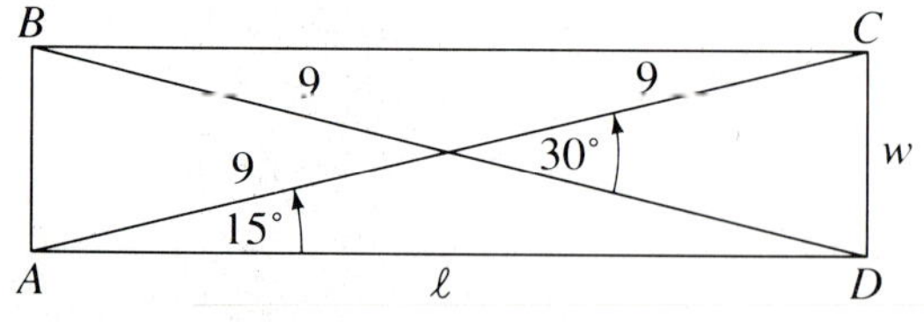

In order to find the surface area we find the area K_S of a side $ADEH$, which is a region formed by a parallelogram with base 17.39 and altitude 8, $K_S \approx 8(17.39) = 139.12$.

The area (K_E) of an end $ABGH$, which is a rectangular face, is found by determining the length of $\overline{AH} = e$. Since $\sin 60° = h/e$, $e = h/(\sin 60°) \approx 8/0.8660 \approx 9.24$ and $K_E = we \approx (4.66)(9.24) \approx 43.06$. Thus the total surface area $S = 2K_B + 2K_S + 2K_E = 2(K_B + K_S + K_E) \approx 2(81.04 + 139.12 + 43.06) = 2(263.22) = 526.44$.

▶ **2.** The base of a prism is a regular pentagonal face with sides 3 cm in length. Face $ABGH$ is perpendicular to the plane of the base. The lateral edges are 5 cm and form an angle of 60° with the base. What is the volume of the prism?

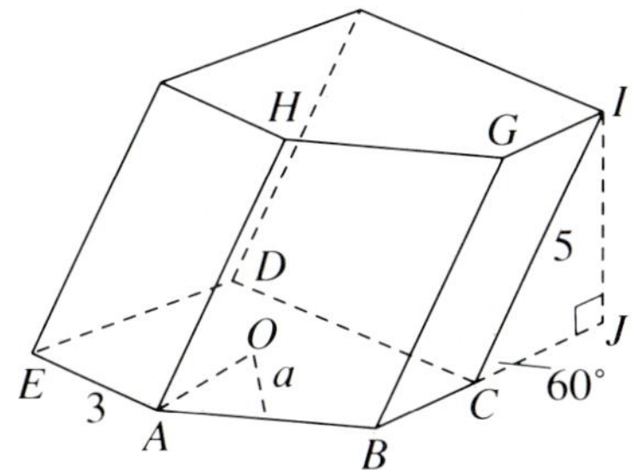

Solution $V = K \cdot h$, so we need the area of the regular pentagon (see page 237). Thus $K = \frac{1}{2}a \cdot p$, and $a = 1.5 \tan 54°$, so $K \approx \frac{1}{2}[(1.5)(1.3764)](25) = 25.74$. The altitude $h = \frac{1}{2} \cdot 5\sqrt{3} \approx 4.33$, since $\angle ICJ$ is a 60° angle and $\overline{CI}$ is the hypotenuse of a right triangle. Thus $V \approx (25.74)(4.33) \approx 111.5 \text{ cm}^3$. ◁

Exercise Set 9.6

1. Find the volume of a cube with an edge of 8 in.
2. A classroom is 38 ft long, 24 ft wide and 10 ft high. Determine the volume and surface area.
3. Find the volume of a right prism if its base is a triangular face with sides 5, 12, and 13 and its altitude is 15.
4. How many (cubic) yards of concrete are required to build a retaining wall 115 ft long, 7 ft high and 18 in. wide?
5. What is the weight of a steel plate 15 ft long, 4 ft wide and $\frac{5}{8}$ in. thick, if steel weighs 475 lb/ft^3?

Definition 9-10 A **pyramid** is a polyhedron that has a polygon for its base and triangles, which meet at a common vertex, for its faces (Fig. 9-22).

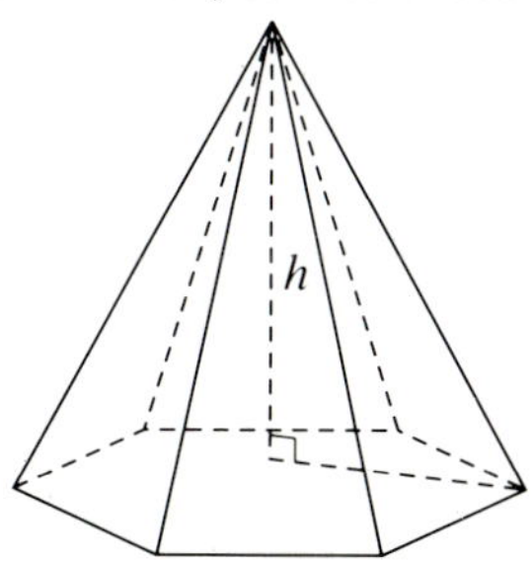

Figure 9-22

Postulate 9-5 The volume (V_P) of a pyramid is equal to one-third the product of the area of its base (K) and the altitude (h) from the vertex to the base. ($V_P = \frac{1}{3}Kh$)

Definition 9-11 The **lateral area** of a pyramid is the sum of the areas of its lateral faces.

Example

The base of a pyramid is a regular hexagon with sides 6 in. long and its altitude measures 10 in. Find its volume and total surface area.

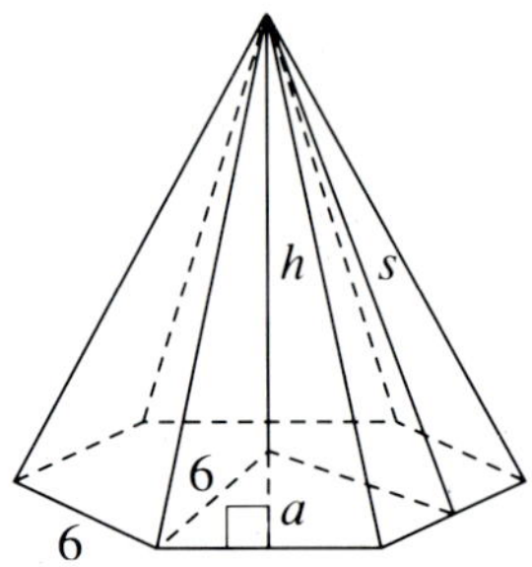

Solution The area of the hexagon may be found by Theorem 8-15, $K = \frac{1}{2}ap$. Since a is opposite the 60° angle of a 30°, 60° right triangle, $a = 3\sqrt{3}$. Thus $K = \frac{1}{2}(3\sqrt{3})(36) = 54\sqrt{3} \approx 93.53$, and $V = \frac{1}{3}(54\sqrt{3})(10) = 180\sqrt{3} \approx 311.76$.

To find the lateral area we need an altitude of one of the triangular faces (this is called the *slant height* of the pyramid). Thus $s^2 = h^2 + a^2 = 10^2 + (3\sqrt{3})^2 = 100 + 27$, and $s = \sqrt{127} \approx 11.27$.

The area of one triangular face is $K_T \approx \frac{1}{2}(6)(11.27) = 33.81$ and its lateral area is $S_P \approx 6(33.81) = 202.86$. The total surface area is $T_P = S_P + K_H \approx 202.86 + 93.53 = 296.39$. ◁

Exercise Set 9.7

1. Find the volume of a solid pyramid with a base area 35 square units and an altitude 9 units in length.
2. Find the lateral area of a pyramid with a square base 6 in. on each side if its altitude measures $3\sqrt{3}$ in.
3. Find the lateral area of a regular triangular pyramid (tetrahedron) if each edge is 12 cm in length.
4. A solid pyramid has a base that is a regular hexagon with sides and height all 8 in. Find its surface area and its volume.
5. Find the altitude of a solid triangular pyramid if each of its edges has a length of 6 in. Find its volume.
6. ▶ Find the surface area of a pyramid with a base 10 in. on each side and lateral edge 13 in. if the base is **(a)** a square, **(b)** a hexagon, and **(c)** an octagon.

Cylinder, Cone and Sphere

The relationships of the cylinder, cone, and sphere are analogous to those of the prism, pyramid, and regular polyhedra. The following are formulas for their areas and volumes.

Cylinder (Fig. 9-23)

Lateral area = circumference of base × altitude

$$S = 2\pi rh$$

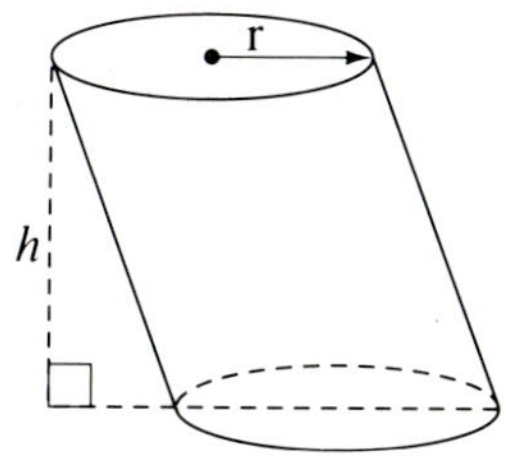

Figure 9-23

Total area = lateral area + area of bases

$$T = 2\pi rh + 2\pi r^2 = 2\pi r(h + r)$$

Volume = Area of base × altitude

$$V = (\pi r^2)h$$

Cone (Fig. 9-24)

Lateral area = $\frac{1}{2}$ circumference of base × slant height

$$S = \tfrac{1}{2}(2\pi r)s = \pi rs$$

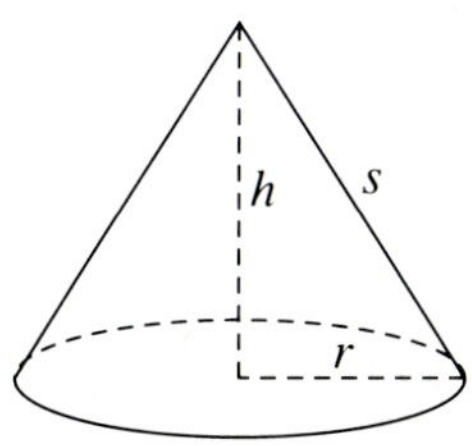

Figure 9-24

Total area = lateral area + area of base

$$T = \pi rs + \pi r^2 = \pi r(r + s)$$

Volume = $\frac{1}{3}$ area of base × altitude

$$V = \tfrac{1}{3}\pi r^2 h$$

Sphere (Fig. 9-25)

Surface area = 4 times area of great circle

$$S = 4\pi r^2$$

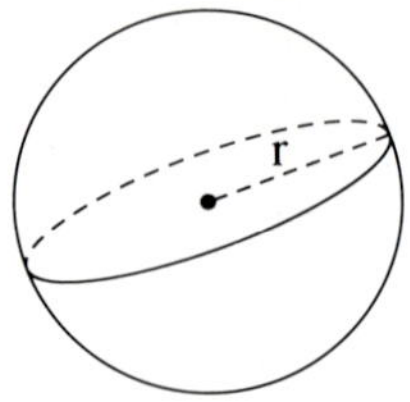

Figure 9-25

Volume $= \frac{4}{3}\pi \times$ cube of radius

$$V = \tfrac{4}{3}\pi r^3$$

Examples

1. A tank is cylindrically shaped with height 60 in. and base 15 in. in diameter.
   (a) Find the lateral area.
   (b) Find the total area if the ends are hemispheres.
   (c) Find the volume of the tank.

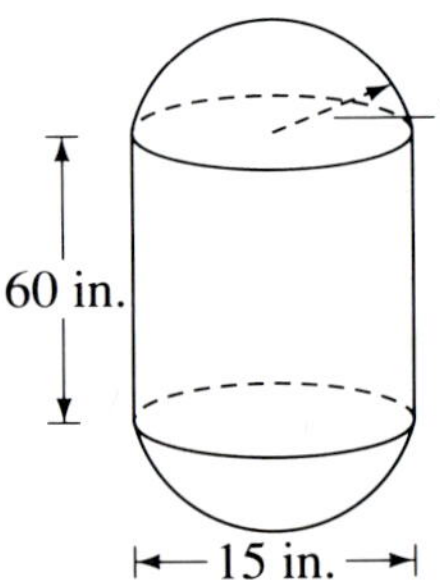

Solution (a) $S = 2\pi rh \approx 2(3.14)(7.5)(60) = 2826$ in.^{2}

(b) $T = S +$ area of bases $\approx 2826 + 4(3.14)(7.15)^2 = 2826 + 706.5 = 3532.5$ in.^{2}

(c) $V = V(\text{cylinder}) + V(\text{sphere})$
$= \pi r^2 h + \frac{4}{3}\pi r^3$
$= \pi r^2(h + \frac{4}{3}r) \approx 176.6(60 + 10)$
$= 12{,}362$ in.^{3}

2. The altitude of a right circular cone is 12 and its slant height is 13. Find its lateral area, total area, and volume in terms of π.

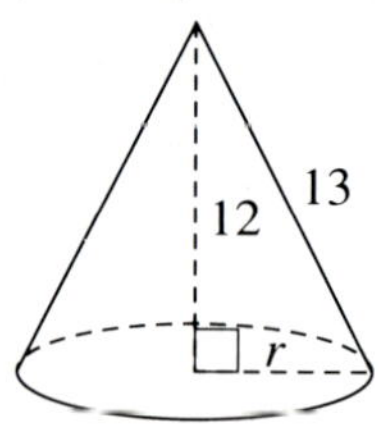

Solution Since $s = 13$ and $h = 12$, $r^2 + 12^2 = 13^2$ by the Pythagorean Theorem and $r = 5$. Thus

$S = \pi rs = \pi(5)(13) = 65\pi$ (square units)
$T = S + \pi r^2 = 65\pi + \pi(5)^2 = 65\pi + 25\pi = 90\pi$ (square units)
$V = \frac{1}{3}\pi r^2 h = \frac{1}{3}(25\pi)(12) = 100\pi$ (cubic units)

3. If the cost of material is \$7.00 per square yard, how much would it cost to construct a spherical balloon 30 ft in diameter?

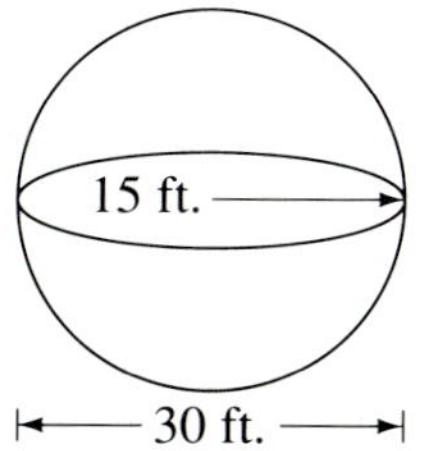

Solution Surface area of a sphere is $S = 4\pi r^2 = 4\pi(15)^2 = 900\pi$ ft^2. Since 9 ft^2 = 1 yd^2, 900π ft^2 = (900π ft^2)/1 × (1 yd^2)/(9 ft^2) = 100π yd^2. The cost of the balloon is (\$7/yd^2)(100π yd^2) = (\$$700\pi$) ≈ \$2198. ◁

Exercise Set 9.8

Give answers in terms of π unless otherwise specified.

1. An equilateral triangular region of side 8 is inscribed in a circular region. If the figure is revolved about one of the altitudes of the triangle, find
   (a) the volume of the cone generated
   (b) the surface area of the sphere
2. If a tub has a bottom with diameter 24 in. and top with diameter 36 in. and its height is 15 in., how much water will it hold? (1 gallon ≈ 231 cubic in.)

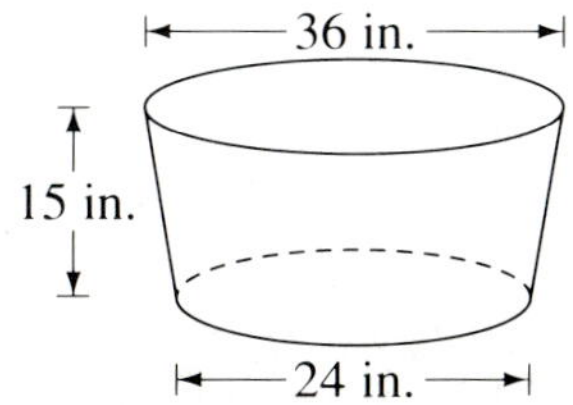

3. A cylindrical chimney of height 50 ft, inside diameter 7 ft, and outside diameter 12 ft is to be built of concrete. If the cost of construction is to be \$65 per cubic yard, what will the chimney cost (to the nearest ten dollars)?
4. A triangle with sides 6, 11, and 15 is rotated about the side of length 11. **(a)** What is the lateral area of the cones which are generated? **(b)** What is the enclosed volume?
5. A hemispherical shell with inside diameter 40 ft is to be constructed of material 8 in. thick. The material weighs 7.2 lb/ft^3. **(a)** What is the volume of the interior of the shell? **(b)** What is the volume of the shell? **(c)** What is the total weight of the shell (to the nearest 10 lb)? **(d)** What is the cost of painting the inside and outside of the shell if labor and material together will cost \$0.68/ft^2?

▶ 6. A cone has a base with an area of 18 square units. A segment of length 8 from the center of the base to the vertex of the cone makes an angle of 55° with a radius. What is the volume of the cone?

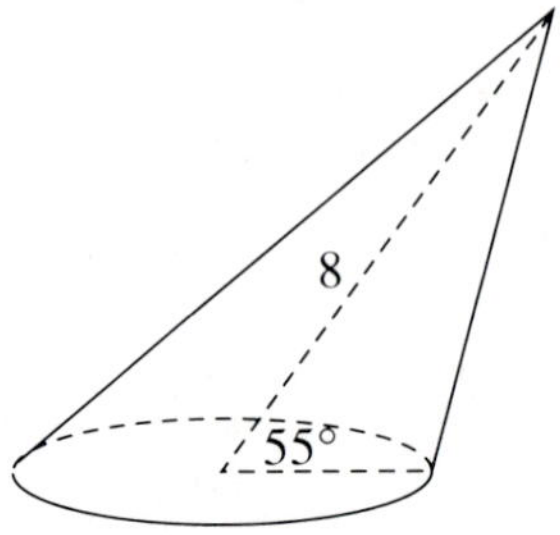

7. If the centerline of a circular cylinder of radius 4 forms an angle of 60° with the base and is twice the length of the radius, find the volume of the cylinder.

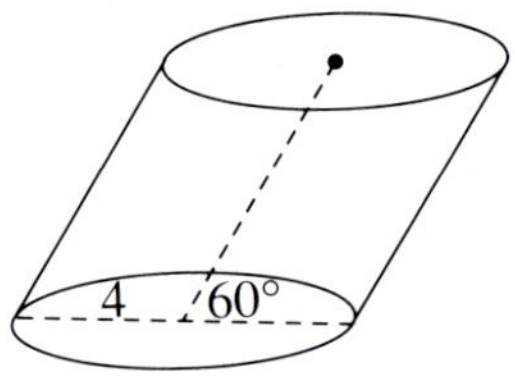

8. A sphere with a diameter of 20 in. is cut by a plane 7 in. from the center of the sphere. Find the area of the circle of intersection.

9. Find the volume of a spherical solid if there are 100π square units in the measure of its surface area.

10. If the volume of a spherical solid is $125\pi/6$ cubic units, what is the length of its radius?

REVIEW EXERCISES

1. A line ℓ_1 is perpendicular to a plane $\mathscr{E}$ at some point P. How many perpendiculars to ℓ_1 at P may be drawn in $\mathscr{E}$? Why?
2. If plane $\mathscr{E}_1$ is perpendicular to ℓ_2 in $\mathscr{E}_2$, is plane $\mathscr{E}_1$ necessarily perpendicular to plane $\mathscr{E}_2$?
3. If two planes $\mathscr{E}_1$ and $\mathscr{E}_2$ are perpendicular to a third plane $\mathscr{E}_3$, are the two planes $\mathscr{E}_1$ and $\mathscr{E}_2$ necessarily parallel? Illustrate.
4. If three planes $\mathscr{E}_1$, $\mathscr{E}_2$ and $\mathscr{E}_3$ are parallel to the same line, are they necessarily parallel to each other? Illustrate.
5. Given parallel planes $\mathscr{E}_1$, $\mathscr{E}_2$ and $\mathscr{E}_3$, with $\mathscr{E}_2$ between $\mathscr{E}_1$ and $\mathscr{E}_3$. From points A and B in $\mathscr{E}_1$ lines are drawn to C and D in $\mathscr{E}_3$, intersecting $\mathscr{E}_2$ at E and F respectively. Prove $m\overline{AE}/m\overline{EC} = m\overline{BF}/m\overline{FD}$.
6. If $\overleftrightarrow{AB}$ and $\overleftrightarrow{CD}$ are skew lines and a plane is perpendicular to $\overleftrightarrow{AB}$ at Q, must $\overleftrightarrow{CD}$ intersect the plane? Illustrate.
7. State whether each of the following statements is sometimes, always, or never true.
   (a) If two planes are perpendicular to the same line, they are parallel.
   (b) A line perpendicular to a plane is perpendicular to every line in that plane.

(c) If two planes are each parallel to a line, they are parallel to each other.
(d) Through a given point not on a plane, there is only one line parallel to the given plane.
(e) A plane containing the bisector of the plane angle of a dihedral angle is the bisecting plane of the dihedral angle.
(f) If a plane is perpendicular to one of two parallel planes, it is perpendicular to the other.
(g) A line segment AB 8 in. in length joins a point on the bisector of the dihedral angle to its edge. The point on the bisector is $4\sqrt{2}$ in. from each of its faces.

8. If the length, width and height of a rectangular solid are, respectively, 8, 6, and 24, find the length of a diagonal.

9. A line segment 18 in. long has parallel planes at its end points. A third plane, parallel to the other two, cuts the segment so that its projections on the third plane are 4 and 5 in. What is the distance between the planes?

10. Prove that the diagonals of a cube are not perpendicular.

11. The surface of a cube with edges of length 6 in. is painted red and then the cube is divided into one in. (length of edge) cubes. Find
(a) the number of inch cubes with one face painted red
(b) the number of inch cubes with two faces painted red
(c) the number of inch cubes with three faces painted red
(d) the number of inch cubes with no faces painted red

12. If ℓ, w, and h denote the length, width, and height of a rectangular solid, find the length of the diagonal when
(a) $\ell = 4$, $w = 6$, $h = 3$ **(b)** $\ell = 10$, $w = 8$, $h = 6$
(c) $\ell = 5$, $w = 7$, $h = 8$ **(d)** $\ell = 2\sqrt{3}$, $w = 2$, $h = 3$
(e) $\ell = 6\sqrt{3}$, $w = 6$, $h = 5$

13. Find the edge of a cube if its diagonal is (a) $5\sqrt{2}$, (b) $3\sqrt{3}$, (c) x.

14. If two spheres have total surface areas of 16π and 36π, respectively, what is the ratio of their volumes?

15. What is the ratio of the volumes of two similar right circular cylinders if the ratio of their radii is $\frac{2}{3}$?

16. The base of a parallelepiped is a rhombus with a diagonal equal in length to its side. If the height is 8 in. and a side of the base is 4 in. find its volume.

17. A cone and a right circular cylinder have bases of equal area and have equal volume. Find the ratio of the lengths of their heights.

18. The volume of a rectangular solid is 1800 cubic ft. Find its dimensions if they are in the ratio of 2 : 3 : 5.

19. Find the surface area of a sphere whose volume is 972π cubic in.

20. A spherical shell 12 in. thick contains the same amount of material as a solid sphere of radius $1\frac{1}{2}$ ft. Find the radius of the outer surface of the shell.

21. If A denotes the total surface area of a cube and V its volume, show that

$$V = \frac{A}{36}\sqrt{6A}$$

22. Find the volume generated by rotating a triangle with sides 3, 4, 5 around **(a)** its shortest side, **(b)** its longest side.

23. By what value must the radius (r) of a sphere be multiplied in order to double its surface area?

24. Find the volume generated by revolving an isosceles right triangle about one of its legs, if the length of each leg is a units.

25. Find the edge of a cube if its total surface area is numerically equal to its volume.

▶ 26. The base of a prism is a regular octagon with all edges 5 in. in length. If one face is perpendicular to the plane of the base but its edge makes an angle of 30° with the base, what is the volume of the solid prism?

Cumulative Review—Chapters 2–9

Complete the following with a word or phrase that makes the statement true.

1. When we write A-X-B, we are indicating that X __?__ A and B.
2. $|a|$ is __?__ when a is positive and it is __?__ when a is negative.
3. If $\angle ABC \cong \angle DEF$ and $\angle DEF \cong \angle XYZ$, then $\angle ABC \cong \angle XYZ$, since congruence of angles is said to be __?__.
4. An angle divides a plane into three distinct sets: (1) __?__ (2) __?__ (3) __?__.
5. If the sum of the measures of two angles is 180, the angles are __?__.
6. The point $(-3, 2)$ is in the __?__ quadrant.
7. The interior of a triangle is a __?__ set.
8. If two sides of a triangle are of measure 5 and 11 and the third side is of measure x, then __?__ $< x <$ __?__.
9. Two points, each equidistant from the endpoints of a segment, determine __?__ of the segment.
10. If two triangles have two angles and a side of one triangle congruent to two angles and a side of the other, the triangles are __?__.
11. If in a triangle $\angle ABC$, $\angle A$ is the largest angle, then __?__ is the longest side.
12. If the diagonals of a parallelogram are perpendicular, the parallelogram is __?__.
13. An angle inscribed in the major arc of a circle has measure a, such that __?__ $< a <$ __?__.
14. The measure of an angle formed by a tangent and a chord is __?__.
15. If the area of a circle is 225π, the measure of its radius is __?__ and the measure of its circumference is __?__.

Problems

16. Find the distance between the following pairs of points on the real number line with coordinates:
    - **(a)** 2, 7
    - **(b)** -3, 5
    - **(c)** $\frac{3}{7}$, 4
    - **(d)** a, b
    - **(e)** p, q, if $q < p$
17. If the measure of $\angle A$ is increased by 30, it is then equal to $\frac{2}{3}$ the measure of its supplement. What is the measure of $\angle A$?
18. If $m\angle AOD = (3x - 5)/5$ and $m\angle BOC = (2x - 7)/3$, what is $m\angle AOB$?

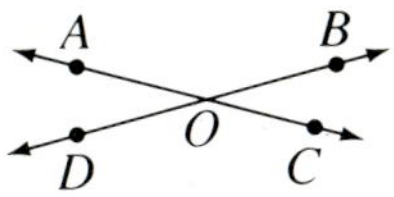

19. $\ell_1 \parallel \ell_2$ and t is a transversal
   (a) If m$\angle 1 = 50$, m$\angle 5 =$ __?__
   (b) If m$\angle 4 = 70$, m$\angle 6 =$ __?__
   (c) If m$\angle 2 = 130$, m$\angle 5 =$ __?__

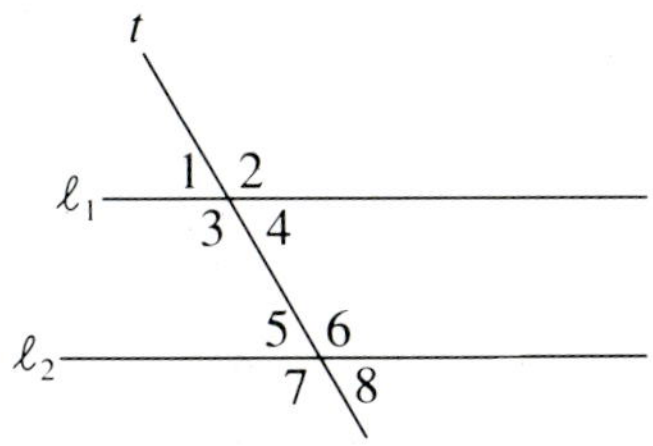

20. Given the points $P(1, 4)$, $Q(5, 8)$ and $R(1, 8)$.
   (a) Find the midpoint of $\overline{PQ}$; $\overline{PR}$; $\overline{QR}$.
   (b) Find the slope of $\overline{PQ}$; $\overline{PR}$; $\overline{QR}$.
   (c) Find the length of $\overline{PQ}$; $\overline{PR}$; $\overline{QR}$.
21. Given $\triangle ABC \sim \triangle DEF$, write the proportionality between the corresponding sides.
22. If the projection of the shortest side of a 30°, 60° right triangle onto the hypotenuse is 12, what is the length of each of the three sides?
23. A 16-ft ladder is placed against a building just touching the roof line. If the base of the ladder is 8 ft from the bottom of the wall, how high is the wall?
24. If the perimeter of a rhombus is 72 in. and its shortest diagonal measures 18 in., find the area of the region.
25. A quadrilateral $ABCD$ is inscribed in a circle. If m°$\widehat{AB} = 120$, m°$\widehat{BC} = 60$, m°$\widehat{CD} = 50$, find
   (a) m$\angle ABC$
   (b) m$\angle DAC$
   (c) m$\angle ADC$
   (d) measure of the angle formed by $\overleftrightarrow{AD}$ and $\overleftrightarrow{BC}$
26. Given a circle of radius 8:
   (a) find the area of the inscribed square
   (b) find the area of the circumscribed square
   (c) find the area of the inscribed regular hexagon
27. With compass and straightedge, inscribe a circle in a square with sides of length 3 in.
28. Find the length of the diagonal of a cube with edge which measures 6 units.
29. Find the volume of the conical solid generated by rotating a 30°, 60° right triangle about its longer leg.
30. How much per mile will it cost to dig a trench 4.5 ft wide at the top, 3 ft wide at the bottom, and 3.5 ft deep, at 15¢/yd^3?

Proofs

31. Prove that the bisectors of a pair of supplementary adjacent angles form a right angle.

32. Prove that any two equilateral triangles are similar.

33. Using coordinates, prove that if two medians of a triangle are congruent, then the triangle is isosceles.

34. Prove that in a circle parallel chords intersect congruent arcs.

35. Prove Construction 9, construction of the perpendicular bisector of a segment. (See page 129.)

Constructions

In the following exercises, use this scale:

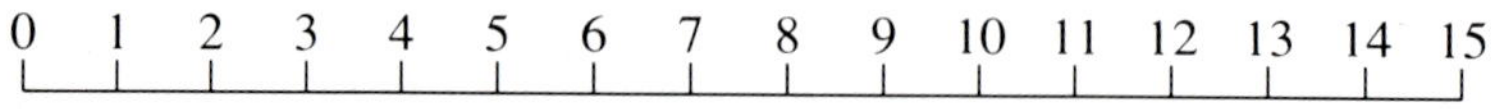

36. Construct a triangle with sides of lengths:
(a) 3, 4, 5 **(b)** 2, 5, 6 **(c)** 4, 4, 4

37. Using the triangle of 1(b), construct the bisectors of its angles.

38. Construct an isosceles triangle with base 5 and congruent sides of length 4.

39. Construct an isosceles triangle with base 5, and vertex angle congruent to angle B:

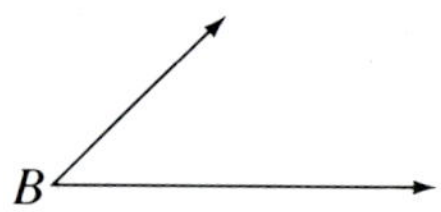

40. Construct an equilateral triangle given its median has length 6.

41. Draw a segment with length 8. By construction, divide it into three congruent parts.

42. Draw a segment with length 9. By construction, divide it into six congruent segments.

43. Circumscribe a circle about a square with 5 unit sides.

44. Construct a parallelogram given one angle measures 60, one side is 4 and the perimeter is 15.

45. The radii of two circles are 2 and 3. Construct them (a) internally tangent, (b) externally tangent.

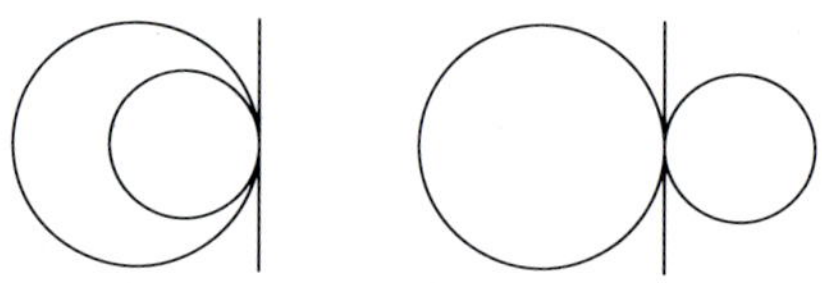

(a) (b)

Appendixes

APPENDIX I. THE REAL-NUMBER SYSTEM

The following laws hold for all real numbers a, b, c, and d.

Axioms of Equality

E-1: (*Reflexive Axiom*) $a = a$.

E-2: (*Symmetric Axiom*) If $a = b$, then $b = a$.

E-3: (*Transitive Axiom*) If $a = b$ and $b = c$, then $a = c$.

E-4: (*Substitution Axiom*) If $a = b$, then a can be substituted for b and b can be substituted for a in any mathematical statement or expression.

The Field Axioms

R-1: (*Closure Axiom for Addition*) $a + b$ is a unique real number.

R-2: (*Commutative Axiom for Addition*) $a + b = b + a$.

R-3: (*Associative Axiom for Addition*) If $a + b + c = (a + b) + c$, then $(a + b) + c = a + (b + c)$.

R-4: (*Identity Element for Addition*) There is a unique real number 0 such that $a + 0 = 0 + a = a$.

R-5: (*Inverse Element for Addition*) There is a unique real number $(-a)$ such that $a + (-a) = (-a) + a = 0$.

R-6: (*Closure Axiom for Multiplication*) ab is a unique real number.

R-7: (*Commutative Axiom for Multiplication*) $ab = ba$.

R-8: (*Associative Axiom for Multiplication*) If $abc = (ab)c$, then $(ab)c = a(bc)$.

R-9: (*Identity Element for Multiplication*) There is a unique real number 1 such that $a \cdot 1 = 1 \cdot a = a$.

R-10: (*Inverse Element for Multiplication*) If $a \neq 0$, there is a unique real number $1/a$ such that $a \cdot (1/a) = (1/a) \cdot a = 1$.

R-11: (*Distributive Axiom*) $a(b + c) = ab + ac$.

Fundamental Theorems

I-1: *Addition Law for Equals*: If $a = b$ and $c = d$, then $a + c = b + d$.

I-2: *Multiplication Law for Equals*: If $a = b$ and $c = d$ then $ac = bd$.

I-3: *Cancellation Law for Equals*: If $a + c = b + d$ and $c = d$, then $a = b$.

I-4: *Division Law for Equals*: If $c \neq 0$ and $ac = bc$, then $a = b$.

I-5: *Zero Law for Multiplication*: $a \cdot 0 = 0 \cdot a = 0$.

Properties of Order

Definition The number a is *less than* the number b ($a < b$) if there is a positive number x such that $b - a = x$, or $b = a + x$. The statement $a < b$ is equivalent to $b > a$, or b is *greater than* a.

0-1: (*Trichotomy Axiom*) For any two numbers a and b exactly one of the following is true: $a < b$, $a = b$, $a > b$.

0-2: (*Transitive Axiom of Order*) If $a < b$ and $b < c$, then $a < c$.

I-6: (Theorem)
(a) $a + c < b + c$ if and only if $a < b$.
(b) If $a < b$ and $c < d$, then $a + c < b + d$.

I-7: (Theorem)
(a) If $c > 0$, $ac < bc$ if and only if $a < b$.
(b) If $c < 0$, $bc < ac$ if and only if $a < b$.

I-8: (Theorem) If $a^2 = b^2$, $a > 0$ and $b > 0$, then $a = b$.

APPENDIX II. SETS

Elements and Subsets

The word set is undefined, but we need to be able to determine whether or not a given *element* or *member* is included in a set. If A is the set of nonzero digits ($A = \{1, 2, 3, 4, 5, 6, 7, 8, 9\}$), we can say that 4 *is* an element of A, ($4 \in A$) and 27 is *not* an element of A ($27 \notin A$). If B is the set of letters of the alphabet, then $2 \notin B$, but $m, n \in B$.

We say that set A is a *subset* of B ($A \subset B$) if and only if every element of A is an element of B. Thus the odd digits form a subset of the set of nonzero digits, i.e., $\{1, 3, 5, 7, 9\} \subset \{1, 2, 3, 4, 5, 6, 7, 8, 9\}$. On the other hand the set $\{1, 2, 3\}$ is not a subset of the set of odd digits, or $\{1, 2, 3\} \not\subset \{1, 3, 5, 7, 9\}$.

Each set discussed thus far has been a *finite* set, containing a countable number of elements. Some sets are *infinite* sets, such as the set of real numbers, or the set of points on a line.

If x is a member of a set, we can sometimes denote the set by using *set-builder* notation, as follows: $K = \{x \mid -3 \leq x \leq 0$ and x is an integer$\}$ is read "K is the set of numbers x such that x lies between -3 and 0, inclusive, and x is an integer." The vertical bar following the first x alerts us to what is being described and is read "such that." Note that $K = \{-3, -2, -1, 0\}$, a finite set. The set $G = \{(x, y)|x + y = 7\}$ is "the set of ordered pairs of real numbers (x, y) such that $x + y = 7$." If no mention is made of the set of numbers intended we will assume the real numbers are meant. It would be impossible to list all of the elements of G, which is an infinite set.

Combinations of Sets

The *union* of sets A and B $(A \cup B)$ is the set of elements contained in A or B, or in both A and B. If $A = \{a, b, c, d\}$ and $B = \{b, d, f, h\}$ then $A \cup B = \{a, b, c, d, f, h\}$.

The *intersection* of sets A and B $(A \cap B)$ is the set of elements contained in both A and B. If $A = \{1, 3, 5, 7, 9\}$ and $B = \{3, 4, 5, 6, 7, 8\}$, then $A \cap B = \{3, 5, 7\}$.

If the intersection of two sets A and B contains no members, then their intersection is the *empty set* $(\emptyset)$ and the sets are said to be *disjoint*. Thus if $A = \{1, 2, 3\}$ and $B = \{5, 7, 9\}$, then $A \cap B = \emptyset$ and A and B are disjoint.

Sets of Points

We will be concerned for the most part with sets of numbers and sets of points. We say that a geometric figure is a nonempty set of points. A line is a set of points. If ℓ is a line, there are at least two points on it. Thus we can consider a point $P \in \ell$ and a point $Q \in \ell$, or a set of points $\{X|X \in \ell\}$.

If two lines ℓ_1 and ℓ_2 are in the same plane, they may either coincide, intersect, or not intersect. Thus if they coincide, $\ell_1 \cap \ell_2 = \ell_1 = \ell_2$; if they intersect $\ell_1 \cap \ell_2 = P$, where P is a point of the plane, or if they do not intersect then $\ell_1 \cap \ell_2 = \emptyset$.

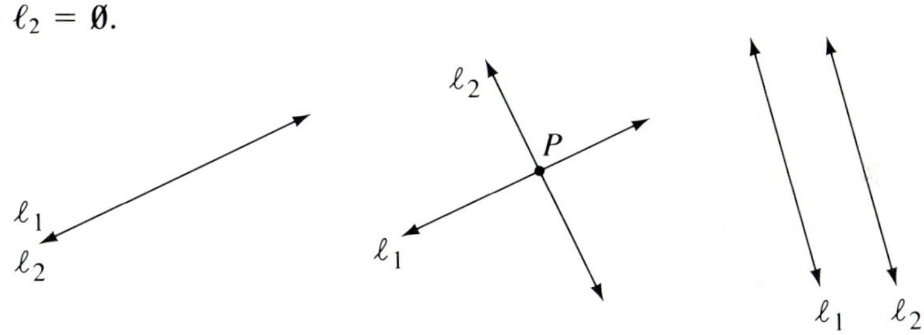

The *union* of two sets of points is the entire set of points contained in either or both of them.

APPENDIX III. PROOFS

The following proofs were omitted from the main body of the text in order not to interrupt its continuity. We present them here so that the interested reader will not have to search elsewhere for them.

Theorem The number $\sqrt{2}$ is irrational.

Proof To say that $\sqrt{2}$ is irrational means that there are no integers p and q such that $\sqrt{2} = p/q$. We will assume that there *are* and then show that this leads to a contradiction. We need to agree beforehand, also, that every rational number has a "simplest form," i.e., if d is the greatest common divisor of integers a and b, then $a/b = (md)/(nd) = m/n$.

Now suppose that $\sqrt{2} = p/q$, with p/q in simplest form. Then $(\sqrt{2})^2 = (p/q)^2$, or $2 = p^2/q^2$, so that $2q^2 = p^2$ which means that p^2 must be even. Thus p is even and can be expressed in the form $p = 2k$, where k an integer. Now $p^2 = (2k)^2 = 2q^2$, so $4k^2 = 2q^2$ or $2k^2 = q^2$. This means q^2 is even, so q is even and can be written in the form $q = 2h$, h an integer. But we originally had said that p/q was in simplest form, and now $p/q = (2k)/(2h)$. The only statement which is subject to doubt was our original one, that $\sqrt{2} = p/q$. Therefore, $\sqrt{2}$ *cannot* be written in the form p/q with p and q integers, so $\sqrt{2}$ is irrational.

Theorem 6-1 If the angles of two triangles are correspondingly congruent, the triangles are similar.

Our proof is for a special case, but it could easily be revised for any pair of positive integers which represent the lengths of $\overline{AC}$ and $\overline{DF}$. If the ratios do not involve rational numbers, however, the proof is somewhat more difficult and we will not attempt it here.

Given: $\triangle ABC$ and $\triangle DEF$ with $\angle A \cong \angle D$, $\angle B \cong \angle E$ and $\angle C \cong \angle F$

Prove: $\triangle ABC \sim \triangle DEF$

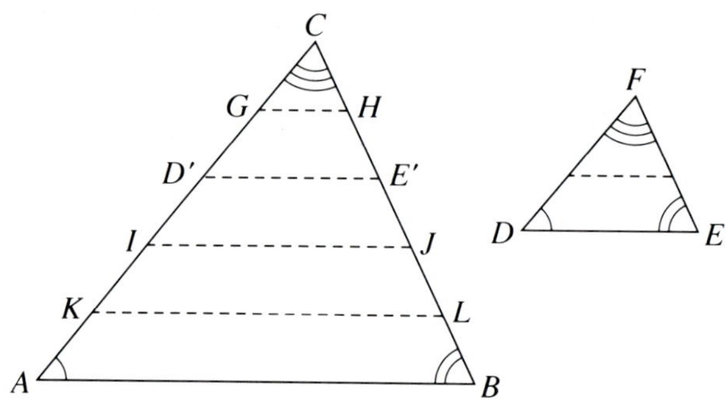

Proof We need to show that $m\overline{AB}/m\overline{DE} = m\overline{BC}/m\overline{EF} = m\overline{AC}/m\overline{DF}$. Let us consider that $m\overline{AC}/m\overline{DF}$ is some rational number, such as $\frac{5}{2}$, and let $m\overline{DF} = 2$ and $m\overline{AC} = 5$. Then we may draw a figure as before.

Statements	*Reasons*
1. There is a point D' on $\overline{AC}$ such that $\overline{CD'} \cong \overline{FD}$ and $m\overline{CD} = 2$	1. Theorem 2-4
2. Through D' there is a line $\overline{D'E'} \parallel \overline{AB}$ and intersecting $\overline{BC}$ at E'	2. Theorem 5-10
3. Thus $\angle CD'E' \cong \angle A \cong \angle D$	3. Theorem 5-15 and Given

Statements	*Reasons*
4. $\triangle D'E'C \cong \triangle DEF$	4. Post. 4-2 (ASA)
5. $\overline{CE'} \cong \overline{FE}$	5. CPCTC
6. There exist points G, I, and K on $\overline{AC}$ such that $m\overline{CG} = m\overline{GD'} = m\overline{D'I} = m\overline{IK} = m\overline{KA} = 1$	6. $m\overline{AC} = 5$ and $m\overline{CD'} = 2$
7. Through G, I, and K there are lines parallel to $\overline{AB}$ intersecting $\overline{CB}$ at H, J, and L, respectively	7. Theorem 5-10
8. $\overline{CH} \cong \overline{HE'} \cong \overline{E'J} \cong \overline{IL} \cong \overline{LB}$	8. Theorem 5-20
9. $m\overline{CE'} = 2m\overline{CH}$ and $m\overline{CB} = 5m\overline{CH}$	9. Def. 2-10
10. $\dfrac{m\overline{CB}}{m\overline{CE'}} = \dfrac{5}{2}$	10. Def. 6-4
11. $\dfrac{m\overline{AC}}{m\overline{DF}} = \dfrac{5}{2}$	11. Given
12. $\dfrac{m\overline{AC}}{m\overline{DF}} = \dfrac{m\overline{BC}}{m\overline{EF}}$	12. Axiom E-3 and Statement 5
13. In like manner we can show $\dfrac{m\overline{AB}}{m\overline{DE}} = \dfrac{m\overline{BC}}{m\overline{EF}}$	13. Statements 1–12
14. $\triangle ABC \sim \triangle DEF$	14. Def. 6-3

APPENDIX IV. TABLES

Table of Trigonometric Ratios

Angle	Sin	Cos	Tan	Angle	Sin	Cos	Tan
1°	0.0175	0.9998	0.0175	11°	0.1908	0.9816	0.1944
2°	0.0349	0.9994	0.0349	12°	0.2079	0.9781	0.2126
3°	0.0523	0.9986	0.0524	13°	0.2250	0.9744	0.2309
4°	0.0698	0.9976	0.0699	14°	0.2419	0.9703	0.2493
5°	0.0872	0.9962	0.0875	15°	0.2588	0.9659	0.2679
6°	0.1045	0.9945	0.1051	16°	0.2756	0.9613	0.2867
7°	0.1219	0.9925	0.1228	17°	0.2924	0.9563	0.3057
8°	0.1392	0.9903	0.1405	18°	0.3090	0.9511	0.3249
9°	0.1564	0.9877	0.1584	19°	0.3256	0.9455	0.3443
10°	0.1736	0.9848	0.1763	20°	0.3420	0.9397	0.3640

Table of Trigonometric Ratios (*cont.*)

Angle	Sin	Cos	Tan	Angle	Sin	Cos	Tan
21°	0.3584	0.9336	0.3839	56°	0.8290	0.5592	1.4826
22°	0.3746	0.9272	0.4040	57°	0.8387	0.5446	1.5399
23°	0.3907	0.9205	0.4245	58°	0.8480	0.5299	1.6003
24°	0.4067	0.9135	0.4452	59°	0.8572	0.5150	1.6643
25°	0.4226	0.9063	0.4663	60°	0.8660	0.5000	1.7321
26°	0.4384	0.8988	0.4877	61°	0.8746	0.4848	1.8040
27°	0.4540	0.8910	0.5095	62°	0.8829	0.4695	1.8807
28°	0.4695	0.8829	0.5317	63°	0.8910	0.4540	1.9626
29°	0.4848	0.8746	0.5543	64°	0.8988	0.4384	2.0503
30°	0.5000	0.8660	0.5774	65°	0.9063	0.4226	2.1445
31°	0.5150	0.8572	0.6009	66°	0.9135	0.4067	2.2460
32°	0.5299	0.8480	0.6249	67°	0.9205	0.3907	2.3559
33°	0.5446	0.8387	0.6494	68°	0.9272	0.3746	2.4751
34°	0.5592	0.8290	0.6745	69°	0.9336	0.3584	2.6051
35°	0.5736	0.8192	0.7002	70°	0.9397	0.3420	2.7475
36°	0.5878	0.8090	0.7265	71°	0.9455	0.3256	2.9042
37°	0.6018	0.7986	0.7536	72°	0.9511	0.3090	3.0777
38°	0.6157	0.7880	0.7813	73°	0.9563	0.2924	3.2709
39°	0.6293	0.7771	0.8098	74°	0.9613	0.2756	3.4874
40°	0.6428	0.7660	0.8391	75°	0.9659	0.2588	3.7321
41°	0.6561	0.7547	0.8693	76°	0.9703	0.2419	4.0108
42°	0.6691	0.7431	0.9004	77°	0.9744	0.2250	4.3315
43°	0.6820	0.7314	0.9325	78°	0.9781	0.2079	4.7046
44°	0.6947	0.7193	0.9657	79°	0.9816	0.1908	5.1446
45°	0.7071	0.7071	1.0000	80°	0.9848	0.1736	5.6713
46°	0.7193	0.6947	1.0355	81°	0.9877	0.1564	6.3138
47°	0.7314	0.6820	1.0724	82°	0.9903	0.1392	7.1154
48°	0.7431	0.6691	1.1106	83°	0.9925	0.1219	8.1443
49°	0.7547	0.6561	1.1504	84°	0.9945	0.1045	9.5144
50°	0.7660	0.6428	1.1918	85°	0.9962	0.0872	11.4301
51°	0.7771	0.6293	1.2349	86°	0.9976	0.0698	14.3007
52°	0.7880	0.6157	1.2799	87°	0.9986	0.0523	19.0811
53°	0.7986	0.6018	1.3270	88°	0.9994	0.0349	28.6363
54°	0.8090	0.5878	1.3764	89°	0.9998	0.0175	57.2900
55°	0.8192	0.5736	1.4281	90°	1.0000	0.0000	

Table of Squares and Square Roots

No.	Square	Square Root	No.	Square	Square Root	No.	Square	Square Root
1	1	1.000	36	1,296	6.000	71	5,041	8.426
2	4	1.414	37	1,369	6.083	72	5,184	8.485
3	9	1.732	38	1,444	6.164	73	5,329	8.544
4	16	2.000	39	1,521	6.245	74	5,476	8.602
5	25	2.236	40	1,600	6.325	75	5,625	8.660
6	36	2.449	41	1,681	6.403	76	5,776	8.718
7	49	2.646	42	1,764	6.481	77	5,929	8.775
8	64	2.828	43	1,849	6.557	78	6,084	8.832
9	81	3.000	44	1,936	6.633	79	6,241	8.888
10	100	3.162	45	2,025	6.708	80	6,400	8.944
11	121	3.317	46	2,116	6.782	81	6,561	9.000
12	144	3.464	47	2,209	6.856	82	6,724	9.055
13	169	3.606	48	2,304	6.928	83	6,889	9.110
14	196	3.742	49	2,401	7.000	84	7,056	9.165
15	225	3.873	50	2,500	7.071	85	7,225	9.220
16	256	4.000	51	2,601	7.141	86	7,396	9.274
17	289	4.123	52	2,704	7.211	87	7,569	9.327
18	324	4.243	53	2,809	7.280	88	7,744	9.381
19	361	4.359	54	2,916	7.348	89	7,921	9.434
20	400	4.472	55	3,025	7.416	90	8,100	9.487
21	441	4.583	56	3,136	7.483	91	8,281	9.539
22	484	4.690	57	3,249	7.550	92	8,464	9.592
23	529	4.796	58	3,364	7.616	93	8,649	9.644
24	576	4.899	59	3,481	7.681	94	8,836	9.695
25	625	5.000	60	3,600	7.746	95	9,025	9.747
26	676	5.099	61	3,721	7.810	96	9,216	9.798
27	729	5.196	62	3,844	7.874	97	9,409	9.849
28	784	5.292	63	3,969	7.937	98	9,604	9.899
29	841	5.385	64	4,096	8.000	99	9,801	9.950
30	900	5.477	65	4,225	8.062	100	10,000	10.000
31	961	5.568	66	4,356	8.124	101	10,201	10.050
32	1,024	5.657	67	4,489	8.185	102	10,404	10.100
33	1,089	5.745	68	4,624	8.246	103	10,609	10.149
34	1,156	5.831	69	4,761	8.307	104	10,816	10.198
35	1,225	5.916	70	4,900	8.367	105	11,025	10.247

Table of Squares and Square Roots (*cont.*)

No.	Square	Square Root	No.	Square	Square Root	No.	Square	Square Root
106	11,236	10.296	121	14,641	11.000	136	18,496	11.662
107	11,449	10.344	122	14,884	11.045	137	18,769	11.705
108	11,664	10.392	123	15,129	11.091	138	19,044	11.747
109	11,881	10.440	124	15,376	11.136	139	19,321	11.790
110	12,100	10.488	125	15,625	11.180	140	19,600	11.832
111	12,321	10.536	126	15,876	11.225	141	19,881	11.874
112	12,544	10.583	127	16,129	11.269	142	20,164	11.916
113	12,769	10.630	128	16,384	11.314	143	20,449	11.958
114	12,996	10.677	129	16,641	11.358	144	20,736	12.000
115	13,225	10.724	130	16,900	11.402	145	21,025	12.042
116	13,456	10.770	131	17,161	11.446	146	21.316	12.083
117	13,689	10.817	132	17,424	11.489	147	21,609	12.124
118	13,924	10.863	133	17,689	11.533	148	21,904	12.166
119	14,161	10.909	134	17,956	11.576	149	22,201	12.207
120	14,400	10.954	135	18,225	11.619	150	22,500	12.247

APPENDIX V.

List of Postulates

Postulate 2-1(a). For any two distinct points there is exactly one line containing them. (p. 13)

Postulate 2-1(b). Every line contains at least two distinct points. (p. 13)

Postulate 2-2. (*Betweenness Postulate*) Given any two distinct points A and B, there exists at least one point between them. (p. 15)

Postulate 2-3. Given any two distinct points R and S, there exists a point T such that R-S-T. (p. 16)

Postulate 2-4. (*Ruler Postulate*) The points on a line can be associated with the set of real numbers so that:
(a) to every point there corresponds a unique real number, and
(b) to every real number there corresponds one and only one point. (p. 19)

Postulate 2-5. (*Ruler Placement Postulate*) For any two points on a line, we may assign zero to one of them and a positive coordinate to the other. (p. 22)

Postulate 2-6. (*Distance Postulate*) To every pair of distinct points we can assign a unique positive number. (p. 23)

Postulate 2-7. If C is a point on $\overline{AB}$ such that A-C-B, then $m\overline{AC} + m\overline{CB} = m\overline{AB}$. (p. 27)

Postulate 2-8. A plane is uniquely determined by three noncollinear points. (p. 32)

Postulate 2-9. (*Plane Separation Postulate*) A line contained in a plane $\mathscr{E}$ separates the plane into three convex sets $\mathscr{H}_1$, $\mathscr{H}_2$, and ℓ such that if $A \in \mathscr{H}_1$ and $B \in \mathscr{H}_2$, then $\overline{AB}$ intersects the line ℓ. (p. 33)

Postulate 3-1. (*Angle Measurement Postulate*) To every angle there corresponds a real number a such that $0 \leq a \leq 180$. (p. 43)

Postulate 3-2. (*Angle Construction Postulate*) Let $\overleftrightarrow{AB}$ be a line in a plane $\mathscr{E}$. For every number a $(0 < a < 180)$ there is exactly one ray AR with R in $\mathscr{H}_1$, such that $m\angle RAB = a$. (p. 43)

Postulate 3-3. (*Angle Addition Postulate*) If P is in the interior of $\angle ABC$, or if P is a point not on a straight angle ABC, then $m\angle ABC = m\angle ABP + m\angle PBC$, or $m\angle APB = m\angle ABC - m\angle PBC$. (p. 44)

Postulate 4-1. (*SAS*) $\triangle ABC \cong \triangle DEF$ if $\overline{AB} \cong \overline{DE}$, $\overline{AC} \cong \overline{DF}$, and $\angle A \cong \angle D$. (p. 90)

Postulate 4-2. (*ASA*) Given triangles ABC and DEF, if $\angle A \cong \angle D$, $\overline{AB} \cong \overline{DE}$ and $\angle B \cong \angle E$, then $\triangle ABC \cong \triangle DEF$. (p. 93)

Postulate 4-3. (*SSS*) Given triangles ABC and DEF, if $\overline{AB} \cong \overline{DE}$, $\overline{BC} \cong \overline{EF}$ and $\overline{AC} \cong \overline{DF}$, then $\triangle ABC \cong \triangle DEF$. (p. 94)

Postulate 5-1. (*The Parallel Postulate*) Through a point not on a given line there is at most one line parallel to the given line. (p. 132)

Postulate 7-1. To every polygonal region there corresponds a unique positive real number. (p. 206)

Postulate 7-2. An area of a polygonal region R is the sum of the areas of the distinct triangular regions $R_1, R_2, \ldots, R_n$ whose union is R. (p. 206)

Postulate 7-3. Congruent triangles determine regions having equal areas. (p. 207)

Postulate 7-4. The area (K) of a right triangular region is one-half the product of the lengths of the legs. (p. 207)

Postulate 8-1. The circumference of a circle is the limit of the perimeters of the inscribed regular polygons. (p. 240)

Postulate 8-2. The ratio of the length of an arc of a circle to its circumference is equal to the ratio of the measure of its central angle to 360. (p. 243)

Postulate 8-3. The area of a circular region is the limit of the areas of the inscribed polygons. (p. 245)

Postulate 8-4. The ratio of the area of a sector of a circle to the area of the circle is equal to the ratio of the measure of the central angle of the sector to 360. (p. 245)

Postulate 9-1. For every plane $\mathscr{E}$ there is at least one point X not on $\mathscr{E}$. (p. 258)

Postulate 9-2. If two distinct points of a line lie in a plane, then all the points of the line lie in the plane. (p. 258)

Postulate 9-3. The volume (V) of a rectangular parallelepiped is equal to the product of its length (ℓ), width (w) and height (h). ($V = \ell wh$). (p. 271)

Postulate 9-4. The volume of a prism is equal to the area of a base (K) times the length of the altitude to the base (h). ($V = K \cdot h$) (p. 272)

Postulate 9-5. The volume (V_P) of a pyramid is equal to one-third the product of the area of its base (K) and the altitude (h) from the vertex to the base. ($V_P = \frac{1}{3}Kh$) (p. 274)

List of Theorems

Theorem 2-1. For every segment AB, CD and EF:
(a) $\overline{AB} \cong \overline{AB}$ (Congruence of segments is reflexive.)
(b) If $\overline{AB} \cong \overline{CD}$, then $\overline{CD} \cong \overline{AB}$. (Congruence of segments is symmetric.)
(c) If $\overline{AB} \cong \overline{CD}$ and $\overline{CD} \cong \overline{EF}$, then $\overline{AB} \cong \overline{EF}$. (Congruence of segments is transitive.) (p. 26)

Theorem 2-2. If A-B-C and R-S-T such that $\overline{AB} \cong \overline{RS}$ and $\overline{BC} \cong \overline{ST}$, then $\overline{AC} \cong \overline{RT}$. (p. 27)

Theorem 2-3. If A-B-C and R-S-T such that $\overline{AC} \cong \overline{RT}$ and $\overline{AB} \cong \overline{RS}$, then $\overline{BC} \cong \overline{ST}$. (p. 27)

Theorem 2-4. (*Segment Construction*) For any line AB and segment CD there exists a point P on ray AB such that $m\overline{CD} = m\overline{AP}$ and $\overline{CD} \cong \overline{AP}$. (p. 27)

Theorem 2-5. Every segment has exactly one midpoint. (p. 29)

Theorem 2-6. P is the midpoint of $\overline{AB}$ if and only if $\overline{AP} \cong \overline{PB}$ and A-P-B. (p. 29)

Theorem 2-7. The segments formed when congruent segments are bisected are congruent. (p. 30)

Theorem 2-8. Two different lines intersect in at most one point. (p. 32)

Theorem 3-1. For any angles A, B, and C:
(a) $\angle A \cong \angle A$. (Congruence of angles is reflexive.)
(b) If $\angle A \cong \angle B$, then $\angle B \cong \angle A$. (Congruence of angles is symmetric.)
(c) If $\angle A \cong \angle B$ and $\angle B \cong \angle C$, then $\angle A \cong \angle C$. (Congruence of angles is transitive.) (p. 52)

Theorem 3-2. If P is in the interior of $\angle ABC$ and Q is in the interior of $\angle RST$, so that $\angle ABP \cong \angle RSQ$ and $\angle PBC \cong \angle QST$, then $\angle ABC \cong \angle RST$. (p. 52)

Theorem 3-3(a). If P is in the interior of $\angle ABC$ and Q is in the interior of $\angle RST$, so that $\angle ABC \cong \angle RST$ and $\angle ABP \cong \angle RSQ$, then $\angle PBC \cong \angle QST$. (p. 53)

Theorem 3-3(b). If $\angle ABC$ and $\angle RST$ are straight angles, P is a point not on

ABC and Q is a point not on RST, and $\angle ABP \cong \angle RSQ$, then $\angle PBC \cong \angle QST$. (p. 53)

Theorem 3-4. Vertical angles are congruent. (p. 54)

Theorem 3-5. If the exterior sides of two adjacent angles form a straight angle, the angles are supplementary. (p. 56)

Theorem 3-6. All right angles are congruent. (p. 56)

Corollary 3-6(a). An angle congruent to a right angle is a right angle. (p. 57)

Corollary 3-6(b). If a pair of angles are equal in measure and supplementary, they are right angles. (p. 57)

Theorem 3-7. Complements of the same angle are congruent. (p. 57)

Theorem 3-8. Complements of congruent angles are congruent. (p. 57)

Theorem 3-9. Supplements of the same angle are congruent. (p. 58)

Theorem 3-10. Supplements of congruent angles are congruent. (p. 58)

Theorem 3-11. If one of the four angles formed by the intersection of two lines is a right angle, the other three are right angles. (p. 62)

Theorem 4-1(a). $\triangle ABC \cong \triangle ABC$. (Triangle congruence is reflexive.) (p. 89)

Theorem 4-1(b). If $\triangle ABC \cong \triangle DEF$, then $\triangle DEF \cong \triangle ABC$. (Triangle congruence is symmetric.) (p. 89)

Theorem 4-1(c). If $\triangle ABC \cong \triangle DEF$ and $\triangle DEF \cong \triangle GHI$, then $\triangle ABC \cong \triangle GHI$. (Triangle congruence is transitive.) (p. 90)

Theorem 4-2. Every angle has exactly one bisector. (p. 101)

Corollary 4-2. The angles formed by bisectors of congruent angles are congruent. (p. 102)

Theorem 4-3. The midpoint of the hypotenuse of a right triangle is equidistant from the vertices. (p. 104)

Theorem 4-4. If two medians of a triangle are congruent, the triangle is isosceles. (p. 104)

Theorem 4-5. (*Isosceles Triangle Theorem*) If two sides of a triangle are congruent, the angles opposite these sides are congruent. (p. 107)

Corollary 4-5. Equilateral triangles are equiangular. (p. 107)

Theorem 4-6. If two angles of a triangle are congruent, the sides opposite these angles are congruent. (p. 108)

Corollary 4-6. An equiangular triangle is equilateral. (p. 108)

Theorem 4-7. (*Exterior Angle Theorem*) An exterior angle of a triangle is greater than either nonadjacent interior angle. (p. 111)

Corollary 4-7. If a triangle has a right angle, the other two angles are acute. (p. 112)

Theorem 4-8. If two sides of a triangle are not congruent, the angle opposite the longer side is the larger angle. (p. 112)

Theorem 4-9. If two angles of a triangle are not congruent, the side opposite the larger angle is the longer side. (p. 113)

Theorem 5-1. In a plane through a given point on a given line, there is one and only one perpendicular to the given line. (p. 117)

Theorem 5-2. Any point on the perpendicular bisector of a segment is equidistant from the endpoints of the segment. (p. 118)

Theorem 5-3. If a point is equidistant from the endpoints of a segment, it lies on the perpendicular bisector of the segment. (p. 118)

Corollary 5-3. If two distinct points are each equidistant from the endpoints of a line segment, they determine the perpendicular bisector of the segment. (p. 119)

Theorem 5-4. Through a point outside a line there is one and only one perpendicular to that line. (p. 123)

Theorem 5-5. The shortest segment joining a point to a line is the perpendicular segment. (p. 125)

Theorem 5-6. A point on the bisector of an angle is equidistant from the sides of the angle. (p. 126)

Theorem 5-7. (*The Triangle Inequality Theorem*) The sum of the lengths of two sides of a triangle is greater than the length of the third side. (p. 126)

Theorem 5-8. If a and b are lengths of two sides of a triangle and $a > b$ then the length of the third side, x, is such that $a - b < x < a + b$. (p. 127)

Theorem 5-9. Two coplanar lines that are perpendicular to a given line in the same plane are parallel. (p. 130)

Theorem 5-10. Through a point P not on a line ℓ_1, there is a line ℓ_2 parallel to ℓ_1. (p. 131)

Theorem 5-11. If two lines are cut by a transversal so that the angles in one pair of alternate-interior angles are congruent, then the angles in the other pair of alternate-interior angles are congruent. (p. 134)

Theorem 5-12. If two lines are cut by a transversal so that a pair of alternate-interior angles are congruent, the lines are parallel. (p. 134)

Theorem 5-13. If two parallel lines are cut by a transversal, then the angles in each pair of alternate-interior angles are congruent. (p. 135)

Theorem 5-14. If two lines are cut by a transversal so that a pair of corresponding angles are congruent, the lines are parallel. (p. 135)

Theorem 5-15. (*Converse of Theorem* 5-14) If two parallel lines are cut by a transversal, then the angles of each pair of corresponding angles are congruent. (p. 135)

Theorem 5-16. If parallel lines are cut by a transversal, the interior angles on the same side of the transversal are supplementary. (p. 135)

Theorem 5-17. In a plane, two lines parallel to a third line are parallel to each other. (p. 135)

Corollary 7-1(b). The opposite angles of a parallelogram are congruent. (p. 199)

Theorem 7-2. The diagonals of a rectangle are congruent. (p. 199)

Theorem 7-3. In a parallelogram any two consecutive angles are supplementary. (p. 199)

Theorem 7-4. The diagonals of a parallelogram bisect each other. (p. 200)

Theorem 7-5. If a quadrilateral has both sets of opposite sides congruent, it is a parallelogram. (p. 202)

Theorem 7-6. If a quadrilateral has one set of opposite sides congruent and parallel, it is a parallelogram. (p. 202)

Theorem 7-7. If the diagonals of a quadrilateral bisect each other, it is a parallelogram. (p. 202)

Theorem 7-8. If a parallelogram has one right angle, then it is a rectangle. (p. 202)

Corollary 7-8. If the diagonals of a parallelogram are congruent, it is a rectangle. (p. 202)

Theorem 7-9. The diagonals of a rhombus are perpendicular to each other. (p. 202)

Theorem 7-10. If the diagonals of a quadrilateral bisect each other and are perpendicular, then the quadrilateral is a rhombus. (p. 202)

Theorem 7-11. The sum of the measures of the angles of a quadrilateral is 360. (p. 205)

Theorem 7-12. The area (K) of a triangle is equal to one-half the product of the length of a base and the length of the altitude to that base. (p. 208)

Corollary 7-12(a). Triangles which have bases of equal length and altitudes of equal length have equal areas. (p. 208)

Corollary 7-12(b). The areas of two triangles having bases of equal length have the same ratio as the lengths of their altitudes. (p. 209)

Corollary 7-12(c). The areas of two triangles having altitudes of equal length have the same ratio as the lengths of their bases. (p. 209)

Corollary 7-12(d). The areas of two triangles have the same ratio as the products of the lengths of their bases and altitudes. (p. 209)

Theorem 7-13. The area (K) of a parallelogram is equal to the product of the lengths of a base (b) and an altitude (a) to that base. ($K = a \cdot b$) (p. 212)

Corollary 7-13(a). The area (K) of a rectangle is equal to the product of the length of its base (b) and the length of its altitude (a). ($K = a \cdot b$) (p. 213)

Corollary 7-13(b). The area (K) of a square with side of length s is s^2. ($K = s^2$) (p. 213)

Theorem 7-14. The area (K) of a trapezoid is equal to one-half the product of the length of its altitude (a) and the sum of the lengths of its bases (b and b'). ($K = \frac{1}{2}a(b + b')$) (p. 213)

Theorem 7-15. The area of a rhombus is equal to one-half the product of the lengths of its diagonals. (p. 213)

Theorem 8-1. A line through the center of a circle perpendicular to a chord bisects the chord. (p. 219)

Theorem 8-2. In the same or congruent circles, chords equidistant from the center of the circle are congruent. (p. 219)

Theorem 8-3. In the same or congruent circles, congruent chords are equidistant from the center of the circle. (p. 219)

Theorem 8-4. A line bisecting a chord that is not the diameter of a circle is perpendicular to the chord. (p. 219)

Theorem 8-5. In the plane of a given circle, the perpendicular bisector of a chord passes through the center of the circle. (p. 219)

Theorem 8-6. If A, B, and C are points on a circle such that ACB, then $m°\overset{\frown}{AC} + m°\overset{\frown}{CB} = m°\overset{\frown}{AB}$. (p. 223)

Theorem 8-7. In the same or congruent circles, if two arcs are congruent, their corresponding chords are congruent. (p. 224)

Theorem 8-8. In the same or congruent circles, if two chords are congruent, their corresponding minor arcs are congruent. (p. 224)

Theorem 8-9. The measure of an angle inscribed in a circle is one-half the degree measure of its intercepted arc. (p. 228)

Corollary 8-9(a). An angle inscribed in a semicircle is a right angle. (p. 229)

Corollary 8-9(b). Angles inscribed in the same arc are congruent. (p. 229)

Theorem 8-10. The measure of the angle formed by two secants intersecting in the interior of a circle is one-half the sum of the degree measures of the arcs intercepted by the angle and its vertical angle. (p. 229)

Theorem 8-11. If a line is tangent to a circle, it is perpendicular to the radius drawn to the point of contact. (p. 232)

Corollary 8-11(a). The two tangent segments from a point of the exterior of a circle are congruent. (p. 232)

Theorem 8-12. The measure of an angle formed by a secant ray and a tangent ray with its vertex on the circle is equal to one-half the degree measure of its intercepted arc. (p. 232)

Theorem 8-13. The measure of an angle formed by two secants intersecting in the exterior of a circle is one-half the difference between the degree measures of the larger and the smaller of the two intercepted arcs. (p. 233)

Theorem 8-14. The measure of a central angle of a regular polygon of n sides is $360/n$. (p. 236)

Theorem 8-15. The area (K) of a regular polygon is equal to one-half the product of its apothem (a) and its perimeter (p). ($K = \frac{1}{2}ap$) (p. 237)

Theorem 8-16. The ratio C/d of the circumference to the diameter is constant for all circles. (p. 240)

Theorem 8-17. The area (K) of a circle of radius r is πr^2. (p. 245)

Theorem 8-18. The bisectors of the angles of a triangle are concurrent at a point equidistant from the sides of the triangle. (p. 248)

Theorem 8-19. The perpendicular bisectors of the sides of a triangle are concurrent at a point equidistant from the vertices of the triangle. (p. 248)

Theorem 8-20. The medians of a triangle are concurrent at a point that is two-thirds of the distance from any vertex to the midpoint of the opposite side. (p. 249)

Theorem 8-21. The lines that contain the three altitudes of a triangle are concurrent. (p. 249)

Theorem 9-1. A plane is uniquely determined by a line and a point not on the line. (p. 258)

Theorem 9-2. A plane is uniquely determined by two intersecting lines. (p. 258)

Theorem 9-3. The intersection of two distinct planes is a straight line. (p. 259)

Theorem 9-4. If a line is perpendicular to two intersecting lines, then it is perpendicular to the plane containing the two intersecting lines. (p. 261)

Theorem 9-5. There is exactly one plane perpendicular to a line at a given point on the line. (p. 261)

Theorem 9-6. Two lines perpendicular to the same plane are parallel. (p. 261)

Theorem 9-7. Two planes perpendicular to the same line are parallel. (p. 262)

Theorem 9-8. The intersections of two parallel planes cut by a third plane are parallel lines. (p. 262)

Theorem 9-9. The plane angles of a dihedral angle are congruent. (p. 265)

Selected Answers

CHAPTER 1

Exercise Set 1.1—Page 3

1. **(f)** $C = 25.12$

5. **(b)** and **(d)** are right triangles.

Exercise Set 1.2—Page 9

1. **(a)** Converse: If the current is on, then the lamp is burning.
Inverse: If the lamp is not burning, then the current is not on.
Contrapositive: If the current is not on, then the lamp is not burning.

3.

P	Q	not (P)	not (P) or Q	$P \to Q$
T	T	F	T	T
T	F	F	F	F
F	T	T	T	T
F	F	T	T	T

6.

P	Q	(Implication) $P \to Q$	not (P)	not (Q)	Inverse not (P) $\to$ not (Q)	Converse $Q \to P$
T	T	T	F	F	T	T
T	F	F	F	T	T	T
F	T	T	T	F	F	F
F	F	T	T	T	T	T

(not equivalent) (equivalent)

Exercise Set 1.3—Page 11

1.

P	Q	P or Q	not (P)	(P or Q) and not (P)	(P or Q) and not (P) $\to Q$
T	T	T	F	F	T
T	F	T	F	F	T
F	T	T	T	T	T
F	F	F	T	F	T

3. Law of Disjunctive Syllogism. P: Max mows the lawn; Q: Max can use his father's car; R: Max will take Cathy to the movies.
5. The card is a diamond. Law of Disjunctive Syllogism; Law of Detachment.

CHAPTER 2

Exercise Set 2.1—Page 14

3. An unlimited (infinite) number.
5. 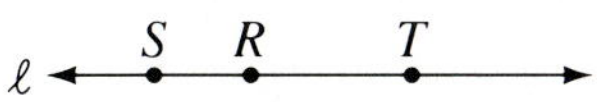

7. 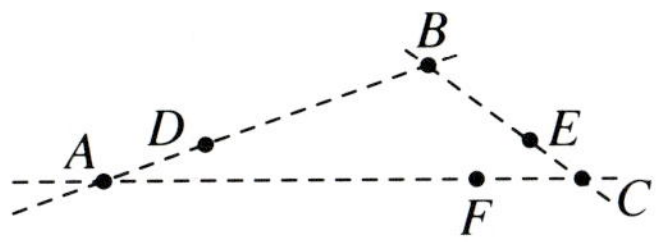

9. **(a)** none **(b)** infinite **(c)** one **(d)** three lines or one line.
11. **(a)** 

Exercise Set 2.2—Page 16

3. **(a), (b)** 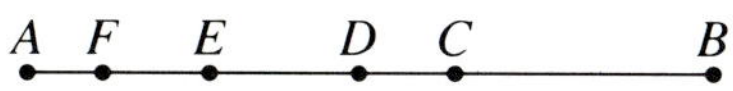

Indefinitely

(c) 15 **(d)** No **(e)** No, because there is no point "next to" A which would then be the endpoint.

Exercise Set 2.3—Page 17

1. 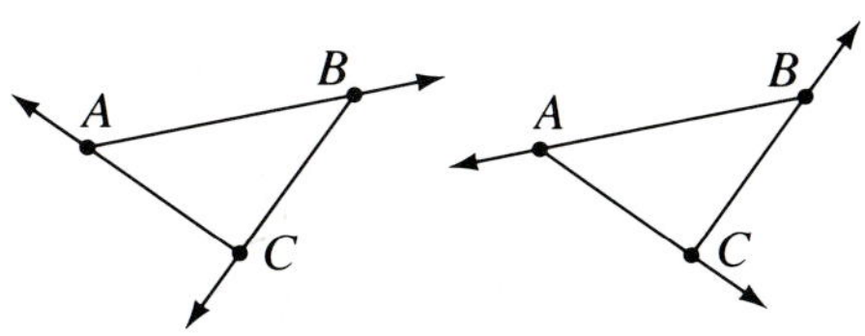

(a) (b)

3. Yes, $\overline{AB} = \overline{BA}$. No, because $\overrightarrow{AB}$ contains points X such that A-B-X, while $\overrightarrow{BA}$ contains points Y such that Y-B-A, and no X is the same as Y.
5. **(a)** No 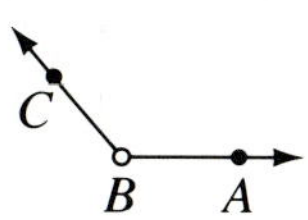

(b) Yes 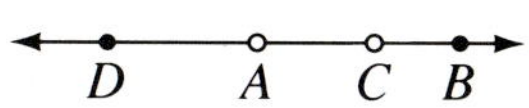

7. **(a)** $\overrightarrow{BC}$, ray **(b)** $\overrightarrow{AB}$, ray
(c) $\overleftrightarrow{AB}$, line **(d)** $\overleftrightarrow{AB}$, line
(e) B, point **(f)** ∅, null set

Exercise Set 2.4—Page 21

3. **(a)** $5 > 3$ **(b)** $-3 > -5$ **(c)** $2 > -1$ **(d)** $-2 < -1$ **(e)** $3 < 5$

5. $\frac{1}{2}, \frac{2}{3}, \frac{3}{4}, \frac{4}{5}, \frac{5}{6}, \frac{6}{7}, \frac{7}{8}, \frac{8}{9}, \frac{9}{10}, \frac{10}{11}$.

7. a is between b and c.

9.

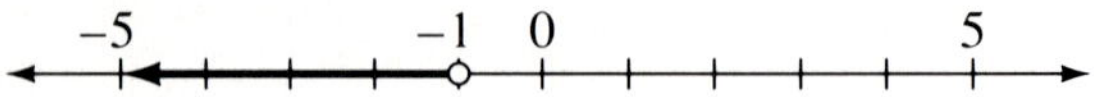

(a)

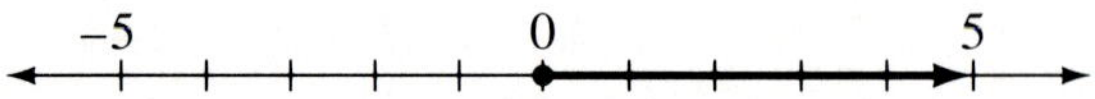

(b)

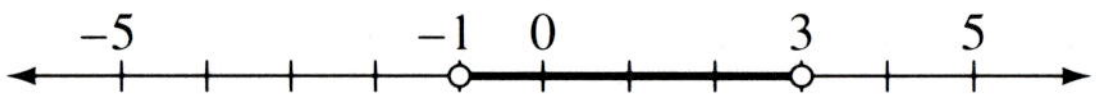

(c)

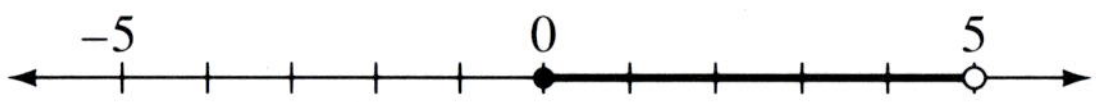

(d)

13.

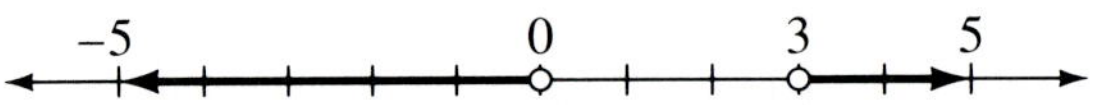

(a)

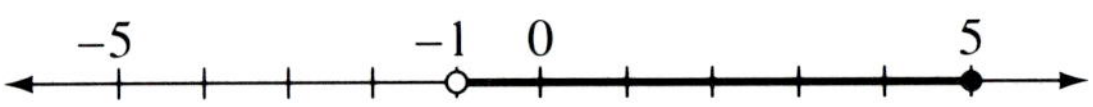

(b)

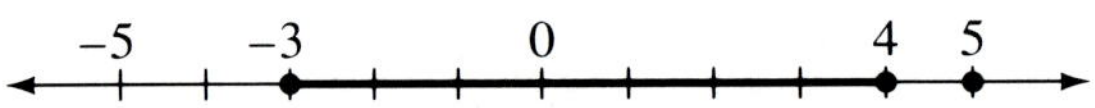

(c)

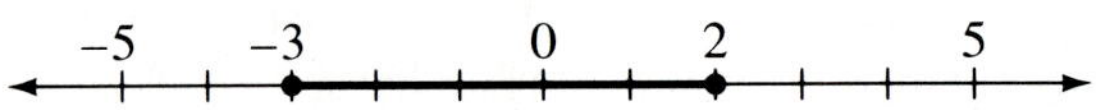

(d)

Exercise Set 2.5—Page 25

1. **(a)**, **(b)**, **(d)**, **(e)**, **(h)**, **(i)**, and **(j)**.

3. 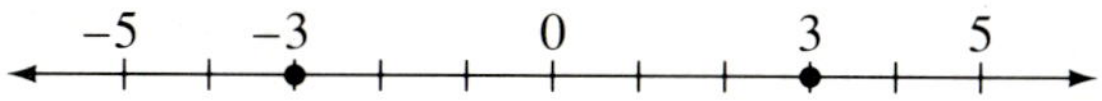

(a)

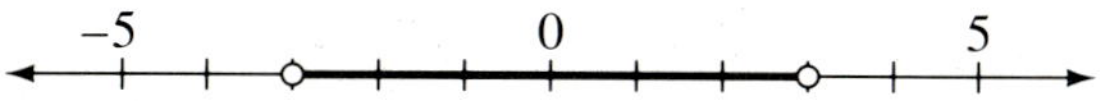

(b)

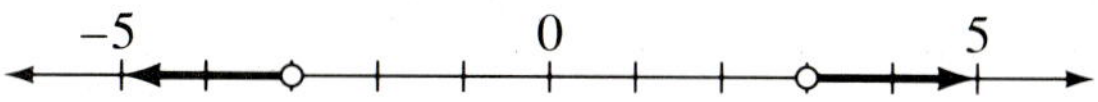

(c)

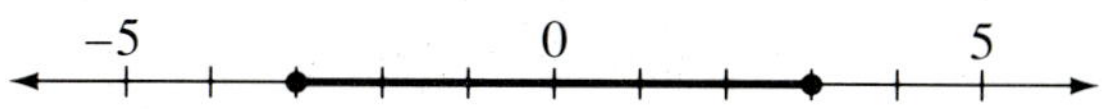

(d)

5. **(a)** 11 **(b)** 5 **(c)** 2 **(d)** 5 **(e)** 1 **(f)** $2|n|$ **(g)** $2|x|$ **(h)** 0

Exercise Set 2.6—Page 30

1. (1) Given (2) Congruent segments (Def. 2-10) (3) Post. 2-7 (4) Substitution axiom (E-4) or Transitive axiom (E-3) (5) Subtraction Law for Equals (6) Congruent segments (Def. 2-10)
3. Yes. Yes: 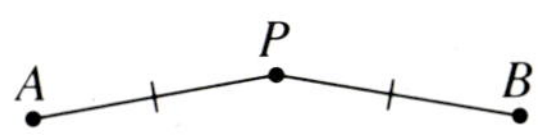
5. Not unless $\overline{RS} \cong \overline{TU}$.

Exercise Set 2.7—Page 34

2. **(a)** Yes. If A and B are any two points on the line, segment AB lies entirely on the line.
   (b) No. If A and B are on opposite sides of the point which is removed, then the set of points in $\overline{AB}$ are not all contained in the given set.
   (c) No. If A is on one line and B is on the other, then $\overline{AB}$ is not entirely in the given set.
   (d) Yes. Any two points in this set can be joined by a segment which is entirely in the given set.
3. **(a)** No. $\mathcal{H}_1 \cup \mathcal{H}_2$ does not contain the edge.
   (b) No. Their edges would form an angle, the interior of which would not contain $\mathcal{H}_1$ or $\mathcal{H}_2$.
5. **(a)** 4

Review Exercises—Page 37

1. True. (Theorem 2-5)
2. False. The length of a ray is infinite.
3. True. (Def. 2-10)

4. False. Congruent segments have the same length but are not necessarily the same set of points.
5. False, unless m$\overline{RS}$ = m$\overline{XY}$.
6. True. m$\overline{AC}$ = $|-2 - 2| = |-4| = 4$, m$\overline{BD}$ = $|1 - 5| = |-4| = 4$
7. False. $|-2| = 2$, and $2 < 5$.
8. False. 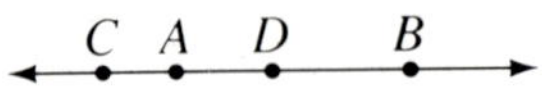

9. False. 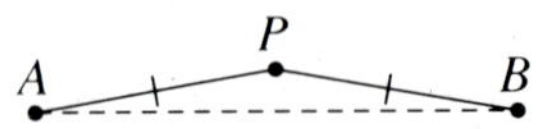

10. True. (Theorem 2-7)
11. Ray
12. $\overrightarrow{AB}$
13. $\overline{AB}$
14. 6
15. Half-line
16. **(e)** (It is a number)
17. **(e)** (Two distinct points are minimum requirement.)
18. **(b)** measure
19. $\{x | x \leq -1\}$ 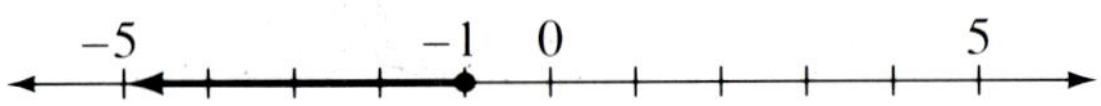

20. $\{x | |x| > 2\}$ 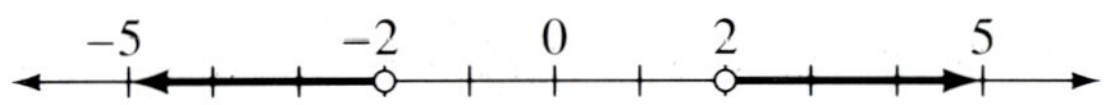

21. $\{x | -2 < x < 3\}$ 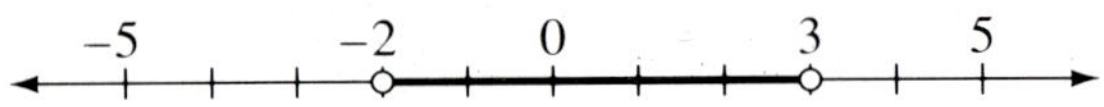

22. **(a)** C, D, E, F **(b)** A, B, C, D
(c) F **(d)** B, C, D
(e) B, C, D

CHAPTER 3

Exercise Set 3.1—Page 41

1. A
3. Interior
5. $\angle ROU$ (only)
7. A, U, B, S, C
9. No. Def. 3-2(a) says P and A are on the same side of $\overleftrightarrow{BC}$. Def. 3-2(b) requires that A, B, and C are not collinear.
11. **(a)** D, E **(b)** A, B, C, G, K **(c)** F, H, I
13. Into two convex sets. Yes, a non-straight angle does not divide the plane into *two* convex sets.
15. No, a portion of $\overline{AB}$ may lie interior to the angle.

Exercise Set 3.2—Page 44

1. (All measurements are approximate.)
(a) 37 **(b)** 22
(c) 33 **(d)** 47
(e) 37

3. **(a)** 20 **(b)** 45
(c) 5 **(d)** 105
(e) 160

7. **(a)** m$\angle RPT$ **(b)** m$\angle RPU$
(c) m$\angle RPT$ **(d)** m$\angle TPU$
(e) m$\angle UPR$ **(f)** m$\angle VPS$
(g) m$\angle VPT$

9. **(c)** 180

Exercise Set 3.3—Page 46

1. 20°, 100°, 150°, 190°, 225°, 300°

2. **(c)** 24°, 36°

Exercise Set 3.4—Page 49

1. **(a)** $\angle AOB$, $\angle BOC$, $\angle DOE$, $\angle EOF$ are acute.
(b) $\angle DOB$, $\angle EOC$, $\angle EOA$, $\angle$FOB are obtuse.
(c) $\angle AOC$, $\angle COD$, $\angle DOF$, $\angle FOA$ are right.
(d) $\angle AOD$, $\angle BOE$, $\angle COF$.

3. 90, since if x is the measure of the angle, $x + x = 180$.

5. **(a)** 150 **(b)** 60 **(c)** 135 **(d)** 25 **(e)** $180 - y$

7. Acute: **(a), (d), (f), (h)**
Right: **(b)**
Obtuse: **(c), (e), (i)**
Straight: **(g)**

9. $51\frac{3}{7}$

11. 75 and 105

13. 135 and 45

Exercise Set 3.5—Page 51

1. 30

3. **(a)** $\angle COD$ is a straight angle.
(b) 180, $\angle AOB$ is a straight angle.
(c) 180, $\angle DOC$ is a straight angle.
(d) 180, $\angle AOB$ is a straight angle.
(e) Transitive axiom, or by substitution.
(f) =, same as **(e)**.

Exercise Set 3.6—Page 53

1. **(a)** Reflexive property.
(b) Theorem 3-3(b).

5. Use Theorem 3-2.

Exercise Set 3.7—Page 54

1. **(a)** $\angle 1$ and $\angle 8$, $\angle 9$ and $\angle 10$, etc.
(b) $\angle 9$ and $\angle 12$, $\angle 12$ and $\angle 11$, etc.
(c) $\angle 9$ and $\angle 11$, $\angle 10$ and $\angle 12$.

3. $m\angle a = m\angle d = 135$, $m\angle c = m\angle b = 45$

5. Use Theorem 3-4 and Theorem 3-1(c).

Exercise Set 3.8—Page 59

3. Use Theorem 3-5 and Theorem 3-10.

9. **(a)** Supplements of the same angle are congruent.
(b) Substitution.
(c) Supplements of congruent angles are congruent.

11. **(a)** Use substitution and the definition of complementary angles.

Exercise Set 3.10—Page 62

1. **(a)** $\perp$ segments
(b) $\perp$ segments
(c) right angles

3. **(a)** $\angle ABR$ and $\angle CBT$
(b) $\overrightarrow{BR} \perp \overrightarrow{BT}$
(c) $\angle ABR$ and $\angle RBC$; $\angle ABT$ and $\angle TBC$

5. 45

Exercise Set 3.11—Page 66

1. Infinitely many on $\overleftrightarrow{OX}$ and $\overleftrightarrow{OY}$.

3. Infinitely many.

Exercise Set 3.12—Page 69

1. $A(2, 2)$; $B(5, 3)$; $C(0, 5)$; $D(-2, 4)$; $E(-3, 3)$; $F(-5, 0)$; $G(-4, -2)$; $H(-2, -3)$; $I(0, -2)$; $J(2, -3)$; $K(4, -3)$; $L(3, -1)$.

3. 5

5. (0, 0)

7. **(a)** (3, 0) **(b)** (2, 0)
(c) (3, 0) **(d)** (−7, 0)
(e) (7, 0) **(f)** (4, 0)

9. (−7, 1), (2, 6), (4, −5), (5, 3), (6, 6)

11. **(a)** I **(b)** IV
(c) IV **(d)** III
(e) II **(f)** None
(g) None **(h)** IV

13. **(a)** II **(b)** III $= \mathcal{H}_2 \cap \mathcal{H}_4$

Exercise Set 3.13—Page 74

1. **(a)** 3, 4 **(b)** 5, 12 **(c)** 4, 3 **(d)** 4, 3 **(e)** 5, 12 **(f)** $\frac{3}{4}$, 1 **(g)** 1, 1 **(h)** 1, 1 **(i)** 2, 2 **(j)** 3, 4 **(k)** 8, 6 **(l)** 4, 0 **(m)** 0, 3 **(n)** 5, 5
3. **(a)** 5 **(b)** 13 **(c)** 5 **(d)** 5 **(e)** 13 **(f)** $\frac{5}{4}$ **(g)** $\sqrt{2}$ **(h)** $\sqrt{2}$ (i) $\sqrt{8} = 2\sqrt{2}$ **(j)** 5 **(k)** 10 **(l)** 4 **(m)** 3 **(n)** $\sqrt{50} = 5\sqrt{2}$

Exercise Set 3.14—Page 77

3. Horizontal: **(b), (c), (e)**
Vertical: **(a), (d), (f)**
5. **(a)** 7 **(b)** -5 **(c)** $y_1 = b$
7. 6
9. 2 and $-\frac{1}{2}$
11. $\frac{2}{3}$ and $-\frac{3}{2}$

Exercise Set 3.15—Page 79

1. **(a)** (3, 4) **(b)** (6, 9) **(c)** (0, 5) **(d)** $(\frac{13}{2}, \frac{13}{2})$ **(e)** $(\frac{5}{16}, \frac{23}{40})$
(f) $\left(\frac{a+c}{2}, \frac{b+d}{2}\right)$
3. (10, 4)
5. $M(4, 3)$, $d(\overline{AM}) = 5\sqrt{2}$, $d(\overline{MB}) = 5\sqrt{2}$

Review Exercises—Page 79

1. Obtuse
2. Vertex
3. Complementary
4. $m\angle ABP + m\angle PBC$
5. 115
6. Congruent.
7. $\angle R \cong \angle T$. Transitivity of congruent angles
8. Vertical or congruent
9. Right angles
10. Perpendicular
11. 12
12. (3, 0)
13. c and e
14. 5

15. (7, −3)
16. 35 and 55
17. $x = 120$, $y = 30$
18. (6, 3)
19. **(a)** (5, −4) **(b)** $\frac{1}{2}$ **(c)** $\sqrt{20} = 2\sqrt{5}$
20. Use vertical angles and Transitive Axiom.
21. **(a)** (−2, −4) **(b)** (−4, 2) **(c)** (−5, 0) **(d)** (3, −5) **(e)** (3, 4) **(f)** (0, −9)
22. **(a)** 40° **(b)** 81° **(c)** 121° **(d)** 171° **(e)** 169° **(f)** 45° **(g)** 22°

Cumulative Review—Chapters 2 and 3—Page 81

1. Straight
2. Geometric
3. *R-T-S*
4. Segment
5. Ray
6. Rational number
7. Coordinate
8. Positive
9. $-x$
10. Symmetric
11. Rays
12. Angle
13. (1) The angle (2) the interior of the angle (3) the exterior of the angle
14. Straight
15. $0 \le a \le 180$
16. Congruent
17. Perpendicular
18. *IV*

Problems

1. Infinite
2. The half line does not contain its initial point.
3. **(a)**

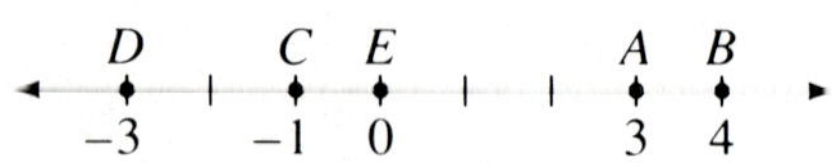

(b) m$\overline{AC} = 4$; m$\overline{BD} = 7$; m$\overline{AD} = 6$; m$\overline{BE} = 4$; m$\overline{CA} = 4$
4. When C is on $\overline{AB}$ and $\overline{AC} \cong \overline{CB}$.
5.

6. **(a)** Half-line **(b)** Segment **(c)** Segment **(d)** Line
(e) Point **(f)** Half-line **(g)** Segment **(h)** Ray

7. **(a)** 6 **(b)** 5 **(c)** 12 **(d)** 13 **(e)** 9 **(f)** -4 **(g)** -7 **(h)** -3
(i) $|-x| = x, x \geq 0; -x, x < 0$ **(j)** $x, x \geq 0; -x, x < 0$

8. **(a)** Yes **(b)** Yes **(c)** Yes **(d)** No, the segment joining the points is not convex. **(e)** Yes, if the point removed is the initial point. No, if any other point is removed. **(f)** Yes **(g)** No, a segment joining two points in the plane is not necessarily convex.

9. **(a)** Yes, satisfies the definition of convex set.
(b) No, does not contain the edge.

10. **(a)**

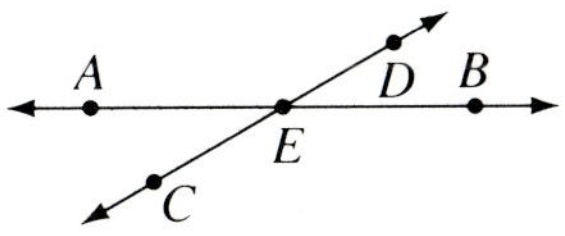

(b) Yes, by definition

11. **(a)** $\angle DBF$ **(b)** $\angle DBC$ **(c)** $\angle DBC$

12. **(a)** 125° **(b)** 75° **(c)** 152.2° **(d)** $(180 - a)°$ **(e)** $a°$ **(f)** $(90 + a)°$

13. 40°

14. **(a)** Yes **(b)** Yes, same angle **(c)** Yes, same angle

15. **(b)** 8; $\sqrt{53}$; 9
(c) (0, 4); $(-\frac{7}{2}, 3)$; $(0, -\frac{1}{2})$
(d) 0; 7; -1

16. **(a)** 94 **(b)** 152 **(c)** 122

CHAPTER 4

Exercise Set 4.1—Page 87

1. No. The points in the interior of the triangle would be in the intersection of the interiors of all three triangles.

3. **(a)** Both separate the plane into three disjoint sets.
(b) The interior of the angle is *open;* i.e., a bisector of the angle can extend infinitely far without intersecting the angle; the triangle is *closed.*

5. **(a)** *BSA*, *BSR*, *BSG* **(b)** *ASB*, *ASR*, *ASG* **(c)** *ARG*, *ASG*, *BSG*, *BRG*; six with *A* as vertex have already been listed.

7. **(a)** Right **(b)** Obtuse **(c)** Acute

9. **(a)** Equal **(b)** Equal

11. **(a)** Equal **(b)** Equal

13. **(a)** (1, 1), (2, 1), (2, 4) **(b)** (1, 1), (1, 2), (1, 3), (2, 1)
(c) (1, 1), (2, 1)

Exercise Set 4.2—Page 92

1. $R \leftrightarrow Q$, $S \leftrightarrow V$, $T \leftrightarrow W$; yes

5. $m\overline{RT} = m\overline{XZ} = 2\frac{1}{2}$ in.
$m\angle STR = m\angle YZX = 49$
$m\angle TRS = m\angle ZXY = 91$

10. $m\angle C = m\angle F = 105$, $m\overline{AC} = m\overline{DF} = 4$ in., $m\overline{BC} = m\overline{EF} = 2\frac{3}{16}$ in.

Exercise Set 4.3—Page 98

1. **(a)** ≅, SAS
(b) ≅, ASA
(c) ≅, SSS
(d) Insufficient information
(e) ≅, ASA
(f) ≅, SAS

5. Show $\overline{DC} \cong \overline{DC}$, then $\triangle ADC \cong \triangle BDC$ by SSS.

7. **(a)** Prove $\overline{AD} \cong \overline{AC}$ by Theorem 2-2, then the triangles are ≅ by SAS.
(b) Prove the triangles are ≅ by SSS.
(c) Use parts (a) and (b) to show that $\angle ABF \cong \angle ACD$ and $\angle BCD \cong \angle CBF$; then use Theorem 3-3(a) to show that $\angle ABC \cong \angle ACB$.

13. Approximately (5.3, 11.3)

Exercise Set 4.4—Page 105

4. **(a)** $m\overline{BC} \approx 1\frac{15}{16}$; $m\overline{AC} \approx 1\frac{15}{16}$; $m\angle C \approx 100$
(b) $m\overline{BC} \approx 1\frac{15}{16}$; $m\angle B \approx 56$; $m\angle C \approx 84$
(c) $m\overline{BC} \approx 2\frac{3}{4}$; $m\overline{AC} \approx 4\frac{1}{4}$; $m\angle C \approx 45$
(d) $m\overline{BC} \approx 2\frac{1}{16}$; $m\angle B \approx 70$; $m\angle C \approx 70$
(e) $m\angle A \approx 48$; $m\angle B \approx 48$; $m\angle C \approx 84$
(f) $m\angle A \approx 60$; $m\overline{AC} \approx 2\frac{1}{2}$; $m\overline{AB} \approx 2\frac{1}{2}$
(g) $m\angle A \approx 37$; $m\angle B \approx 53$; $m\angle C \approx 90$

9. Refer to Theorem 4-3.

Exercise Set 4.5—Page 108

1. Prove $\angle R \cong \angle S$; then use ASA.

3. Using Theorem 4-5, the triangles are congruent by SAS.

5. Show that $\angle FDE \cong \angle FED$, then that $\angle 1 \cong \angle 4$. Next show that $\overline{DJ} \cong \overline{EJ}$ (Theorem 4-6) and that $\angle GJD \cong \angle HJE$, hence, the triangles are congruent by ASA.

7. Prove $\triangle ABE \cong \triangle CBE$. Then by using CPCTC and reflexive property of congruent segments, prove $\triangle ADE \cong \triangle CDE$. $\overline{AD} \cong \overline{CD}$ by CPCTC.

9. Show that $\overline{AE} \cong \overline{BD}$, $\angle CAB \cong \angle CBA$ and $\overline{AB} \cong \overline{BA}$, then $\triangle ABD \cong \triangle BAE$ and $\overline{AD} \cong \overline{BE}$.

11. Prove $\triangle ADC \cong \triangle BDC$ by ASA; then
(a) $\angle ADC \cong \triangle BDC$ by CPCTC; hence $\overline{AB} \perp \overline{CD}$
(b) CPCTC

13. Use the Midpoint Formula and Distance Formula.
Median to $\overline{TR}$ is $4\sqrt{2}$;
median to $\overline{RS}$ is $\sqrt{26}$;
median to $\overline{TS}$ is $\sqrt{26}$.

Exercise Set 4.6—Page 113

1. **(a)** 1, 2, 4, 5, 7, 9 **(b)** 6 & 8, 3 & 8, 3 & 6 **(c)** $\angle 8$

2. **(a)** $<$ **(b)** $<$ **(c)** $>$ **(d)** $<$

4. $\overline{AB}$; $\overline{BC}$
5. $\overline{RT} < \overline{ST} < \overline{RS}$

Review Exercises—Page 114

1. 11 inches
5. **(a)** They should all equal 60° **(b)** 60°
7. **(a)** SSS or SAS **(b)** SSS or SAS
   (c) ASA **(d)** No
   (e) SSS or SAS **(f)** No
   (g) No **(h)** No

Chapter 5

Exercise Set 5.1—Page 121

3. **(a)** 6 **(b)** 4 **(c)** 5
5. **(a)** 3 **(b)** $5\sqrt{2}$, $5\sqrt{2}$ **(c)** (1, 3) **(d)** $\frac{1}{3}$ **(e)** -3 **(f)** -1

Exercise Set 5.2—Page 127

1. **(a)** Can be drawn as stated
   (b) Not enough conditions
   (c) Can be drawn as stated
   (d) Too many conditions
   (e) Too many conditions
3. $\overline{TQ} < \overline{TR} < \overline{TS}$
5. **(a)** and **(e)** if $1 < e < 9$
11. $2 < x < 20$

Exercise Set 5.4—Page 133

1. **(a)** ∠HDA and ∠*CAB*
   ∠*FEB* and ∠*ABE*
   (b) ∠*EDA* and ∠*DAB*
   ∠*FEB* and ∠*EBG*
   (c) ∠*CDE* and ∠*DAB*
   ∠*CEF* and ∠*EBG*

Exercise Set 5.5—Page 136

1. **(a)** Alternate-exterior angles
   (b) Alternate-interior angles
   (c) Same as (a)
   (d) Same as (b)
   (e) Interior angles on the same side of the transversal
   (f) Corresponding angles
   (g) Same as (f)

(h) Same as (f)
(i) Same as (e)

3. **(a)** Only if $\overline{AD} \parallel \overline{BC}$
   (b) Only if $\angle 2 \cong \angle 6$
   (c) Yes

13. **(a)** $\overline{RU}$ and $\overline{ST}$ **(b)** $\overline{RU}$ and $\overline{ST}$ **(c)** $\overline{RS}$ and $\overline{UT}$

17. 4.2

19. **(a)** 2 **(b)** 2.4 **(c)** 3 **(d)** 6 **(e)** 4 **(f)** 7.2

Exercise Set 5.6—Page 141

1. **(a)** 80 **(b)** 50 **(c)** 6 **(d)** $180 - 3x$ **(e)** 60

5. If angles have their sides $\parallel$, left side to left side and right side to right side, the angles are $\cong$. (See Prob. 4(b) on page 137) Also $\text{m}\overline{AB} = \text{m}\overline{DE}$, hence triangles are congruent by ASA.

9. $\text{m}\angle A = 60$, $\text{m}\angle ACB = 50$

11. **(a)** $\frac{2}{3}$ **(b)** $-\frac{4}{3}$ **(c)** (3, 4)

13. **(a)** 50 **(b)** 180 **(c)** 180 **(d)** 130
    (e) Their sum equals $\text{m}\angle CBD$.

Review Exercises—Page 145

1. **(a)** $\angle 1$ and $\angle 5$, $\angle 3$ and $\angle 7$, $\angle 2$ and $\angle 6$, $\angle 4$ and $\angle 8$
   (b) $\angle 3$ and $\angle 6$, $\angle 4$ and $\angle 5$

2. $\overline{XY}$, $\overline{YZ}$

3. **(a)** 100 **(b)** 70 **(c)** 75

4. 40, 70

5. 15

6. 50

7. $\overline{PA}$ (Theorem 5-5)

8. L does, since $d(\overline{LU}) = d(\overline{LS})$ (Theorem 5-3)

9. $\overline{AD} \parallel \overline{RK}$, since $m_{\overline{AD}} = m_{\overline{RK}}$ (Theorem 5-23)

11. **(a)** SAS
    (b) If two lines (segments) intersect to form congruent adjacent angles, the lines are $\perp$.

12. Prove $\triangle ADC \cong \triangle BFE$ by SAS; then $\overline{CD} \parallel \overline{EF}$ by corresponding angles $\cong$.

CHAPTER 6

Exercise Set 6.1—Page 149

1. **(a)** 3 **(b)** 6 **(c)** $5\frac{1}{4}$ **(d)** $4\frac{4}{5}$ **(e)** $2\frac{1}{3}$ **(f)** 5 **(g)** $\dfrac{bc}{d}$ **(h)** $\frac{1}{2}$

3. **(a)** $\frac{4}{3}$ **(b)** $\frac{2}{a}$ **(c)** $\frac{3}{1}$ **(d)** $\frac{1}{5}$ **(e)** $\frac{3}{1}$ **(f)** $\frac{3}{1}$

5. b, c, and e are true.

7. a and d, b and f, b and i, f and i, a and g, d and g, h and j
9. 1360.80 g
11. $22\frac{4}{5}$ in.

Exercise Set 6.2—Page 151

1. 18
3. $m\overline{EF} = 9\frac{3}{5}$, $m\overline{DF} = 16\frac{4}{5}$
5. **(a)** $\frac{a}{r} = \frac{b}{s} = \frac{c}{t}$

   (b) $\frac{p}{a} = \frac{q}{m} = \frac{r}{t}$
7. **(a)** $x = 24$, $y = 6$
   (b) $x = 19.2$, $y = 20.8$
   (c) $x = 37\frac{1}{2}$, $y = 32$

Exercise Set 6.3—Page 157

1. Yes. Angles are congruent and sides are in proportion (1 to 1).
3. Not necessarily. They need not have their corresponding angles congruent nor their corresponding sides proportional.
5. No. To be parallel the segment must divide the sides proportionally; $\frac{3}{4} \neq \frac{4}{6}$.
7. **(a)** 70 **(b)** 8
9. Using alternate-interior angles and vertical angles, prove by AAA Similarity Theorem.
11. **(a)** 12, Theorem 6-6
    (b) 30
    (c) $\overline{UV} \parallel \overline{RS}$ by Theorem 6-6; hence $\angle TUV \cong \angle TRS$ by Theorem 5-15
13. Prove the triangles formed are similar to the given triangles.
15. Prove similarity to be reflexive, symmetric, and transitive.
17. If the triangles are ABC and DEF with vertex angles A and D such that $m\angle A = m\angle D = x$ then $m\angle B = m\angle C = m\angle E = m\angle F = 90 - x/2$; hence are similar by AAA Similarity Theorem.
19. $5\frac{85}{168}$ ft

Exercise Set 6.4—Page 162

1. **(a)** 0, 2 **(c)** -5, 3 **(e)** $\frac{1}{2}$, $\frac{1}{4}$
   (g) -3, -7 **(i)** ± 1
2. **(a)** 6 **(b)** 2 **(c)** 18
3. $y = 27$, $b = 3\sqrt{10}$, $a = 9\sqrt{10}$
4. $a = 4\sqrt{3}$, $b = 4$, $x = 2\sqrt{3}$
5. $y = 4$, $b = 5\sqrt{21}$, $x = 2\sqrt{21}$
6. **(a)** $ac/(b + c)$ **(b)** 4
7. **(a)** $a = 9$, $p = 3\sqrt{10}$, $q = 9\sqrt{10}$ **(b)** $m = 16$, $q = 20$

Exercise Set 6.5—Page 167

1. a, c, d
3. 6 or $1\frac{1}{2}$
5. $5\sqrt{2}$
7. 10.82
9. $3\sqrt{2}$

Exercise Set 6.6—Page 171

1. **(a)** 6, $12\sqrt{3}$ **(b)** $6\sqrt{3}$, 12 **(c)** $6\sqrt{2}$, $12\sqrt{2}$
3. **(a)** $a = 60$, $x = 3\sqrt{3}$ **(c)** $a = 45$, $y = 5$, $c = 5\sqrt{2}$
5.

	A	a	b	c	m	n	h_c
(a)		3	**4**	**5**	$\frac{9}{5}$	$\frac{16}{5}$	$\frac{12}{5}$
(b)		**5**	**12**	13	$\frac{25}{13}$	$\frac{144}{13}$	$\frac{60}{13}$
(c)		**7**	24	**25**	$\frac{49}{25}$	$\frac{576}{25}$	$\frac{168}{25}$
(d)		**10**	7.5	12.5	8	4.5	**6**
(e)		6	$12\sqrt{2}$	**18**	**2**	16	$4\sqrt{2}$
(f)	**30°**	4	$4\sqrt{3}$	**8**	2	6	$2\sqrt{3}$
(g)	**45°**	$4\sqrt{2}$	$4\sqrt{2}$	**8**	4	4	4
(h)	**60°**	$4\sqrt{3}$	4	**8**	6	2	$2\sqrt{3}$

7. **(a)** $6\sqrt{3}$ **(b)** $6\sqrt{3}$ **(c)** 12
11. 2.8 mi

Exercise Set 6.7—Page 174

1. -3
3. **(a)** $d(\overline{AB})^2 = (4)^2 + (-7)^2 = 16 + 49 = 65$
   $d(\overline{BC})^2 = (6)^2 + (3)^2 = 36 + 9 = 45$
   $d(\overline{AC})^2 = (2)^2 + (-4)^2 = 4 + 16 = 20$
   $65 = 45 + 20$
   $\therefore$ a right triangle

Exercise Set 6.8—Page 178

1. **(a)** X: $\frac{3}{5}$, $\frac{4}{5}$, $\frac{3}{4}$; Y: $\frac{4}{5}$, $\frac{3}{5}$, $\frac{4}{3}$
3. $\frac{1}{2}\sqrt{3}$, $\frac{1}{2}$, $\sqrt{3}$

Exercise Set 6.9—Page 182

1. **(a)** sin 20° = 0.3420
cos 20° = 0.9397
tan 20° = 0.3640
(b) sin 37° = 0.6018
cos 37° = 0.7986
tan 37° = 0.7536
(c) sin 64° = 0.8988
cos 64° = 0.4384
tan 64° = 2.0503

3. **(a)** 39° **(b)** 72° **(c)** 27° **(d)** 31° **(e)** 82°

5. 68°, 4.5, 11.1

Exercise Set 6.10—Page 185

1. $b = 5$, $c = 7.1$

3. $a = 7.5$

5. m$\angle A = 41$, $b = 9.6$

7. 71°

9. 34°

11. **(a)** 21°; 8.5
(b) (−7.3, 6.1)

13. 39°

15. 29°, 7.8 in.

Review Exercises—Page 187

1. $N = 95$

2. $N = 117$

3. $N = \frac{1}{6}$

4. $N = 14$

5. $N = \frac{13}{2} = 6\frac{1}{2}$

6. $N = 9$

7. $\frac{6}{40} = \frac{7}{x} = \frac{8}{y}$, $x = 46\frac{2}{3}$, $y = 53\frac{1}{3}$

8. $s = \frac{2}{3}h\sqrt{3}$

9. 26 ft

10. 120

11. 135 ft

12. 21.5 ft

13. $m = \frac{16\sqrt{13}}{13}$, $n = \frac{36\sqrt{13}}{13}$, $h = \frac{24\sqrt{13}}{13}$

14. **(a)** $h = 3\sqrt{3}$ **(b)** median has length 6 **(c)** $x = 30$

16. 11.9 ft

17. m$\overline{RV} = 4\frac{1}{11}$, m$\overline{UV} = 4\frac{10}{11}$

18. m$\overline{XZ} = 15.8$, m$\overline{YZ} = 12.3$

19. $\approx 27°$
20. $m_{\overline{PQ}} = \frac{1}{3}$, $m_{\overline{QR}} = -3$; hence $\angle PQR$ is a right angle (Theorem 6-14)

Cumulative Review—Chapters 2–6—Page 189

1. $0 < a < 90$
2. Side, vertex, empty
3. Supplementary
4. Congruent
5. Complementary
6. A, C, transitive
7. Adjacent or vertical (supplementary or congruent)
8. Hypothesis, conclusion
9. Hypothesis, conclusion, figure, statements, reasons
10. Equiangular
11. Scalene triangle
12. Obtuse triangle
13. Perpendicular lines
14. Auxiliary lines
15. Parallel
16. 120
17. d/b
18. b/d
19. z/y
20. 4/3
21. 2.7
22. 60
23. $\sqrt{3} \approx 1.732$
24. $\sqrt{2}/2 \approx 0.7071$
25. Right triangle

Problems

1. $m\angle ABF = 122$, $m\angle DBC = 90$
2. $\angle ABF$, $\angle DBF$, $\angle DBC$
3. $\angle ABD$, $\angle FBC$
4. $\angle EBC$, $\angle DBC$
5. **(a)** 66 **(b)** 30 **(c)** 72.6 **(d)** $90 - x$ **(e)** x **(f)** $50 + x$
6. $102\frac{1}{2}$, $77\frac{1}{2}$
7. 65, 25
8. 65, 65, 130, 50, 115

9. **(a)** $\angle EAF$ and $\angle FAB$, $\angle AEB$ and $\angle BEC$, $\angle BEC$ and $\angle CED$
(b) $\angle AFE$ and $\angle BFD$, $\angle AFB$ and $\angle EFD$
(c) $\angle AGE$ and $\angle EGD$, $\angle EGD$ and $\angle DGC$, $\angle DGC$ and $\angle CGA$, $\angle CGA$ and $\angle AGE$

10. **(a)** Isosceles **(b)** Right triangle (scalene) **(c)** Equiangular (equilateral)
(d) Isosceles, right **(e)** Equilateral

12. $90 - x$

13. **(a)** Isosceles right ($\sqrt{10}$, $\sqrt{10}$, $2\sqrt{5}$)
(b) Scalene (5, 2, $\sqrt{13}$)
(c) Isosceles ($4\sqrt{5}$, $4\sqrt{5}$, $4\sqrt{2}$)

14. 70

15. If a point is equidistant from the end points of a segment it lies on the perpendicular bisector of the segment.

16. 29

17. Show $\triangle ABC \sim \triangle EFD$, then the corresponding sides are proportional.

19. **(a)** $m\overline{AB} = 25$, $m = 9$, $n = 16$ **(b)** 6
(c) 12 **(d)** 16

20. No. $\frac{\sqrt{3}}{2} + \frac{1}{2} \neq 0$

21. **(a)** 1 **(b)** 1 **(c)** 1

CHAPTER 7

Exercise Set 7.1—Page 196

1. Concave
(a) 5 **(b)** 5 **(c)** $\overline{AB}$, $\overline{BC}$, $\overline{CD}$, $\overline{DE}$ and $\overline{EA}$ **(d)** $\overline{AB}$ and $\overline{BC}$, $\overline{BC}$ and $\overline{CD}$, $\overline{CD}$ and $\overline{DE}$, $\overline{DE}$ and $\overline{EA}$ **(e)** $\overline{AB}$ and $\overline{CD}$, $\overline{BC}$ and $\overline{DE}$, etc. **(f)** $\overline{EB}$, $\overline{EC}$

3. **(a)** None **(b)** Two **(c)** 5 **(d)** 9 **(e)** 27 **(f)** $\frac{n(n-3)}{2}$ **(g)** 2144

Exercise Set 7.2—Page 200

1. Rhombus

3. Draw $\overline{AC}$. By Theorem 6-6 $\overline{EF} \parallel \overline{AC}$ and $m\overline{EF} = \frac{1}{2}m\overline{AC}$, also $\overline{HG} \parallel \overline{AC}$ and $m\overline{HG} = \frac{1}{2}m\overline{AC}$.

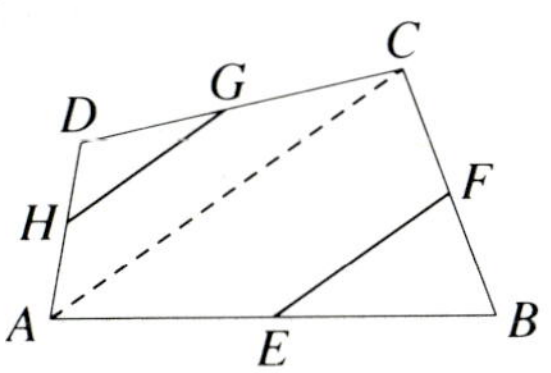

7. Prove $\triangle ABD$ and $\triangle BAC$ are congruent (SAS); then $\overline{AC} \cong \overline{BD}$ by CPCTC.

11. Draw the altitudes from the upper vertices to the base and prove the right triangles formed are congruent.

Exercise Set 7.3—Page 204

1. By Theorem 6-3, $\overline{AF} \parallel \overline{DF}$ and $\overline{AD} \parallel \overline{EF}$, also $\overline{FB} \parallel \overline{DE}$, and $\overline{DF} \parallel \overline{BE}$; hence $AFED$ and $BFDE$ are parallelograms.

5. Use Theorem 7-3 and Corollary 7-1(b) to show all four angles are right angles, hence, by definition the parallelogram is a rectangle.

7. Prove that $\triangle ABE \cong \triangle BCE \cong \triangle CDE \cong \triangle DAE$ by SAS; then $\overline{AB} \cong \overline{BC} \cong \overline{CD} \cong \overline{DA}$ by CPCTC.

9. **(a)** $R(6, 2)$, $S(9, 6)$, $T(5, 9)$, $U(2, 5)$
(b) $m\overline{RS} = m\overline{ST} = m\overline{TU} = m\overline{UR} = 5$
(c) $m_{\overline{RT}} = -\frac{7}{1}$, $m_{\overline{SU}} = \frac{1}{7}$, $m_{\overline{RT}} \cdot m_{\overline{SU}} = -1$
(d) $m_{\overline{RS}} = \frac{4}{3}$, $m_{\overline{ST}} = -\frac{3}{4}$, $m_{\overline{RS}} \cdot m_{\overline{ST}} = -1$; $\therefore \angle RST$ is a right angle.

10. $U(5.5, 9.6)$, $V(11.5, 9.6)$

11. Place a coordinate system on the parallelogram as indicated. Given $\overline{AC} \cong \overline{BD}$, hence $d(\overline{AC}) = d(\overline{BD})$. Show that b must equal 0 and $\angle A$ is a right angle, then by Theorem 7-8 it is a rectangle.

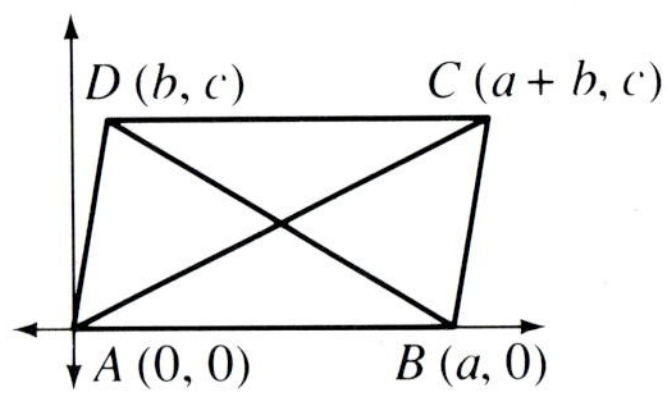

Exercise Set 7.4—Page 205

1. Draw the polygons, then draw all the diagonals possible from one vertex. Observe the number of triangles formed.

3. **(a)** 30 **(b)** 30

5. 125 ft

Exercise Set 7.5—Page 210

3. 20, $14\frac{14}{29}$

5. 24

7. $\frac{9}{14}$

9. Show that the altitudes remain constant and the measures of the bases are one-half the measure of the original base.

11. 16

▶ **13.** 3.42, 25.65

▶ **17.** 993 in.^{2}

Exercise Set 7.6—Page 214

1. **(a)** 80 **(c)** $150\sqrt{2}$ **(d)** $120\sqrt{3}$ **(f)** 123.42

3. 160

7. The area of the parallelogram is twice the area of the triangle.

11. 7, 13

13. 60 cm

Review Exercises—Page 215

2. 4, 5, 6

3. EFGH is a parallelogram, since $\overline{HG}$ is parallel and congruent to $\overline{EF}$. Also $\overline{HE} \parallel \overline{FG}$, hence $\angle GHE$ is a right angle. ($\angle CIB$ is a right angle.)

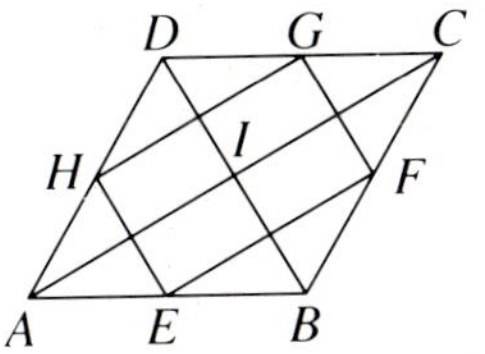

4. $\angle DAB \cong \angle CBA$ and $\overline{AD} \cong \overline{BC}$ and $\overline{AB} \cong \overline{BA} \rightarrow \triangle DAB \cong \triangle CBA \rightarrow \overline{AC} \cong \overline{BD} \rightarrow \overline{HG} \cong \overline{FG}$, and $EFGH$ is a parallelogram.

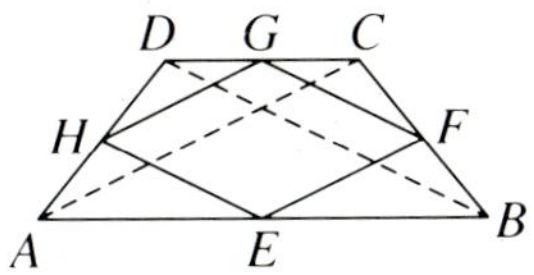

5. $\angle ABE$ and $\angle BED$ are supplementary, $\angle CBE$ and $\angle BEF$ are supplementary; hence $\angle GBE$ and $\angle BEG$ are complementary and $\angle BGE$ is a right angle. $\angle GBE \cong \angle BEH$ and $\angle GEB \cong \angle HBE$, therefore $\overline{GB} \parallel \overline{EH}$ and $\overline{GE} \parallel \overline{HB}$, hence $BGHE$ is a parallelogram with one right angle and therefore is a rectangle.

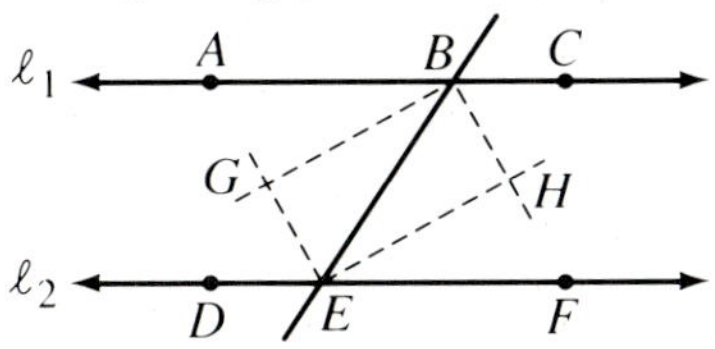

6. Draw altitudes $\overline{DE}$ and $\overline{CF}$, and $\overline{DE} \cong \overline{CF}$. Thus $\triangle ADE \cong \triangle BCF$ by ASA, and $\overline{AD} \cong \overline{BC}$.

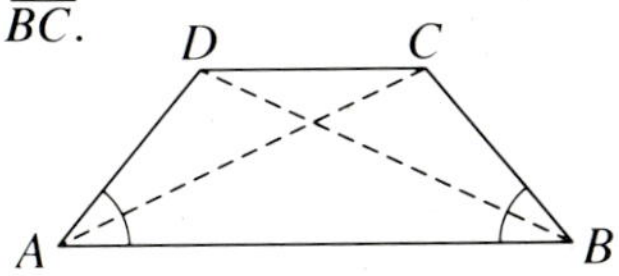

7. $\dfrac{240}{580} = \dfrac{48}{x}$, $x = \dfrac{48(580)}{240} = 116$

8. $K_{ADE} = K_{EDC}$ and $K_{ABE} = K_{EBC}$
$K_{ADE} + K_{ABE} = K_{EDC} + K_{EBC}$

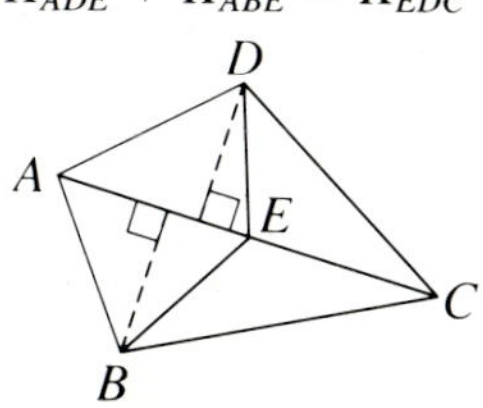

9. 6 in. and 9 in.
10. $16\frac{7}{8}$ ft^2
11. 288 ft^2
12. $\frac{h_1}{h_2} = \frac{1}{3}$
13. 100 ft^2
14. $16\sqrt{3}$ in.^{2}
15. If s is the side, $\frac{1}{2}s\sqrt{3}$ is the altitude and $K = \frac{1}{2}s \cdot \frac{1}{2}s\sqrt{3} = \frac{1}{4}s^2\sqrt{3}$
16. 150 ft^2
17. $h = 20$
18. $122.40
19. $4650.00
20. Right triangle (Postulate 7-4) 10, 15, $5\sqrt{13}$

CHAPTER 8

Exercise Set 8.1—Page 220

1. **(a)** False **(b)** True **(c)** False **(d)** False (only two) **(e)** False **(f)** True
7. $\sqrt{13}$
9. $6\sqrt{3}$
*11. $(x - 3)^2 + (y - 4)^2 = 17$ or $x^2 - 6x + y^2 - 8y + 8 = 0$

Exercise Set 8.2—Page 225

1. **(a)** 125 **(b)** 270 **(c)** 125
3. Using Theorem 8-7, prove the opposite pairs of sides are congruent.
5. Draw radii $\overline{OC}$ and $\overline{OD}$, then show that $\angle COB \cong \angle DOB$; hence $\angle COA \cong \angle DOA$ and then $\triangle AOC \cong \triangle AOD$ by SAS; $\therefore \overline{AC} \cong \overline{AD}$.
9. **(a)** $9\sqrt{3}$ **(b)** $2\sqrt{3}$ **(c)** $2\sqrt{2}$ **(d)** 11.5
13. **(a)** 20, 20, 20
    (b) 30, 30, 30
    (c) 35, 35, 35
    (d) 45, 45, 45
    (e) 50, 50, 50
    (f) The measure of the angle is one-half the measure of its intercepted arc.
15. **(a)** 40, 20
    (b) 100, 50
    (c) 170, 85
    (d) 190, 85

Exercise Set 8.3—Page 230

1. **(a)** 120 **(b)** 140 **(c)** 90 **(d)** 90 **(e)** 100

3. Use Theorem 8-9 and the fact that the sum of the degree measures of the arcs of a circle is 360.

5. Using the Exterior Angle Theorem, we have $m\angle AED = m\angle EAD + m\angle EDA = \frac{1}{2}m°\widehat{BD} + \frac{1}{2}m°\widehat{AC}$.

7. $m°\widehat{AD} = 140$; $m°\widehat{AB} = 140$; $m°\widehat{BC} = 40$; $m\angle ADC = 90$; $m\angle ACD = 70$; $m\angle BAC = 20$; $m\angle ABC = 90$.

9. Show that $\triangle ABE \sim \triangle DCE$; then, from the proportionality $a/d = c/b$, we have $a(b) = c(d)$.

11. **(a)** Reflexive **(b)** Inscribed angles that intersect the same arc are congruent. **(c)** AAA Similarity Theorem **(d)** Corresponding sides of similar triangles are proportional. **(e)** Multiplication Law for Equals

Exercise Set 8.4—Page 233

1. $\triangle AOP \cong \triangle BOP$ by Corollary 6-8(b) hence $\overline{AP} \cong \overline{BP}$.

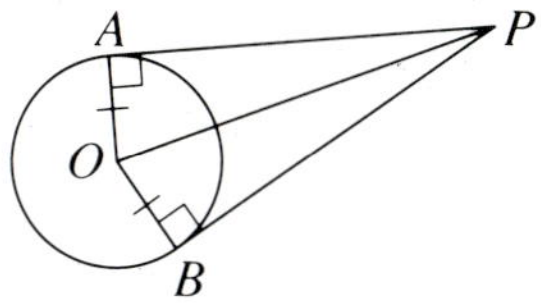

3. **(a)** 4 **(b)** $\frac{16\sqrt{3}}{3}$

5. $m\angle FAD = m\angle ADB + m\angle B$ or $m\angle B = m\angle FAD - m\angle ADB = \frac{1}{2}m°\widehat{AD} - \frac{1}{2}m°\widehat{AC}$

7. $\triangle ACB \sim \triangle DAB$ by AAA Similarity Theorem; hence $m\overline{AB}/m\overline{BC} = m\overline{DB}/m\overline{AB}$ by definition of similar triangles.

9. $\overline{AB} \cong \overline{AC}$ and $\overline{AC} \cong \overline{AD}$ by Corollary 8-9(a), hence $\overline{AB} \cong \overline{AD}$ by Transitive property of congruent segments.

13. 110

Exercise Set 8.5—Page 238

5. The radii to the opposite vertices form a central angle of measure 90 (Theorem 8-14); hence a side is $r\sqrt{2}$ and the area $2r^2$.

7. **(a)** 128 square units **(b)** 256 square units

9. **(a)** 4 **(b)** 24 **(c)** $24\sqrt{3}$ square units

11. **(a)** 108 **(b)** 120 **(c)** 140 **(d)** 144 **(e)** 156 **(f)** 160 **(g)** 162

13. **(a)** 15 **(b)** 10 **(d)** 5

▶ 15. 1120

19. 0.2165

Exercise Set 8.6—Page 242

3. 28π ft

5. 12 ft

7. 2π, 10π

9. 16π
11. 26π
13. **(a)** 654 ft^2 **(b)** 22.3 ft

Exercise Set 8.7—Page 243

1. **(a)** $\frac{5\pi}{3}$ **(b)** $\frac{10\pi}{3}$ **(d)** $\frac{15\pi}{2}$
3. 5π
5. $\frac{15\pi}{2}$
7. ≈ 8.53 in.
9. 2.2 in.

Exercise Set 8.8—Page 246

1. **(a)** 8π **(b)** 10π **(c)** 16π **(d)** 20π
3. **(a)** 16π **(b)** 20π **(c)** 9π **(d)** $\frac{12}{\pi}$ **(e)** $\frac{196}{15}\pi$
5. 2 to 1
7. **(a)** $\frac{16}{3}\pi$ **(b)** 8π **(c)** $\frac{64}{5}\pi$ **(d)** $\frac{64}{3}\pi$
9. **(a)** 50 **(b)** 160
11. 6
13. 360 $in.^2$

Review Exercises—Page 251

1. **(a)** Radius **(b)** Chord **(c)** Tangent **(d)** Secant
2. They have the same center.
3. $m\overline{AB} > m\overline{CP}$
4. Right
5. **(a)** 50 **(b)** 40 **(c)** 20 **(d)** 120 **(e)** 90 **(f)** 30
6. **(a)** 40 **(b)** 40 **(c)** 120
7. $m\overline{BD} = 20$
8. $12/\sqrt{3}$ in.
9. ABC is a right triangle. $\overline{BD}$ is an altitude to the hypotenuse.
10. $\pi R^2 - \pi r^2 - 225\pi, \therefore R^2 - r^2 = 225 = x^2, x = 15$, thus $m\overline{RS} = 30$.
11. They must be $180 - 48 = 132$ and $180 - 83 = 97$.
12. $\overline{AD} \cong \overline{BC}$, then $\widehat{AD} \cong \widehat{BC}$. If $\angle CDB \cong \angle DBA$, then $\overline{DC} \parallel \overline{AB}$ and $ABCD$ is a trapezoid.
13. $m\angle BAC = \frac{1}{2}(75) = 37\frac{1}{2}$
$m\angle BCA = \frac{1}{2}(75) = 37\frac{1}{2}$
$m\angle B = 180 - 75 = 105$
$m\angle OAC = 90 - 37\frac{1}{2} = 52\frac{1}{2}$
14. $m\angle DAC = \frac{1}{2}m°\widehat{DC}$, $m\angle BAD = \frac{1}{2}m°\widehat{AD}$. Since $\widehat{AD} \cong \widehat{DC}$, $m\angle DAC = m\angle BAD$.
15. $d^2 = 12^2 - 4^2 = 144 - 16 = 128$, $d = \sqrt{128} = 8\sqrt{2}$

16. $\overline{AD} \cong \overline{DC}$ and $\overline{DC} \cong \overline{DB}$, by Corollary 8-11.
17. $2\pi(4) = 2C$, $C = \pi(4)$, $d = 4$
18. $\pi(r + 2)^2 = 81\pi$
$r^2 + 4r + 4 = 81 \rightarrow r^2 + 4r - 77 = 0 \rightarrow (r - 7)(r + 11) = 0$
$\therefore r = 7$
19. **(a)** 300 square units **(b)** 293 square units
20. **(a)** 4π in.
(b) 30π in.^{2}
21. $\sqrt{2}$ in. each side

Cumulative Review—Chapters 2–8—Page 253

1. Divides the triangle proportionally
2. $a\sqrt{2}$
3. 50
4. Rectangle
5. 5
6. Rhombus
7. 25 sq. in.
8. 1 to 16
9. Median of a triangle
10. Supplementary
11. Midpoint
12. Collinear
13. Plane
14. $\overline{AB}$
15. Intersection
16. $\sqrt{(c - a)^2 + (d - b)^2}$
17. 1
18. $(-1, 3)$
19. 40
20. Sector

Problems

1. 65, 50
2. $\triangle AEB \sim \triangle BDC$ by Corollary 6-1(b); therefore the proportion holds.
3. $\triangle ABC \sim \triangle EBD$
(a) 60
(b) $m\overline{BE} = 10\frac{2}{3}$
▶ 4. $h = 12 \sin 40° \approx 7.72$
$b = 2\,(12 \cos 40°) \approx 18.4$
6. Use Theorem 2-3 to prove $\overline{AC} \cong \overline{ED}$, then triangles $\cong$ by SAS, hence $\angle FDE \cong \angle BCA$ and $\overline{BC} \parallel \overline{FD}$.

7. **(a)** $m_{\overline{RS}} = -2$, $m_{\overline{ST}} = \frac{1}{2}$
9. **(b)** $M_{\overline{GT}} = (\frac{1}{2}, \frac{13}{2})$, $M_{\overline{TO}} = (-2, 4)$, $M_{\overline{OG}} = (\frac{5}{2}, \frac{5}{2})$
11. 15
12. $W = 7$; $L = 13$
13. **(a)** $x = 11$ **(b)** $m\angle D = 125$ **(c)** $m\angle C = 55$
14. $K = 30$ sq. in.
15. (3, 4)
16. $b = 15$ *ft*
17. $56\sqrt{3}$ sq. in.
18. 18
19. ▶ 64 cos 20° sin 20°
20. 28 ft^2
21. 20
22. **(a)** 35, 35 **(b)** similar **(c)** 9
23. 1225π $ft^2 \approx 3848.5$ ft^2
24. **(a)** 48π in.$^2 \approx 150.8$ in.^{2}
    (b) $(48\pi - 36\sqrt{3})$ in.$^2 \approx 26.1$ in.^{2}
25. **(a)** 60 **(b)** 30 **(c)** 70
26. 20 in.
27. $75\pi + 100 \approx 335.6$ in.^{2}
28. $\pi(8\frac{1}{2})^2 - 72\frac{1}{4}\pi \approx 227.0$ in.^{2}
29. 16 in.
30. $81\pi \approx 254.5$ in.^{2}

CHAPTER 9

Exercise Set 9.1—Page 260

1. **(a)** None
   (b) None
   (c) One (Post. 2-8)
3. Line *AB* lies entirely in $\mathscr{E}$ (Post. 9-2).
5. $\overleftrightarrow{AB}$ lies in the plane $\mathscr{Q}$ (Post. 9-2).
7. **(a)** 3: *ADC*, *BDC*, *ADB*
   (b) 6: *FEG*, *FEH*, *FEI*, *GEH*, *GEI*, *HEI*
9. **(a)** $m\overline{AB} < m\overline{AC}$
   (b) $m\angle ABC > m\angle ACB$

Exercise Set 9.2—Page 263

1. Yes; as in the corner of a room.
3. No.
5. Yes. $\overleftrightarrow{AB}$ and $\overleftrightarrow{CD}$ in parallel planes.

Exercise Set 9.3—Page 265

1. $\angle A\text{-}XY\text{-}D$, $\angle A\text{-}XY\text{-}E$, $\angle A\text{-}XY\text{-}F$, $\angle D\text{-}XY\text{-}E$, $\angle D\text{-}XY\text{-}F$, $\angle E\text{-}XY\text{-}F$

▶ **9.** 49°

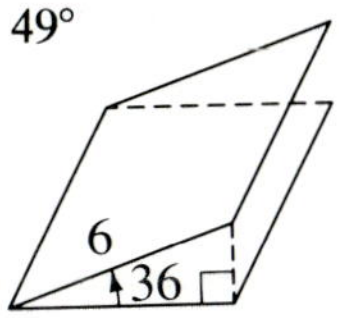

Exercise Set 9.4—Page 268

1. **(a)** No **(b)** No **(c)** Yes

3. 12 edges, 6 vertices (See Fig. 9-17)

Exercise Set 9.5—Page 270

1. **(b)** 6 faces, 12 edges, 8 vertices

3. $A_T = 2(5^2 + 5^2 + 5^2) = 2(75) = 150 \text{ in.}^2$

5. **(a)** $A_L = 18(40) = 720 \text{ in.}^2$
(b) $A_T = 2(48 + 180 + 180) = 816 \text{ in.}^2$

7. 1540 square units

9. 2, 3, 4

11. $d = \sqrt{12} = 2\sqrt{3}$

Exercise Set 9.6—Page 273

1. $V = 512 \text{ in.}^3$

3. $K = \frac{1}{2}(5)(12) = 30$
$V = Kh = 30(15) = 450$ cubic units.

5. $V = (15)(4)(\frac{5}{96}) = \frac{25}{8} = 3\frac{1}{8}$
$W = V \cdot 475 = (\frac{25}{8})475 = 1484\frac{3}{8} \approx 1484$

Exercise Set 9.7—Page 274

1. $V = 105$ cubic units

3. $K_T = \frac{1}{2}(12)(6\sqrt{3}) = 36\sqrt{3} \approx 62.4$
$S_P = 3(36\sqrt{3}) = 108\sqrt{3} \approx 187.1$

5. $h^2 = (3\sqrt{3})^2 - (\sqrt{3})^2 = 27 - 3 = 24$
$h = 2\sqrt{6} \approx 4.9$
$V = \frac{1}{3}[\frac{1}{4}(6)^2\sqrt{3}][2\sqrt{6}] = 6\sqrt{18} = 18\sqrt{2} \approx 25.5$

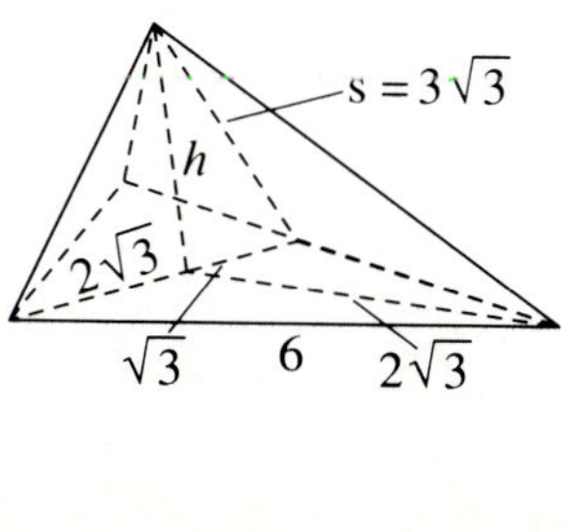

Exercise Set 9.8—Page 277

1. **(a)** $h = 4\sqrt{3}$, $V = \frac{1}{3}\pi(4)^2(4\sqrt{3}) = \frac{64}{3}\pi\sqrt{3} \approx 116.1$

(b) $S = 4\pi\left(\frac{8}{\sqrt{3}}\right)^2 = \frac{256\pi}{3} \approx 268.1$

3. \$8980

5. **(a)** $V_i \approx 16{,}755.2\ \text{ft}^3$
(b) $V_s \approx 1732.0\ \text{ft}^3$
(c) $W \approx 12{,}470$
(d) Cost = \$3,534

7. $h = 4\sqrt{3}$, $V \approx 348.2$

9. $V \approx 524\ \text{in.}^3$

Review Exercises—Page 278

2. Yes

3. No

4. No

5. If ℓ_1 and ℓ_2 intersect at some point P, then a triangle CDP is formed with $\overline{AB}$ and $\overline{EF} \parallel \overline{CD}$ and if a line is parallel to a side of a triangle, it divides the other two sides proportionally. If $\ell_1 \parallel \ell_2$, the proof is similar.

6. No, $\overleftrightarrow{CD}$ could be parallel to the plane.

7. **(a)** Always true **(b)** False **(c)** Sometimes true **(D)** Never true **(e)** Sometimes true **(f)** Always true **(g)** Sometimes true

8. $d = \sqrt{8^2 + 6^2 + 24^2} = \sqrt{676} = 26$

9. $5\sqrt{3}$ and $4\sqrt{3}$

10. The diagonals of the cube will be in the plane of the diagonals of a pair of sides of the cube, hence in a rectangle with sides e and $e\sqrt{2}$ the diagonals will not be perpendicular.

11. **(a)** 96 **(b)** 48 **(c)** 8 **(d)** 64

12. **(a)** $\sqrt{61}$ **(b)** $10\sqrt{2}$ **(c)** $\sqrt{138}$ **(d)** 5 **(e)** 13

13. **(a)** $\frac{5\sqrt{6}}{3}$ **(b)** 3 **(c)** $\frac{x\sqrt{3}}{3}$

14. $\frac{V_1}{V_2} = \frac{8}{27}$

15. $\frac{V_1}{V_2} = \frac{8}{27}$

16. $V = 64\sqrt{3}\ \text{in.}^3$

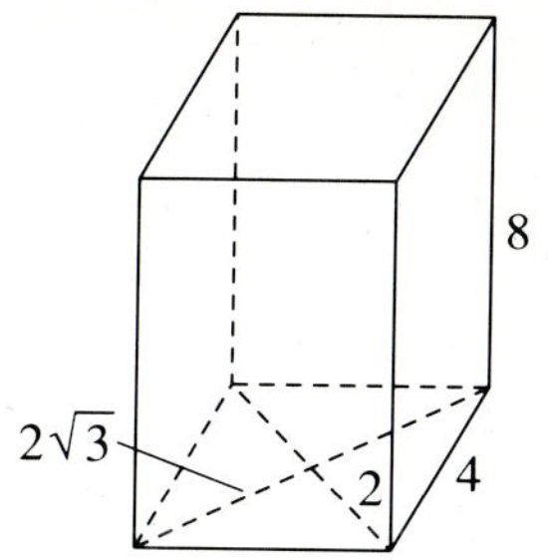

17. $\frac{3}{1}$

18. $l = 2\sqrt[3]{60}$, $w = 3\sqrt[3]{60}$, $h = 5\sqrt[3]{60}$

19. $S = 324\pi$

20. $r = 4$

22. **(a)** 16π
(b) $\frac{48}{5}\pi$

23. $x = \sqrt{2}$

24. $V = \frac{a^3}{3}\pi$

25. $e = 6$

▶ **26.** $V \approx 322$ in.^{3}

Cumulative Review—Chapters 2–9—Page 280

1. Lies between
2. a, $-a$
3. Transitive
4. Its interior, the angle, its exterior
5. Supplementary
6. Second
7. Convex
8. 6, 16
9. Perpendicular bisector
10. Similar
11. $\overline{BC}$
12. A rhombus
13. 0, 90
14. One-half the degree measure of the intercepted arc
15. 15, 30π
16. 5, 8, $3\frac{4}{7}$, $|b - a|$, $p - q$
17. 54
18. 169
19. **(a)** 50 **(b)** 110 **(c)** 50

20. **(a)** (3, 6), (1, 6), (3, 8)
(b) 1, undefined, 0
(c) $4\sqrt{2}$, 4, 4

21. $\frac{m\overline{AB}}{m\overline{DE}} = \frac{m\overline{BC}}{m\overline{EF}} = \frac{m\overline{AC}}{m\overline{DF}}$

22. 24, $24\sqrt{3}$, 48

23. $8\sqrt{3}$

24. $162\sqrt{3}$

25. **(a)** 90 **(b)** 25 **(c)** 90 **(d)** 35

26. **(a)** 128 **(b)** 256 **(c)** $96\sqrt{3}$

28. $6\sqrt{3}$

29. $\frac{\sqrt{3}}{24}\pi s^3$ square units

30. $1,156

Index

A

B

C

D

E

F

G

H

I

L

M

N

O

P

Q

R

S

T

U

V

X

Y

Z